About the cover:

The graphic on the cover, depicting Moiré patterns, is generated by the Mathematica™ command:
DensityPlot(Sin(x/y), {x,-5,5},{y,-1.8,1.8}, PlotPoints->800, Mesh->False).

Source: Theodore W. Gray and Jerry Glynn, *Exploring Mathematics with Mathematica*, Addison Wesley, 1991.

A. Fässler
E. Stiefel
with English translation by
Baoswan Dzung Wong

Group Theoretical Methods
and Their Applications

Birkhäuser
Boston • Basel • Berlin

Albert Fässler
Ingenieurschule Biel
Ecole d'Ingénieurs Bienne
Postfach CP 1180
CH-2500 Biel-Bienne 1

Baoswan Dzung Wong
(Translator)
Wegächerstrasse 3
CH-5417 Untersiggenthal
Switzerland

Eduard Stiefel
Late Professor at the
Department of Mathematics
Eidgenössische Technische
Hochschule ETH
CH-8092 Zürich, Switzerland
(died in 1978)

Library of Congress Cataloging-in-Publication Data
Gruppentheoretische Methoden und ihre Anwendung. English
    Group theoretical methods and their applications / edited by
Albert Fässler, Eduard Stiefel ; translated by Baoswan Dzung Wong.
        p.   cm.
    Rev. translation of: Gruppentheoretische Methoden und ihre
Anwendung / Eduard Stiefel, Albert Fässler
    Includes bibliographical references and index
    ISBN 0-8176-3527-0 (alk. paper) : ISBN 3-7643-3527-0 (alk. paper)
    1. Representations of groups   2. Linear operators.   3. Lie
algebras.   I. Fässler, Albert, 1943-   .   II. Stiefel, Eduard L.,
1909-1978.   III. Title.
QA176.G7813   1992   91-47544
512 ' .2--dc20   CIP

Printed on acid-free paper.
Original German Edition: Gruppentheoretische Methoden und ihre
Anwendung; B. G. Teubner, Stuttgart; first edition: 1979
© Birkhäuser Boston, 1992
ISBN 0-8176-3527-0
ISBN 3-7643-3527-0

Camera-ready copy prepared by the Authors in TeX.
Printed and bound by Quinn-Woodbine, Woodbine, New Jersey.
Printed in the USA.

9 8 7 6 5 4 3 2 1

Contents

1 Preliminaries   **1**
  1.1 The Concept of Groups 1
      1.1.1 Transformation Groups 3
  1.2 Price Index in Economics 5
  1.3 The Realization of Groups 8
  1.4 Representation of Groups 11
  1.5 Equivalence of Representations 14
  1.6 Reducibility of Representations 16
  1.7 Complete Reducibility 17
  1.8 Basic Conclusions 21
  1.9 Representations of Special Finite Groups 22
      1.9.1 The Cyclic Group C_g 22
      1.9.2 The Dihedral Group D_h 23
  1.10 Kronecker Products 25
  1.11 Unitary Representations 27
      1.11.1 Unitary Representations and Unitary Matrices . . . 29
  Problems 30

2 Linear Operators with Symmetries   **33**
  2.1 Schur's Lemma 33
  2.2 Symmetry of a Matrix 35
      2.2.1 Representations of Abelian Groups 36
  2.3 The Fundamental Theorem 36
  Problems 47

3 Symmetry Adapted Basis Functions   **51**
  3.1 Illustration by Dihedral Groups 51
      3.1.1 The Representation ϑ_{per} 52
      3.1.2 The Notion of the Orbit 52
      3.1.3 Instructions on How to Use Table (3.7) 54
      3.1.4 Real Orthonormal Symmetry Adapted Basis 56
  3.2 Application in Quantum Physics 57
  3.3 Application to Finite Element Method 59
      3.3.1 Discretization and Symmetry of the Problem 59
      3.3.2 Elliptic Boundary Value Problem 60

		3.3.3	Heat conduction	63
	3.4		Perturbed Problems with Symmetry	64
	3.5		Fast Fourier Transform on Finite Groups	65
		3.5.1	Definitions and Properties	65
		3.5.2	Direct and Fast Algorithm	66
		3.5.3	The Classical Fast Fourier Transform (FFT)	68
		3.5.4	Applications and Remarks	70

4 Continuous Groups And Representations — **73**

	4.1		Continuous Matrix Groups	73
		4.1.1	Comments on U(n)	74
		4.1.2	Comments on SU(n)	75
		4.1.3	Connectedness of Continuous Groups	75
	4.2		Relationship Between Some Groups	76
		4.2.1	Relationship between SL(2, C) and the Lorentz group	77
		4.2.2	Relationship Between SU(2) and SO(3)	81
	4.3		Constructing Representations	83
		4.3.1	Irreducible Representations of SU(2)	84
		4.3.2	Irreducible Representations of SO(3)	85
		4.3.3	Complete Reduction of Representations of SU(2) and SO(3)	86
	4.4		Clebsch-Gordan Coefficients	89
		4.4.1	Spherical Functions	90
		4.4.2	The Kronecker Product $\vartheta_\ell \otimes \vartheta_m$	91
		4.4.3	Spherical Functions and Laplace Operator	95
	4.5		The Lorentz group and SL(2,**C**)	98
		4.5.1	Irreducible Representations	99
		4.5.2	Complete Reduction	99
	Problems			103

5 Symmetry Ad. Vectors, Characters — **105**

	5.1		Orthogonality of Representations	105
	5.2		Algorithm for Symmetry Adapted Bases	108
	5.3		Applications	117
	5.4		Similarity Classes of Groups	133
		5.4.1	Subgroups of SO(3)	134
		5.4.2	Permutation Groups	136
	5.5		Characters	139
		5.5.1	General Properties	139
		5.5.2	Orthogonality Relations of Characters	142
		5.5.3	A Composition Formula for Characters	143
		5.5.4	Fundamental Theorems about Characters	143
		5.5.5	Projectors	146
		5.5.6	Summary	147
	5.6		Representation Theory of Finite Groups	148
		5.6.1	The Regular Representation	148
		5.6.2	The Character Matrix	149

        5.6.3   Completeness 151
  5.7   Extension to Compact Lie Groups 151
  Problems 157

6 Various Topics of Application   159
  6.1   Bifurcation and A New Technique 159
        6.1.1   Introduction 159
        6.1.2   The Hopf Bifurcation 160
        6.1.3   Symmetry-Breaking 163
        6.1.4   A New Approach 166
        6.1.5   The Brusselator 168
  6.2   A Diffusion Model in Probability Theory 170
        6.2.1   Introduction 170
        6.2.2   General Considerations 172
        6.2.3   The Bernoulli-Laplace Model 176
        6.2.4   Discussion and Applications 179
  Problems 179

7 Lie Algebras   181
  7.1   Infinitesimal Operator and Exponential Map 181
        7.1.1   Infinitesimal Operators 181
        7.1.2   The Exponential Mapping 185
  7.2   Lie Algebra of a Continuous Group 186
  7.3   Representation of Lie Algebras 190
  7.4   Representations of SU(2) and SO(3) 197
        7.4.1   Infinitesimal Aspects of the Kronecker Product . . . 200
        7.4.2   Clebsch-Gordan Coefficients 201
  7.5   Examples from Quantum Mechanics 202
        7.5.1   Energy and Angular Momentum 203
        7.5.2   Spin-Orbit Coupling 204
  Problems 206

8 Applications to Solid State Physics   209
  8.1   Lattices 209
        8.1.1   Reciprocal Lattice 211
        8.1.2   Brillouin Zone, Stabilizer 212
  8.2   Point Groups and Representations 214
        8.2.1   The List of All (+)-Groups 214
        8.2.2   Representations and Characters of (+)-Groups . . . 215
        8.2.3   (±)-Groups of the First Kind 216
        8.2.4   (±)-Groups of the Second Kind 217
        8.2.5   Review 218
  8.3   The 32 Crystal Classes 219
  8.4   Symmetries and the Ritz Method 220
  8.5   Examples of Applications 223
  8.6   Crystallographic Space Groups 232
  Problems 236

9 Unitary and Orthogonal Groups   **239**
   9.1   The Groups U(n) and SU(n)   239
           9.1.1   Similarity Classes, Diagram   240
           9.1.2   Characters of Irreducible Representations   242
           9.1.3   Division Algorithm   244
           9.1.4   Weights   247
           9.1.5   Algorithm for the Complete Reduction   248
   9.2   The Special Orthogonal Group SO(n)   250
           9.2.1   Weyl Group and Diagram of SO(n)   252
           9.2.2   Algorithm for the Computation of Weights   253
   9.3   Subspaces of Representations of SU(3)   254

A   **265**
   Answers to Selected Problems   265
   Bibliography   280
   Index   286

Preface to the English edition

Group theory as a mathematical tool to study symmetries has an abundance of applications from various fields.

It was my intention to add the following topics to the extended and renewed edition of the German version:

1. **An introductory example from economics**, Section 1.2.

2. **The algorithm "Fast Fourier Transforms on Finite Groups"**
   as a contribution to numerical analysis, Section 3.5.

3. **An application to the Bernoulli-Laplace diffusion model from**
   the field of stochastics, Section 6.2.

4. **Bifurcation as a field of application with non-linear problems,**
   Section 6.1.

The topics in items 2 and 3 were undoubtedly initiated by an original seminar talk by Professor Persi Diaconis of Harvard University given in 1988 at the University of Minnesota, Minneapolis, where I spent my sabbatical year. I wish to express my thanks for this and the subsequent correspondence I had with him.

The topic under item 4 was brought about by scientific contacts with Professor Bodo Werner of Hamburg University. Many thanks also for his reading the manuscript.

I owe the example under item 1 to Dr. Arthur Vogt in Bern, and I also wish to express my appreciation for his reading the manuscript.

In Dr. Baoswan Dzung Wong I had the privilege of not only finding an excellent translator but also an excellent mathematician. We had the pleasure of studying mathematics together at ETH, Zürich before she wrote her dissertation in functional analysis at the University of California, Berkeley under the guidance of Professor Tosio Kato.

With her intensive, precise style of working and her abilities she improved this edition from the linguistic and the mathematical point of view. Moreover, the present typesetting is the direct result of a computer output, generated by a code, which she wrote by use of the document preparation system LaTeX. I wish to thank her for the beautiful work and the numerous discussions on the contents of this book.

I am indebted to Peter Fässler, Neu-Technikum Buchs, Switzerland, for drafting the figures, to my students Kurt Rothermann and Stefan Strähl for computer enhancing and labeling the graphics, to Pascal Felder and Markus Wittwer for a simulation program that generated the figures in the stochastics sections.

My thanks go to my new colleague at work, Daniel Neuenschwander, for the inspiring discussions related to the section in stochastics and for reading the manuscript to it.

I am also grateful to Dacfey Dzung for reading the whole manuscript.

Thanks go especially to Professor Walter Gander of ETH, Zürich, who at the finishing stage and as an expert of TeX generously invested numerous hours to assist us in solving software as well as hardware problems; thanks go also to Martin Müller, Ingenieurschule Biel, who made the final layout of this book on the NeXT computer.

Thanks are also due to Helmut Kopka of the Max Planck Institute, for solving software problems, and to Professor Burchard Kaup of the University of Fribourg, Switzerland for adding some useful software; also to Birkhäuser Boston Inc. for the pleasant co-operation.

Finally, let me be reminiscent of Professor E. Stiefel (deceased 1978) with whom I had many interesting discussions and true co-operation when writing the book in German.

I wish to express my thanks to our school, the Engineering College Biel/Bienne, especially to Director Dr. Fredy Sidler for making available the hardware so pleasant to work with, namely that of Macintosh and NeXT.

Last, but not least, I express my gratitudes to my wife Carmen for her moral support; also to Wingdzi(9), Manuel(9), Dominik(8), and Daniel(6) who have borne with our standard phrase (a proposition which Baoswan and I found independently of each other): "We'll do this and that with you as soon as the book's finished!"

Biel/Bienne, Fall 1991   A. Fässler

Preface to the German edition

Numerous problems of the natural and technical sciences possess certain symmetry properties. If we manage to recognize and to understand them, then a mathematical treatment adjusted to the symmetry properties may lead to considerable simplification.

A basic tool in this connection, as will be shown in this book, will be supplied by the **representation theory of groups** and particularly, in view of the applications, by **Schur's lemma**.

Namely, let a linear operator of a given problem in a vector space V be given such that it commutes with a group G of linear transformations of V. If now V decomposes into subspaces that are irreducible and invariant under G, then Schur's lemma enables us to split the problem into simpler partial problems.

Our main concern thus consists in establishing algorithms to determine the invariant subspaces (Chapters 3, 5, 7, 9) and their symmetry adapted basis vectors or basis functions. (We note here that less emphasis is put upon constructing the irreducible representations and characters of a group, as for applications tables are often available.)

For pedagogical reasons, we develop the theory in such a way that, as we proceed, applications of progressive importance from various fields will quickly become accessible. We shall mention here the following applications:

Chapter 3 is dedicated to the group theoretic treatment of certain partial differential equations (as occurring in the **engineering sciences**), whose underlying domains have symmetries. However, the boundary conditions need not possess any symmetries. Furthermore, Chapter 3 contains a simple example from quantum physics of molecules.

In Chapter 5, after constructing an algorithm of central interest, an application about **molecular oscillations** is given.

Chapter 7 gives an introduction to the theory of Lie algebras as an essential device of **quantum mechanics**. As an example, the spin orbit coupling of a particle in a central field is worked out thouroughly.

After an introduction to crystallography, Chapter 8 treats applications from **solid state physics**.

Chapter 9 treats the reduction of representations of unitary and orthogonal groups. In particular, an algorithm (which can be generalized) to determine the invariant irreducible subspaces is formulated for the case of the unimodular unitary 3-dimensional group. Two examples from **elementary particle physics** related to the quark model will be considered.

The first two chapters are prerequisite to understanding the subsequent chapters of this book. Chapters 3, 5, and 8 may be read independently from one another, but after Chapter 2; while Chapters 4, 7, 9 to some extent are based one upon another.

The book is addressed to mathematicians and physicists on the one hand, but on the other hand also to, in particular, engineers, chemists, and natural scientists. We assume that the reader is familiar with the basics of linear algebra.

Messrs Heinz Ungricht, Gabor Groh, and Sandro Pelloni have helped us with their suggestions and with preparing examples. Messrs Karl Flöscher, Dieter Issler, Wolfgang Reim, and Hansueli Schwarzenbach have contributed both in terms of content and editing. Ms Carla Blum did the typewriting. We wish to thank all our collaborators.

We are also grateful to the Swiss Federal Institute of Technology (ETH) in Zürich for supporting the preparatory research work and the Teubner publishing house for the pleasing layout.

Zürich, Summer 1978   E. Stiefel, A. Fässler

Chapter 1

Preliminaries

1.1 The Concept of Groups

Niels Henrik Abel (1802-1829) and Evariste Galois (1811-1832) were the true pioneers in the theory of abstract **groups**.

Abel at an early stage believed he had found a general method to solve a quintic equation, but soon discovered his error and proved that a solution was impossible (1824). He worked in Berlin with Crelle, then in Paris with Cauchy.

About a farewell letter that the 21-year-old Galois left behind on the eve of his fatal duel and whose content marked the beginning of modern algebra and geometry, Hermann Weyl (1885-1955) wrote in his beautiful book "Symmetry" (Weyl, 1952):

"This letter, if judged by the novelty and profundity of ideas it contains, is perhaps the most substantial piece of writing in the whole literature of mankind."

We consider a set G of finitely or infinitely many objects, individuals, or simply, **elements**

$$a, b, c, \ldots, s, \ldots$$

in which we define a law of combination that assigns to every **ordered** pair of elements $a, b \in G$ a **unique** element $c = a \cdot b$ in G. Here the **products** $a \cdot b$ and $b \cdot a$ need not be identical. We often denote the operation by a dot or sometimes just omit the symbol. In certain cases it is favorable to use the symbol $+$.

Example 1.1 $N =$ the set of all natural numbers 1,2,3,... with multiplication as the law of combination.

Example 1.2 $R =$ the set of all positive real numbers with multiplication as the law of combination. This set is uncountable.

Example 1.3 $\mathbf{Z}$ = the set of integers $\ldots, -2, -1, 0, 1, 2, \ldots$ with subtraction as the law of combination. This set is countably infinite.

Example 1.4 G = the set of all rotations in the plane that map a regular hexagon onto itself. Here the law of combination is the composition of two rotations. This set is finite. If we include the rotation through $0°$ (that is, the identity e), it consists of six elements, namely

$$a, a \cdot a = a^2, a^3, a^4, a^5, a^6 = e$$

where a is the rotation through $60°$.

Example 1.5 G = the set of all proper rotations in space that leave a sphere invariant, i.e., rotations about any axis passing through the center of the sphere through any angle of rotation that are not reflections. Here again the law of combination is the composition of two rotations. G is continuous (see Chapter 4). In general, $ab \neq ba$, because if we let a be the rotation about the x_1-axis through $90°$ mapping the north pole

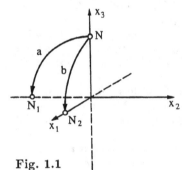

Fig. 1.1

N into N_1

and b be the rotation about the x_2-axis through $90°$ mapping the north pole

N into N_2 (see Fig. 1.1)

then $b \cdot a$ means: first perform a, then perform b. Under this composition N is mapped into N_1, as N_1 is invariant under b. Under the composition $a \cdot b$, however, N is mapped into N_2.

We analyze these examples for the following properties:

I. There exists a **neutral element** $e \in G$, also called the **unit element** or the **identity element**, such that

$$s \cdot e = e \cdot s = s \qquad \forall s \in G \qquad (1.1)$$

Examples 1.1 and 1.2 $e = 1$.

Example 1.3 There is no element e such that $e - x = x$ $\forall x \in G$.

Examples 1.4 and 1.5 e is the rotation through $0°$.

II. To every $s \in G$ there exists an **inverse** $s^{-1} \in G$ such that

$$s \cdot s^{-1} = s^{-1} \cdot s = e \ \forall s \in G \qquad (1.2)$$

Example 1.1 The set $\mathbf{N}$ does not have this property.

Example 1.2 $s^{-1} = \frac{1}{s}$.

Example 1.3 There is no unit e.

Example 1.4 For $s = a^k$ we have $s^{-1} = a^{6-k}$.

Example 1.5 s^{-1} and s are rotations about the same axis, the rotation angles are equal in magnitude but of opposite sign.

III. The composition is **associative**, i.e.,

$$(ab)c = a(bc) \qquad \forall a, b, c \in G \qquad (1.3)$$

Convention: If the associative law (1.3) holds, it is customary to write $abc = (ab)c = a(bc)$, since the way of bracketing does not matter.

In all these examples the associative law holds, except in Example 1.3, where for instance $(6 - 3) - 1 \neq 6 - (3 - 1)$. Only Example 1.5 requires explaining. To this end, we consider a point P on the surface of the sphere. Hereafter, *multiple products of operators shall always be performed from right to left.*

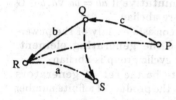

Fig. 1.2

Figure 1.2 shows the three transformations a, b, and c, where $c : P \rightarrow Q$, $b : Q \rightarrow R$, $a : R \rightarrow S$. The left side of (1.3) is represented by the dashed line, the right side by the dot dash line. These considerations give rise to the following

Definition. A set G of elements with a law of combination that assigns to every ordered pair a, b of elements of G a unique element $c = ab$ of G and that has the properties I, II, and III is called a **group**.

The Examples 1.2, 1.4, and 1.5 are groups, but not the Examples 1.1 and 1.3.

Note: $(ab)^{-1} = b^{-1}a^{-1}$, because $(ab)(b^{-1}a^{-1}) = a(bb^{-1})a^{-1} = aea^{-1} = e$.

Definition. A subset of a group is called a **subgroup** provided that its elements by themselves constitute a group under the operation law of G.

In the additive group of integers $\ldots, -2, -1, 0, 1, 2, \ldots$ the even numbers constitute a subgroup, but not the odd numbers.

1.1.1 Transformation Groups

Definition. A **transformation** is a one-to-one[1] mapping $f : x \rightarrow f(x)$ of some set W onto[2] itself. W is called the **range**.

---

[1] A mapping f is said to be one-to-one if $x \neq y$ implies $f(x) \neq f(y)$.

[2] A mapping f is said to be onto if for all b in W there exists an a such that $f(a) = b$.

The set of all transformations with the operation law

$$fg(x) = f(g(x)) \ \forall x \in W \tag{1.4}$$

(i.e., the composition of two transformations g and f, where g is followed by f) forms a group, the so-called **complete transformation group** of W.

The reason is that there exists an identity element (namely the identity transformation), and for each element there exists an inverse. The associative law follows directly from the definition of the composition law (1.4); that is, the argument given in III for Example 1.5 applies to this case.

Subgroups of the complete transformation group of W are called **transformation groups** of W.

Examples 1.4 and 1.5 are transformation groups. The range W consists of all points of the plane or the space, respectively.

Further definitions:

A group G is said to be **abelian** or **commutative** if $ab = ba \ \forall a, b \in G$ (commutative law). Examples 1.2 and 1.4 are abelian.

A group is said to be **cyclic** provided it consists exactly of the powers a^k (k integer) of some element a; a is called **the generating element**. Example 1.4 describes a cyclic group. Every cyclic group is abelian.

A set of elements of a group G is said to be the **set of generators** provided every element of G can be written as the product of a finite number of elements of this generating set and of their inverses.

The set of all rotations about two coordinate axes in Example 1.5 is a set of generators for the following reason: Every rotation d described by the three Eulerian angles φ, ψ, ϑ (see Figure 1.3) may be factored into a product of three rotations in the plane

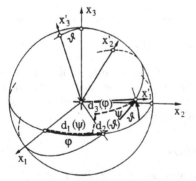

$$d = d_3(\varphi) \cdot d_2(\vartheta) \cdot d_1(\psi)$$

where

$d_1(\psi)$ = rotation about the x_3-axis through angle ψ;

$d_2(\vartheta)$ = rotation about the x_1-axis through angle ϑ;

$d_3(\varphi)$ = rotation about the x_3-axis through angle φ.

Fig. 1.3

In general, the set of generators is not unique, not even for a prescribed number of generators (compare Section 1.9.2).

By the **order** $|G|$ or g of the group G we mean the number of its elements. In Example 1.4, the order $|G| = 6$.

1.2 Price Index in Economics

The problem of constructing an adequate price index in economics has had a long history. Galileo Galilei (1564-1642) had already attempted it and is mentioned by the economist, engineer, and statistician F. Divisia (Divisia, 1927). The price index P for a single commodity is defined by $P = p^1/p^0$; that is, simply by the quotient of the prices p^1 in the observed situation and p^0 in the base situation. However, establishing a price index for more than one commodity presents greater difficulty:

For example, suppose that the price of bread doubled from $p_1^0 = \$1$ per *kg* in the base situation to $p_1^1 = \$2$ per *kg* in the observed situation and that the price for fabric tripled from $p_2^0 = \$3$ per *ft* to $p_2^1 = \$9$ per *ft*. What value should then be attributed to a price index for the two items?

In 1738 Dutot proposed taking the sum of the prices in the observed situation divided by the sum of the prices in the base situation:

$$P_{\text{Dutot}} = \frac{p_1^1 + p_2^1}{p_1^0 + p_2^0} = \frac{2 + 27}{1 + 9} = 2.900 \qquad (1.5)$$

In the general case of n commodities, the price index is a real-valued positive function P of the n quantities $\vec{q^0} = (q_1^0, q_2^0, \ldots, q_n^0)$ and the n prices $\vec{p^0} = (p_1^0, p_2^0, \ldots, p_n^0)$ in the base situation as well as of the n quantities $\vec{q^1} = (q_1^1, q_2^1, \ldots, q_n^1)$ and the n prices $\vec{p^1} = (p_1^1, p_2^1, \ldots, p_n^1)$ in the observed situation. Thus P is a function of $4n$ independent variables:

$$P = P(\vec{q^0}, \vec{p^0}, \vec{q^1}, \vec{p^1}) \qquad (1.6)$$

A good price index must necessarily satisfy the following axioms 1 through 4 and should preferably meet the tests 5 through 8; see (Fisher, 1927), (Eichhorn, 1978):

1. **Identity axiom:** If none of the prices alter from the base to the observed situation, then the price index is equal to 1.

2. **Commodity reversal axiom:** The price index does not depend upon the ordering of the n commodities.

3. **Commensurability axiom:** A change in the units of measure of the commodities does not change the value of the price index.

   Remark: Dutot's price index (1.5) violates this axiom: If in the example just mentioned the unit of measure *feet* for fabric is replaced by *meters*, then its price multiplies by 3.28 (1 *meter* = 3.28 *ft*) and we obtain

   $$P_{\text{Dutot}} = \frac{2 + 3.28 \cdot 9}{1 + 3.28 \cdot 3} = 2.815$$

   which obviously differs from the value given in (1.5).

4. **Proportionality axiom:**

$$P(\vec{q^0}, \vec{p^0}, \vec{q^1}, \lambda \vec{p^0}) = \lambda \tag{1.7}$$

5. **Time reversal test:**

$$P(\vec{q^0}, \vec{p^0}, \vec{q^1}, \vec{p^1}) = \frac{1}{P(\vec{q^1}, \vec{p^1}, \vec{q^0}, \vec{p^0})} \tag{1.8}$$

This states that interchanging the base and the observed situations in the price index function yields the reciprocal value.

6. **Factor reversal test:**

$$P(\vec{q^0}, \vec{p^0}, \vec{q^1}, \vec{p^1}) = \frac{V}{P(\vec{p^0}, \vec{q^0}, \vec{p^1}, \vec{q^1})} \tag{1.9}$$

where the so-called **value index** V is defined by

$$V = \frac{\displaystyle\sum_{i=1}^{n} q_i^1 \cdot p_i^1}{\displaystyle\sum_{i=1}^{n} q_i^0 \cdot p_i^0}$$

Hence the product of the original price index and the price index with prices and quantities interchanged equals V.

In addition to the properties 1 through 6, Vogt introduced the two following tests, which are of economic significance (Vogt, 1987):

7. **Price reversal test:**

$$P(\vec{q^0}, \vec{p^0}, \vec{q^1}, \vec{p^1}) = \frac{1}{P(\vec{q^0}, \vec{p^1}, \vec{q^1}, \vec{p^0})} \tag{1.10}$$

8. **Quantity reversal test:**

$$P(\vec{q^0}, \vec{p^0}, \vec{q^1}, \vec{p^1}) = P(\vec{q^1}, \vec{p^0}, \vec{q^0}, \vec{p^1}) \tag{1.11}$$

He also postulates that combining properties makes sense: For example, let us take the composition *factor reversal (1.9)* followed by *time reversal (1.8)*. Then

$$P(\vec{q^0}, \vec{p^0}, \vec{q^1}, \vec{p^1}) = \frac{V}{P(\vec{p^0}, \vec{q^0}, \vec{p^1}, \vec{q^1})} = V \cdot P(\vec{p^1}, \vec{q^1}, \vec{p^0}, \vec{q^0}) \tag{1.12}$$

The composite of (1.10) followed by (1.9) yields

$$P(\vec{q^0},\vec{p^0},\vec{q^1},\vec{p^1}) = \frac{1}{P(\vec{q^0},\vec{p^1},\vec{q^1},\vec{p^0})} \cdot \frac{\sum\limits_{i=1}^{n} q_i^0 p_i^1}{\sum\limits_{i=1}^{n} q_i^1 p_i^0} \cdot P(\vec{p^1},\vec{q^0},\vec{p^0},\vec{q^1}) \quad (1.13)$$

If the order in the previous composition is reversed, i.e., (1.9) followed by (1.10), then

$$P(\vec{q^0},\vec{p^0},\vec{q^1},\vec{p^1}) = \frac{1}{P(\vec{p^0},\vec{q^0},\vec{p^1},\vec{q^1})} = V \cdot P(\vec{p^0},\vec{q^1},\vec{p^1},\vec{q^0}) \quad (1.14)$$

Let us take a closer look at the **permutations of the 4 vectors** $\vec{q^0}$, $\vec{p^0}$, $\vec{q^1}$, $\vec{p^1}$ described by (1.8) through (1.14). We then realize that, if the identity is included, these elements **form a group** of order 8.

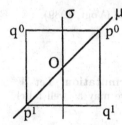

Next let us assign to each of the 4 vectors one of the vertices of a square in the manner of Fig. 1.4. We introduce the so-called **dihedral group** D_4 consisting of all linear transformations of the plane $\mathbf{R}^2$ that leave the square invariant. This group is also of order 8.

Fig. 1.4

In fact, we find that D_4 **is isomorphic with the group of permutations of the 4 vectors** $\vec{q^0}$, $\vec{p^0}$, $\vec{q^1}$, $\vec{p^1}$ described by (1.8) through (1.14). Namely, (1.8) corresponds to the rotation about O through angle π; (1.13) and (1.14) correspond to the rotations about O through angles $\pm\pi/2$; (1.10) and (1.11) correspond to the reflections about the diagonals; (1.9) and (1.12) correspond to the reflections about the vertical and the horizontal lines.

The fact that the order of composing (1.8) and (1.9) does not matter justifies naming the composite (1.12) the **simultaneous time and factor reversal test**.

We conclude this section with a few remarks. The not unusual construction of price indices introduced by Laspeyres and Paasche

$$P_{\text{Laspeyres}} = \frac{\sum\limits_{i=1}^{n} q_i^0 p_i^1}{\sum\limits_{i=1}^{n} q_i^0 p_i^0}, \qquad P_{\text{Paasche}} = \frac{\sum\limits_{i=1}^{n} q_i^1 p_i^1}{\sum\limits_{i=1}^{n} q_i^1 p_i^0} \quad (1.15)$$

both fail the time reversal and factor reversal tests (1.8) and (1.9). And so

Fisher in 1922 recommended taking the geometric mean of the two:

$$P_{\text{Fisher}} = \sqrt{P_{\text{Laspeyres}} \cdot P_{\text{Paasche}}} \qquad (1.16)$$

The point is that P_{Fisher} has each of the desired afore-mentioned quali-
ties (identity axiom through (1.14)). The verification is left as an exercise.
(Funke, Voeller, 1978) provides a proof that P_{Fisher} is the only index satisfying (1.8), (1.9), and (1.11). Thus Fisher had good reasons to call it the
"ideal index".

Economic considerations led Fisher to the time reversal and the factor
reversal test; he did not use the vector notation presented here. Inves-
tigation of the group theoretical structure of the four argument vectors
$\vec{q}^0, \vec{p}^0, \vec{q}^1, \vec{p}^1$, however, leads to the tests 7 and 8.

Nevertheless, the index most commonly used for calculating consumer
price index numbers is the one due to Laspeyres. It has the practical
advantage that only the quantities of the base situation need to be known.

This example illustrates that purely mathematical arguments may yield
results that are relevant in applied sciences.

For a historical review on the price index we refer to (Vogt, 1989).

1.3   The Realization of Groups

A transformation of a **finite** range W is called a **permutation**. Let W
consist of N objects. As their nature is immaterial we may as well label
them by the numbers 1, 2, ..., N.

We write a permutation in the following way ($N = 5$):

$$\left\downarrow \begin{array}{ccccc} 1 & 2 & 3 & 4 & 5 \\ 3 & 2 & 4 & 5 & 1 \end{array} \right.$$

The group of all permutations of N objects is called the **symmetric group**
S_N and has the order $1 \cdot 2 \cdot \ldots \cdot N = N!$

Definition. Subgroups of S_N are called **permutation groups**.

Example 1.6 The symmetric group on 3 objects $W = \{1, 2, 3\}$. The
group elements are the permutations:

$$e = \left\downarrow \begin{array}{ccc} 1 & 2 & 3 \\ 1 & 2 & 3 \end{array}\right. \qquad a = \left\downarrow \begin{array}{ccc} 1 & 2 & 3 \\ 2 & 3 & 1 \end{array}\right. \qquad b = \left\downarrow \begin{array}{ccc} 1 & 2 & 3 \\ 3 & 1 & 2 \end{array}\right.$$

$$c = \left\downarrow \begin{array}{ccc} 1 & 2 & 3 \\ 1 & 3 & 2 \end{array}\right. \qquad d = \left\downarrow \begin{array}{ccc} 1 & 2 & 3 \\ 3 & 2 & 1 \end{array}\right. \qquad f = \left\downarrow \begin{array}{ccc} 1 & 2 & 3 \\ 2 & 1 & 3 \end{array}\right.$$

The following is true: $b = a^2$, $d = ca = bc$, $f = cb = ac$, $e = a^3$. This group
is not abelian, because $ca \neq ac$. The first three elements form a cyclic
subgroup.

Generally, a permutation of N objects is called a **transposition** if it interchanges 2 objects and leaves the remaining $(N-2)$ unchanged.

It should be noted that $(\text{transposition})^2 = e$.

In the example above, the last three permutations c, d, and f are transpositions. They do not form a subgroup of S_3; the elements a and c form a set of generators of the entire group S_3.

Here is an opportunity to introduce the **group table** of a finite group G. This really is the multiplication table of the group. It is obtained as follows:

In the head row, write the elements

$$e, a, b, \ldots, s, \ldots \tag{1.17}$$

of the group. An arbitrary row corresponding to a fixed element $u \in G$ contains all products

$$ue = u, ua, ub, \ldots, us, \ldots \tag{1.18}$$

In other words, the head row is multiplied from the left by u.

In Example 1.6, the table for the group S_3 has the form (1.19). It completely describes the **abstract structure** of S_3 without referring to the numbers 1, 2, and 3.

$\diagdown$	e	a	b	c	d	f
e	e	a	b	c	d	f
a	a	b	e	f	c	d
b	b	e	a	d	f	c
c	c	d	f	e	a	b
d	d	f	c	b	e	a
f	f	c	d	a	b	e
Inverse	e	b	a	c	d	f

$$\tag{1.19}$$

We now give a general proof of

Theorem 1.1 *If all elements s of a finite group G are multiplied from the left by a fixed element $u \in G$, then* $\left\downarrow \begin{array}{c} s \\ us \end{array}\right.$ *is the permutation $P(u)$ assigned to the element u of the group.*

Proof. The elements in (1.18) are pairwise distinct. For, if $us = ut$ for some two elements $s \neq t$, then multiplying the equation from the left by u^{-1} would yield the result $s = t$, contradicting the hypothesis. $\square$

By virtue of Theorem 1.1, each $u \in G$ is associated with some permutation $P(u)$ of the group elements $e, a, b, \ldots, s, \ldots$ For instance, in the group S_3 Table (1.19) yields

$$P(c) = \left\downarrow \begin{array}{cccccc} e & a & b & c & d & f \\ c & d & f & e & a & b \end{array}\right.$$

Theorem 1.2 *The composite of permutation $P(a)$ followed by permutation $P(b)$ is the permutation $P(ba)$; that is,*

$$P(b)P(a) = P(ba) \tag{1.20}$$

Proof. Let s again be an arbitrary element of the group. Then

$$P(a) = \left\lfloor \begin{matrix} s \\ as \end{matrix} \right. \quad ; \quad P(b) = \left\lfloor \begin{matrix} s \\ bs \end{matrix} \right. = \left\lfloor \begin{matrix} as \\ b(as) \end{matrix} \right.$$

The associative law yields

$$P(b)P(a) = \left\lfloor \begin{matrix} s \\ b(as) \end{matrix} \right. = \left\lfloor \begin{matrix} s \\ (ba)s \end{matrix} \right. = P(ba)$$

$\square$

Definitions. More generally, if to each element a of an arbitrary group we assign a transformation $T(a)$ of some range W such that

$$T(b)T(a) = T(ba) \qquad \forall a, b \in G \tag{1.21}$$

then this is called a **realization** of G.

This again is a special case of a still more general situation: To each element a of a group G we assign an element a' of a group G' such that

$$(ba)' = b'a' \qquad \forall a, b \in G \tag{1.22}$$

Such a map

$$a \in G \to a' \in G' \tag{1.23}$$

is called a **homomorphism**.

If, in addition, the map (1.23) is one-to-one and onto, then it is said to be an **isomorphism**. We also say that G and G' are isomorphic. The two groups are distinct only by name. They have the **same abstract structure** and—in the case of finite groups—the same group table.

Theorem 1.3 *The realization of a finite group G by permutations as given in Theorem 1.1 is an isomorphism.*

Proof. Property (1.21) has already been shown. The map $s \to P(s)$ is one-to-one, because the two permutations $P(a)$ and $P(b)$, where $a \neq b$, are distinct, since they map the identity into a or b, respectively. $\square$

Remark. Theorems 1.1, 1.2, and 1.3 are also valid for non-finite groups in the sense that the word "permutation" is replaced by the word "transformation". The same is true for the proofs of the three theorems. The range of the transformations is the group G.

1.4 Representation of Groups

Let G be an arbitrary group and V be a **complex** n-dimensional vector space. Here $n = 1, 2, \ldots$, but also $n = \infty$ is admitted.

Definition. A **representation** ϑ of G assigns to each element s of the group G a unique linear transformation $D(s)$ of V such that the product $st \in G$ corresponds to the product $D(s)D(t)$ of the corresponding transformations, i.e.,

$$D(s)D(t) = D(st) \qquad \forall s, t \in G \tag{1.24}$$

In particular, (1.24) implies

$$D(e) = E = \text{identity}, \quad D(s^{-1}) = D(s)^{-1} \tag{1.25}$$

since $D(a) = D(ae) = D(a)D(e)$ and $D(s^{-1}s) = D(e) = E = D(s^{-1})D(s)$.

If V has finite dimension, then upon introducing a basis in V the linear transformations can be described by non-singular $n \times n$ matrices. We usually denote them by $D(s)$ again if there is no ambiguity. (1.24) and (1.25) also hold for the corresponding matrices where the identity E is described by the identity matrix.

V is called the **representation space** and n is the **dimension** (or the **degree**) of ϑ. Thus the range consists of all vectors of the representation space.

The representation is said to be **faithful** if the map $s \to D(s)$ is one-to-one; that is, if two distinct elements of the group always correspond to distinct matrices.

Note that faithful is equivalent to saying that G and $\vartheta(G)$ are **isomorphic groups**.

Unless otherwise stated, we always deal with representations ϑ of finite dimensions.

A particularly simple representation is the so-called **unit representation**. This representation sends each element of a group into the number 1, whence it transforms the 1-dimensional representation space identically.

Remarks. In constructing representations, we may occasionally arrive at the result

$$D(t)D(s) = D(st) \qquad \forall s, t \in G$$

rather than (1.24). But often the order will turn out to be irrelevant for applications.

In the following we shall repeatedly make use of the fact that the coordinates of the image of the ith basis vector is the ith column of a matrix.

Example 1.7 (The symmetric group S_3) (see Example 1.6). We consider the three basis vectors b_1, b_2, b_3 of the 3-dimensional complex vector space. To each permutation of the numbers 1, 2, 3 we assign the linear transformation that permutes the basis vectors correspondingly.

Hence the transformation corresponding to the element

$$a = \begin{vmatrix} 1 & 2 & 3 \\ 2 & 3 & 1 \end{vmatrix}$$

maps b_1 into the vector b_2, whose coordinates $0, 1, 0$ appear in the first column of the matrix $D(a)$. Handling the other two basis vectors in the same way leads to

$$a \to D(a) = \begin{bmatrix} 0 & 0 & 1 \\ 1 & 0 & 0 \\ 0 & 1 & 0 \end{bmatrix} \tag{1.26}$$

The analog is obtained for the remaining five elements of S_3. For example,

$$c \to D(c) = \begin{bmatrix} 1 & 0 & 0 \\ 0 & 0 & 1 \\ 0 & 1 & 0 \end{bmatrix} \tag{1.27}$$

The correspondence described above between the 6 group elements and the 6 matrices is a representation of dimension 3.

This is conceptually plausible if one considers the fact that it does not matter whether the three permuting objects are numbers or vectors. But one can also verify that, for instance,

$$D(c)D(a) = \begin{bmatrix} 0 & 0 & 1 \\ 0 & 1 & 0 \\ 1 & 0 & 0 \end{bmatrix} = D(ca) = D(d) \tag{1.28}$$

Definitions. A representation of an arbitrary permutation group on N objects is called the **natural representation** ϑ_{nat} if it is constructed in a way analogous to Example 1.7 by permuting the N basis vectors $b_1, b_2, \ldots, b_N$.

Representing matrices of a natural representation have the property that each row and each column has exactly one entry 1, the remaining entries being 0. They are called **permutation matrices**.

Remark. Natural representations are faithful, i.e., G is isomorphic with the group $\vartheta(G)$ of the representing permutation matrices.

Definition. A **permutation** of N objects is said to be **even or odd** according as the determinant of the corresponding $N \times N$ permutation matrix of ϑ_{nat} equals $+1$ or -1.

Remarks. The determinant det of a permutation matrix can be determined by successively interchanging two column vectors each until the matrix is eventually carried into the identity matrix. Let a be the total number of changes that are needed, then

$$\det = (-1)^a$$

Successively interchanging columns also means that each permutation can be written as a product of transpositions. The even permutations of S_N obviously form a subgroup of order $N!/2$, the so-called **alternating group** A_N.

Example 1.8 In order to obtain another representation ϑ_2 of the symmetric group S_3, we choose the vertices of an equilateral triangle as the objects to be permuted.

Each permutation of the vertices can be realized by a reflection or a rotation of the plane of the paper carrying the triangle into itself.

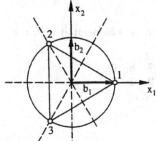

In order to obtain the corresponding matrices, we describe these transformations relative to the basis b_1, b_2 (Figure 1.5). The three transpositions correspond to the reflection about the dashed axes. The element a is associated with the 120° rotation.

Fig. 1.5

In this coordinate system, the following matrices are assigned to the two generators a and c (compare Example 1.6):

$$D(a) = \begin{bmatrix} -\frac{1}{2} & -\frac{\sqrt{3}}{2} \\ \frac{\sqrt{3}}{2} & -\frac{1}{2} \end{bmatrix}, \qquad D(c) = \begin{bmatrix} 1 & 0 \\ 0 & -1 \end{bmatrix} \qquad (1.29)$$

The images of the basis vectors are read from the columns. We leave it to the reader to verify by aid of Example 1.6 that thus an even permutation corresponds to a proper rotation (det $= +1$), while an odd permutation corresponds to an improper rotation (det $= -1$).

Example 1.9 In order to numerically solve a **partial differential equation** of a certain type on the square region D of Figure 1.6, we use a lattice made up of $n = k^2$ points arranged in a square. The desired solution function is approximated by values on these n points (the lattice function).

The figure is carried into itself by rotations through multiples of 90° and by the four reflections about midlines and diagonals of the square. These elements form the so-called **dihedral group** D_4 (see Section 1.9) or the **group of symmetries of the square.**

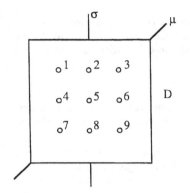

The two axial reflections σ and μ in Figure 1.6 are an example of a generating set for this group. For instance, the rotation through 90° is the composite of these reflections. If we label the lattice points ($n = 9$) in some arbitrary way, e.g., as indicated in Figure 1.6, then each element α of D_4 is associated with a permutation $p(\alpha)$ of the numbers 1, 2, ..., 9.

Fig. 1.6

For instance, the reflection σ about the vertical axis is associated with the permutation

$$p(\sigma) = \begin{array}{ccccccccc} 1 & 2 & 3 & 4 & 5 & 6 & 7 & 8 & 9 \\ 3 & 2 & 1 & 6 & 5 & 4 & 9 & 8 & 7 \end{array}$$

Thus we obtain a representation ϑ of D_4

$$\alpha \rightarrow p(\alpha) \xrightarrow{\vartheta_{\mathrm{nat}}} P(\alpha) \tag{1.30}$$
$$\underset{\vartheta}{\underline{\hspace{3cm}}}$$

which will appear to be of interest when further investigating this problem (Examples 1.12, 2.2, and 2.3). Here ϑ_{nat} is the natural representation by permutation matrices $P(\alpha)$. The basis vectors e_i of the 9-dimensional representation space are those lattice functions that have value 1 at lattice point no. i and value 0 elsewhere.

Remark. The foregoing Examples 1.7, 1.8, and 1.9 all dealt with faithful representations. In such cases the multiplicative group of the representating matrices $\{D(s) \mid s \in G\}$ can replace the respective abstract group without loss of information.

1.5   Equivalence of Representations

We discuss this fundamental concept of the theory of representations by aid of the well-established example of the symmetric group S_3. The representation ϑ_2 in (1.29) resulted from describing the rotations and reflections in the coordinate system of Fig. 1.5. Now we describe **the same** rotations and reflections in the skew coordinate system shown in Fig. 1.7.

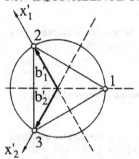

$$a \xrightarrow{\vartheta'_2} D'(a) = \begin{bmatrix} 0 & -1 \\ 1 & -1 \end{bmatrix}$$

$$c \xrightarrow{\vartheta'_2} D'(c) = \begin{bmatrix} 0 & 1 \\ 1 & 0 \end{bmatrix}$$

The coordinates of the images of the basis vectors b'_1 and b'_2 are seen in the columns.

Fig. 1.7

Unlike the matrices of (1.29), those of ϑ'_2 are not all orthogonal. However, the matrix entries are all integers, which plays a role in crystallography (see Chapter 8). The two representations ϑ_2 and ϑ'_2 describe **the same** linear transformations (in different coordinate systems). Therefore, they are said to be **equivalent**.

More generally, we can formulate the following

Definition. Two n-dimensional representations $\vartheta : s \rightarrow D(s)$ and $\vartheta' : s \rightarrow D'(s)$ of a group G are said to be **equivalent** provided there exists a fixed non-singular $n \times n$ matrix T such that

$$D'(s) = T^{-1}D(s)T \qquad \forall s \in G \qquad (1.31)$$

To further illustrate this, we regard the matrices $D(s)$ as linear transformations of the vector space V and the matrices $D'(s)$ to be linear transformations of the vector space V' relative to suitable coordinate systems in each of the two representation spaces.

Hence

$$\begin{aligned} y &= D(s)x & x, y \in V & \qquad (1.32) \\ y' &= D'(s)x' & x', y' \in V' & \qquad (1.33) \end{aligned}$$

By introducing a new coordinate system in V by means of

$$x = Tu, \qquad y = Tv \qquad (1.34)$$

where T is the matrix of the coordinate transformation, and substituting (1.34) into (1.32), we obtain

$$v = (T^{-1}D(s)T)u \qquad (1.35)$$

Comparison of (1.33) with (1.35), together with the definition, yields the following **statement**:

ϑ and ϑ' are equivalent if and only if there exists a basis in V such that the representing matrices of ϑ relative to this new coordinate system are identical with those of ϑ'; that is, if

$$v = D'(s)u, \qquad y' = D'(s)x' \qquad \forall s \in G \qquad (1.36)$$

In other words, (1.36) says that the appropriately chosen coordinates of the vectors in V transform in **exactly the same way** *as do those in V'.*

Remarks.

1. In the following we shall often denote equivalent representations the same ($\vartheta = \vartheta'$).

2. If ϑ and ϑ' are two equivalent 1-dimensional representations, then the representing numbers satisfy $D(s) = D'(s)$ $\forall s \in G$.

1.6   Reducibility of Representations

Definitions. A representation ϑ of an arbitrary group G with the finite- or infinite-dimensional representation space V is said to be **reducible** provided there exists a proper linear subspace V_1 of V with the property that **every** transformation of ϑ maps **every** vector of V_1 into V_1. Proper means that $V_1 \neq 0$ and $V_1 \neq V$.

A non-reducible representation is said to be **irreducible**. We also use the terms **reducible** or **irreducible representation space**.

We illustrate this through the natural representation ϑ_{nat} of the symmetric group S_3 described in Example 1.7. The representing matrices all permute the basis vectors b_1, b_2, b_3; whence the vector sum $v = b_1 + b_2 + b_3$ spans a one-dimensional invariant subspace V_1.

Let us go back to the general finite-dimensional case. If the coordinate system in V is chosen such that the first m basis vectors span a proper invariant subspace V_1, then the matrices of the representation $\vartheta : s \to F(s)$ have the following reduced form:

$$F(s) = \left. \begin{array}{|c|c|} \hline Q_1(s) & R(s) \\ \hline 0 & Q_2(s) \\ \hline \end{array} \right\} n \qquad (1.37)$$

Here $Q_1(s)$ and $Q_2(s)$ are square matrices with m and $m - n$ rows, respectively. $R(s)$ is a rectangular matrix, and 0 stands for the null matrix.

The $Q_1(s)$ describe a representation ϑ_1, which denotes the restriction of ϑ to the invariant subspace V_1. The representation ϑ_1 itself may be reducible or irreducible.

In particular, if in the example above v is the first basis vector, then $m = 1$ and $Q_1(s) = 1$ $\forall s$. Hence ϑ_1 is the unit representation.

Remarks.

1. The explicit determination of invariant subspaces of a representation is an important tool employed in applied representation theory. This problem can also be posed as follows: Find a non-singular matrix T such that the matrices

$$F(s) = T^{-1}D(s)T \qquad \forall s \in G \qquad (1.38)$$

have the reduced form (1.37).

2. Each 1-dimensional representation is irreducible.

Example 1.10 Let us investigate whether or not the 2-dimensional representation ϑ_2 of the symmetric group S_3 in (1.29), Example 1.8, is reducible.

Suppose it were reducible. Then there would exist a 1-dimensional invariant subspace of ϑ_2, i.e., a vector that is an eigenvector for all representing matrices (possibly with different eigenvalues). The matrix

$$D(a) = \begin{bmatrix} -1/2 & -\sqrt{3}/2 \\ \sqrt{3}/2 & -1/2 \end{bmatrix}$$

has two complex eigenvectors $\begin{bmatrix} 1 \\ \pm i \end{bmatrix}$, where $i = \sqrt{-1}$.

But $D(c)$ satisfies

$$\begin{bmatrix} 1 & 0 \\ 0 & -1 \end{bmatrix} \begin{bmatrix} 1 \\ \pm i \end{bmatrix} = \begin{bmatrix} 1 \\ \mp i \end{bmatrix}$$

i.e., $D(c)$ interchanges the two eigenvectors. Hence there is no proper subspace that is simultaneously invariant for all matrices. Therefore, ϑ_2 is irreducible.

Consider the representation obtained by restricting ϑ_2 to the cyclic subgroup e, a, and a^2 of S_3. This, however, is reducible: As $D(a^2) = D(a)^2$, the two eigenvectors of $D(a)$, being eigenvectors of $D(a)^2$ likewise, each span a one-dimensional invariant subspace of this representation.

This gives rise to the following

Remark. If an irreducible representation of a group is restricted to a subgroup, then it need not necessarily remain irreducible. However, it is obvious that irreducibility of a representation with respect to a subgroup implies irreducibility of the representation of the entire group.

1.7 Complete Reducibility

If a representation ϑ is reducible and if there is a coordinate transformation T such that for the reduced form (1.37)

$$R(s) = \text{null matrix} \qquad \forall s \in G \qquad (1.39)$$

then we say ϑ **reduces** to the representations

$$\vartheta_1 : s \rightarrow Q_1(s) \qquad \vartheta_2 : s \rightarrow Q_2(s) \qquad (1.40)$$

because then the last $(n - m)$ basis vectors also span an invariant subspace V_2. The vector space V is the direct sum of the two subspaces and is denoted as follows:

$$V = V_1 \oplus V_2$$

In order to express the reduction of the representation ϑ into the two representations ϑ_1 and ϑ_2, we use the following notation:

$$\vartheta = \vartheta_1 \oplus \vartheta_2 \qquad (1.41)$$

Let us return to the introductory example at the beginning of Section 1.6. There we stated that the vector sum $v = b_1 + b_2 + b_3$ remains fixed. In the following, we shall make our considerations on the basis of a cartesian coordinate system. Now ϑ_{nat} is a representation by orthogonal matrices. Hence every transformation sends orthogonal vectors of the 3-dimensional real vector space (which is embedded in the representation space) into other such vectors. Consequently, the plane orthogonal to v

$$\varepsilon : x_1 + x_2 + x_3 = 0$$

is a 2-dimensional invariant subspace (Fig. 1.8). Here the coordinates x_i relate to the orthonormal basis b_1, b_2, b_3. We choose a new coordinate system adapted to the reduction by taking the column vectors of the matrix

$$T = \begin{bmatrix} \frac{1}{\sqrt{3}} & \frac{-2}{\sqrt{6}} & 0 \\ \frac{1}{\sqrt{3}} & \frac{1}{\sqrt{6}} & \frac{-1}{\sqrt{2}} \\ \frac{1}{\sqrt{3}} & \frac{1}{\sqrt{6}} & \frac{1}{\sqrt{2}} \end{bmatrix} \qquad (1.42)$$

as the new basis vectors g_1, g_2, g_3.

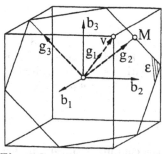

ε = invariant plane;
M = midpoint of a side;
g_1 = unit vector in direction of v, the two remaining columns g_2 and g_3 are orthogonal unit vectors in the plane ε. Thus T is an orthogonal matrix.

Fig. 1.8

Therefore, the representation equivalent to ϑ_{nat} given by $F(s) = T^{-1}D(s)T$ also consists of such orthogonal matrices. Using (1.26), (1.27), (1.42) and observing that the inverse T^{-1} equals the transpose of T, calculations specific to the generating set yield

$$F(a) = \begin{array}{|c|c|c|} \hline 1 & 0 & 0 \\ \hline 0 & -\frac{1}{2} & -\frac{\sqrt{3}}{2} \\ \hline 0 & \frac{\sqrt{3}}{2} & -\frac{1}{2} \\ \hline \end{array} \qquad F(c) = \begin{array}{|c|c|c|} \hline 1 & 0 & 0 \\ \hline 0 & 1 & 0 \\ \hline 0 & 0 & -1 \\ \hline \end{array} \tag{1.43}$$

as the new representing matrices. The special form of the two matrices confirms that the representation reduces into

$$\vartheta_{\text{nat}} = \vartheta_1 \oplus \vartheta_2 \tag{1.44}$$

where ϑ_1 is the unit representation and ϑ_2 the representation (1.29) from Example 1.8.

Now let us continue the general theory: Proceeding in this way, the method of reduction may be iterated such that in a suitable coordinate system the matrices of a representation ϑ decompose further:

$$s \xrightarrow{\vartheta} \qquad \vartheta = \vartheta_1 \oplus \vartheta_2 \oplus \ldots \oplus \vartheta_J \tag{1.45}$$

The matrices of the representation take the form of square blocks of possibly different sizes arranged along the main diagonal. The rest consists of zeros. In other words, the representing matrices $D(s)$ form a direct sum of matrices D_j of dimension n_j $(1 \leq j \leq J)$.

Definition. A representation that is irreducible or that reduces into nothing but irreducible representations is said to be **completely reducible**.

Example 1.11 We consider the representation ϑ_{nat} of S_3 in the form (1.43), (1.44). ϑ_1 is a 1-dimensional representation, hence irreducible, and ϑ_2 was found to be irreducible in Example 1.10. Therefore, ϑ_{nat} of S_3 is completely reducible.

Remark. Not every representation is completely reducible, as is exemplified by the following: Let G be the multiplicative group of the nonzero real numbers x. The mapping

$$x \to \begin{bmatrix} 1 & \log|x| \\ 0 & 1 \end{bmatrix} \tag{1.46}$$

is a representation, as can be readily verified. It is reducible, since the first basis vector spans a 1-dimensional invariant subspace. If it were completely reducible, then there would exist a coordinate transformation such that all matrices would be diagonal. But this is impossible, because the matrix $\begin{bmatrix} 1 & 1 \\ 0 & 1 \end{bmatrix}$ is not diagonalizable for lack of a basis of eigenvectors that correspond to the single eigenvalue 1.

Example 1.12 In order to extend our considerations to the applications, we wish to find **all irreducible subspaces** of the completely reducible[3] 9-dimensional representation ϑ (see (1.30)) of Example 1.9.

In order to find these subspaces, it will prove to be extremely useful to label the vectors in the lattice according to Fig. 1.6. Let the ith component be written at the position of lattice point no. i. We have tabulated the irreducible subspaces below.

subspace	V_1^1	V_1^2	V_1^3	V_2	V_3	V_4^1	V_4^2
basis	x_1	x_2	x_3	y	z	u_1, v_1	u_2, v_2

$$(1.47)$$

The representation space decomposes into 5 subspaces of dimension 1 and 2 subspaces of dimension 2. The notations in (1.47) will later appear to be useful.

Table of basis vectors:

$$
x_1 = \begin{matrix} . & . & . \\ . & 1 & . \\ . & . & . \end{matrix} \quad
y = \begin{matrix} . & 1 & . \\ -1 & . & -1 \\ . & 1 & . \end{matrix} \quad
u_1 = \begin{matrix} . & . & . \\ -1 & . & 1 \\ . & . & . \end{matrix} \quad
v_1 = \begin{matrix} . & 1 & . \\ . & . & . \\ . & -1 & . \end{matrix}
$$

$$
x_2 = \begin{matrix} . & 1 & . \\ 1 & . & 1 \\ . & 1 & . \end{matrix} \quad
z = \begin{matrix} -1 & . & 1 \\ . & . & . \\ 1 & . & -1 \end{matrix} \quad
u_2 = \begin{matrix} -1 & . & 1 \\ . & . & . \\ -1 & . & 1 \end{matrix} \quad
v_2 = \begin{matrix} 1 & . & 1 \\ . & . & . \\ -1 & . & -1 \end{matrix}
$$

$$
x_3 = \begin{matrix} 1 & . & 1 \\ . & . & . \\ 1 & . & 1 \end{matrix}
$$

$$(1.48)$$

Dots stand for zero entries.

Owing to the lattice notation, it is easily seen that these subspaces are invariant under ϑ, namely by applying the two generating permutation matrices $P(\sigma), P(\mu)$ to the various basis vectors (σ and μ are reflections about the vertical line or the 45°-line, respectively). Hence, by "reflecting" u_1 about the 45°-line we obtain, for example,

$$
P(\mu)u_1 = P(\mu) \begin{matrix} . & . & . \\ -1 & . & 1 \\ . & . & . \end{matrix} = \begin{matrix} . & 1 & . \\ . & . & . \\ . & -1 & . \end{matrix} = v_1
$$

---

[3] The proof of complete reducibility will be given in Theorem 1.5.

If we work the same way with, say, the basis vectors of the 2-dimensional invariant subspaces V_4^1 and V_4^2, then we obtain the following transformations:

$$P(\sigma)u_i = -u_i \qquad P(\sigma)v_i = v_i$$
$$P(\mu)u_i = v_i \qquad P(\mu)v_i = u_i \qquad i = 1, 2 \qquad (1.49)$$

In the completely reduced form

$P(\sigma)$ contains exactly two blocks of the form $\begin{bmatrix} -1 & 0 \\ 0 & 1 \end{bmatrix}$

and $P(\mu)$ contains exactly two blocks of the form $\begin{bmatrix} 0 & 1 \\ 1 & 0 \end{bmatrix}$.

The reader will be able to verify that by (1.49) the subspaces V_4^1 and V_4^2 are transformed irreducibly.

Remark. The vectors in Table (1.48) have certain "**symmetries**". This concept is closely linked to the theory of representations. An algorithmic computation of the basis vectors (1.48) will follow in Section 5.2.

1.8 Basic Conclusions

Our introductory considerations suggest that two main problems arise:

  I Determination of all invariant irreducible subspaces of a given completely reducible representation.

 II Construction of all of the irreducible inequivalent representations of a given group. Thereby we obtain a complete survey of all of the completely reducible representations of G, since they are equivalent to some direct sum of irreducible constituents.

In view of the applications, **I** is of central importance as will be shown in the next Chapter 2. Compared with that, **II** is rather a mathematical technicality.

To meet the purpose of this book we shall give more attention to the aspects connected with **I** than to the construction of irreducible representations, which we shall often merely record without proof. Nevertheless, it is desirable to have a criterion to decide whether or not a list of irreducible representations of a group is complete. In the case of finite groups, this is done by the following

Theorem 1.4 *A finite group G of order $|G|$ has only a finite number of inequivalent irreducible representations ϑ_j. If K is their total number and n_j the dimension of ϑ_j, then*

$$\sum_{j=1}^{K} n_j^2 = |G| \qquad (1.50)$$

The proof will follow in Theorem 5.13.

Example 1.13 (A survey of the irreducible representations of the group S_3) $(g = 3! = 6)$. So far, we have introduced the unit representation and the 2-dimensional representation ϑ_2 in Example 1.8. Equation (1.50) says that there exists precisely one more irreducible representation. Because $1^2 + 2^2 = 5$, this representation must be of dimension 1. It is the so-called **alternating representation** ϑ_{alt}. It assigns to each even (odd) permutation the number $+1$ (-1). Hence for the two elements a and c of the generating set

$$\vartheta_{alt} : a \to 1, \qquad c \to -1 \tag{1.51}$$

1.9 Representations of Special Finite Groups

1.9.1   The Cyclic Group C_g

This group is generated by an element a and consists of the powers a, a^2, ..., $a^g = e$. It can be realized by rotations of the plane (or the space) around a fixed point (or a fixed axis) through the rotation angles

$$k \cdot \frac{2\pi}{g} \qquad (k = 0, 1, \ldots, g - 1).$$

The representation is completely determined if the representing matrix of the generating element is known. Let now

$$\omega_k = e^{\frac{2\pi i}{g} k} \qquad k = 0, 1, \ldots, g - 1$$

be the gth roots of unity. Then the map

$$a \to \omega_k \qquad k = 0, 1, \ldots, g - 1 \tag{1.52}$$

furnishes a 1-dimensional representation for each fixed k. Hence we obtain a total of g inequivalent (why?) irreducible 1-dimensional representations. Their contribution to the left side of (1.50) is $\sum_{j=1}^{g} 1^2 = g$. Therefore, (1.52) describes **all** irreducible representations.

Because of complete reducibility[4], every arbitrary representation of the group C_g is equivalent to some representation consisting entirely of diagonal matrices. Their entries are the gth roots of unity. Hence completely reducing a representation of C_g is equivalent to diagonalizing the representing matrix of the generating element. Representation theory guarantees the existence of such a transformation.

---

[4] see Theorem 1.6.

1.9.2   The Dihedral Group D_h

This is the group of h rotations and h reflections of the plane that carry a regular h-sided polygon onto itself. Its order is thus $2h$. The rotation angles are integral multiples of

$$\varphi = \frac{2\pi}{h} \qquad (1.53)$$

The rotations form a subgroup of D_h which is isomorphic with the cyclic group C_h. In particular, the group D_2 consists of 2 reflections, the rotation through π, and the identity. It is abelian, but not cyclic. This group is also called the **Klein**[5] **four group**.

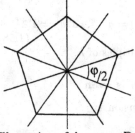

A possible set of generators of the group D_h are the two reflections about axes forming an angle of $\varphi/2$ (Fig. 1.9).

Illustration of the group D_5

Fig. 1.9

In the following, let d be the rotation through φ and s be an arbitrary reflection about some fixed axis. Then the rotations of the group are the elements

$$d^k \qquad (k = 0, 1, \ldots, h) \qquad (1.54)$$

and the reflections about the various axes are the elements

$$sd^k \qquad (k = 0, 1, \ldots, h) \qquad (1.55)$$

Thus the two elements s and d also form a set of generators. Along with the two relations

$$s^2 = e, \qquad d^h = e \qquad (1.56)$$

we also obtain from Fig. 1.10 that

$$d^{-k}s = sd^k \qquad (1.57)$$

---

[5]Klein, Felix (1849-1925). Created the so-called Erlangen Program in geometry. He was not only involved in non-Euclidean geometry but was also strongly concerned with the teaching of applied mathematics.

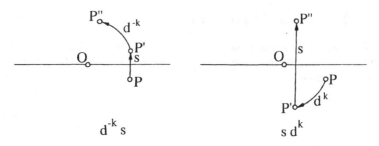

Fig. 1.10

This is true because both $d^{-k}s$ and sd^k carry a general point P into P''. Equations (1.56) and (1.57) imply

$$d^k(sd^\ell) = sd^{\ell-k}, \quad (sd^k)d^\ell = sd^{k+\ell}, \quad (sd^k)(sd^\ell) = d^{\ell-k}, \quad d^k d^\ell = d^{k+\ell}$$

$$\text{(1.58)}$$

The exponents are understood up to integral multiples of h.

Irreducible Representations of D_h, where h is even

There are 4 inequivalent representations of dimension 1:

	d^k	sd^k
ϑ_1^1	1	1
ϑ_2^1	1	-1
ϑ_3^1	$(-1)^k$	$(-1)^k$
ϑ_4^1	$(-1)^k$	$(-1)^{k+1}$

$$\text{(1.59)}$$

Moreover, there are $\frac{h}{2} - 1$ representations of dimension 2 that are constructed via the hth root of unity $\omega = e^{i\varphi}$; namely,

$$d^k \xrightarrow{\vartheta_j^2} \begin{bmatrix} \omega^{jk} & 0 \\ 0 & \omega^{-jk} \end{bmatrix} \quad sd^k \xrightarrow{\vartheta_j^2} \begin{bmatrix} 0 & \omega^{jk} \\ \omega^{-jk} & 0 \end{bmatrix} \quad j = 1, 2, \ldots, \frac{h}{2} - 1$$

$$\text{(1.60)}$$

Straightforward computations show that both (1.59) and (1.60) are indeed representations if (1.58) is used.

Irreducibility: This is clear for the 1-dimensional representations. If a 2-dimensional representation were reducible, then there would exist a 1-dimensional subspace simultaneously invariant for all matrices. But by (1.60)

$$\left\{ \begin{array}{l} \begin{bmatrix} 1 \\ 0 \end{bmatrix} \text{ and } \begin{bmatrix} 0 \\ 1 \end{bmatrix} \quad \text{are eigenvectors of the representing matrix of } d \text{ corresponding to distinct eigenvalues.} \\ \\ \begin{bmatrix} 1 \\ 1 \end{bmatrix} \text{ and } \begin{bmatrix} 1 \\ -1 \end{bmatrix} \quad \text{are eigenvectors of the representing matrix of } s \text{ corresponding to distinct eigenvalues.} \end{array} \right.$$

Hence there is no common eigenspace.

Inequivalence: Again, this is clear for 1-dimensional representations. If ϑ_j^2 were equivalent to ϑ_ℓ^2 for some $j \neq \ell$, then for, say, $k = 1$ the two matrices on the right side of (1.60) associated with the two representations would have the same eigenvalues. But this is only possible if $j = \ell$, contradicting the hypothesis.

Completeness: Our list contributes

$$4 \cdot 1^2 + (\frac{h}{2} - 1) \cdot 2^2 = 2h$$

to the sum in (1.50). Hence our catalog is complete.

Irreducible Representations of D_h, where h is odd

There is a total of 2 representations of dimension 1:

	d^k	sd^k
ϑ_1^1	1	1
ϑ_2^1	1	-1

(1.61)

Moreover, there are $\frac{h-1}{2}$ representations of dimension 2 that are constructed according to (1.60) with $j = 1, 2, \ldots, \frac{h-1}{2}$. The subsequent investigations are conducted in the same way.

1.10   Kronecker Products

Given two representations ϑ_1 and ϑ_2 of a group G, let V_1 and V_2 be the representation spaces and m and n be their dimensions. Furthermore, let $e_1^{(1)}, e_2^{(1)}, \ldots, e_m^{(1)}$ be the basis vectors of V_1 and correspondingly, $e_1^{(2)}, e_2^{(2)}, \ldots, e_n^{(2)}$ be the basis vectors of V_2. The images of the basis vectors under ϑ_1 and ϑ_2 are

$$e_i^{(1)} \xrightarrow{\vartheta_1} \sum_{\lambda=1}^{m} d_{\lambda i}^{(1)}(s) e_\lambda^{(1)}, \qquad e_k^{(2)} \xrightarrow{\vartheta_2} \sum_{\mu=1}^{n} d_{\mu k}^{(2)}(s) e_\mu^{(2)}; \qquad s \in G \qquad (1.62)$$

Hence the representing matrices are

$$D_1(s) = (d_{\lambda\nu}^{(1)}(s)), \qquad D_2(s) = (d_{\mu\kappa}^{(2)}(s))$$

We now assume that the $m \cdot n$ formal products $e_i^{(1)} \cdot e_k^{(2)}$ are basis vectors of an $m \cdot n$-dimensional vector space V and define in V a representation ϑ by

$$e_i^{(1)} e_k^{(2)} \xrightarrow{\vartheta} \sum_{(\lambda,\mu)} d_{\lambda i}^{(1)}(s) d_{\mu k}^{(2)}(s) (e_\lambda^{(1)} \cdot e_\mu^{(2)}) \qquad (1.63)$$

The right side is produced by expanding the product of the right sides of
(1.62). In other words, an arbitrary product $v \cdot w \in V$, where $v \in V_1$, $w \in V_2$,
is linear in both v and w. The transformations of this representation are
denoted by $D(s) = D_1(s) \otimes D_2(s)$, and the representation itself is called
the **Kronecker product**[6]

$$\vartheta = \vartheta_1 \otimes \vartheta_2 \tag{1.64}$$

of ϑ_1 and ϑ_2.

The transformation (1.63) can be written explicitly in the form of a
matrix if we define an order for the basis vectors $e_i^{(1)} \cdot e_j^{(2)}$. If, for instance,
we take the following ordering for the pairs of subscripts (i, j)

$$(1,1); (1,2); \ldots; (1,n); (2,1); (2,2); \ldots; (2,n); \ldots; (m,n)$$

then the representing matrices $D(s)$, written in block notation, with the
argument s omitted for the sake of simplicity, are

$$D(s) = D_1 \otimes D_2 = \begin{bmatrix} d_{11}^{(1)} D_2 & d_{12}^{(1)} D_2 & \ldots & d_{1m}^{(1)} D_2 \\ d_{21}^{(1)} D_2 & d_{22}^{(1)} D_2 & \ldots & d_{2m}^{(1)} D_2 \\ \vdots & \vdots & & \vdots \\ d_{m1}^{(1)} D_2 & d_{m2}^{(1)} D_2 & \ldots & d_{mm}^{(1)} D_2 \end{bmatrix} \tag{1.65}$$

Thus blockwise multiplication of the two representing matrices $D(s)$
and $D(t)$ of the form (1.65) and the representation property of ϑ_1 and ϑ_2

$$\begin{aligned} D(s) \cdot D(t) &= [D_1(s) \otimes D_2(s)] \cdot [D_1(t) \otimes D_2(t)] \\ &= [D_1(s) \cdot D_1(t)] \otimes [D_2(s) \cdot D_2(t)] \\ &= [D_1(st) \otimes D_2(st)] = D(st) \end{aligned} \tag{1.66}$$

imply the representation property for ϑ.

In a way similar to (1.63), we can construct Kronecker products $\vartheta = \vartheta_1 \otimes \vartheta_2 \otimes \ldots \otimes \vartheta_J$ of several representations $\vartheta_1, \vartheta_2, \ldots, \vartheta_J$ of G. If n_j is
the dimension of ϑ_j $(1 \leq j \leq J)$, then ϑ is a representation of dimension
$(n_1 \cdot n_2 \cdot \ldots \cdot n_J)$. Correspondingly, the basis vectors of the representation
space V of ϑ have the form

$$e_i^{(1)} \cdot e_j^{(2)} \cdot \ldots \cdot e_r^{(J)}$$

An important special case. Let the representation space V_1 of ϑ_1 consist
of functions $\varphi(x_1, x_2, \ldots, x_p)$ that are spanned by the basis functions φ_1,
φ_2, ..., φ_m and let V_2 of ϑ_2 consist of functions $\psi(y_1, y_2, \ldots, y_q)$ that are
spanned by the basis functions ψ_1, ψ_2, ..., ψ_n.

Then the products $\varphi_i \cdot \psi_j$ (usual multiplication of functions) form a basis
of the $m \cdot n$-dimensional vector space V consisting of functions of the $p + q$
variables $x_1, x_2, \ldots, x_p, y_1, y_2, \ldots, y_q$. Using these, we can then construct a
Kronecker product $\vartheta = \vartheta_1 \otimes \vartheta_2$.

---

[6]Kronecker, Leopold (1823-1891). He gave lectures at the University of Berlin from
1861 until 1883 and was a professor there from 1883 on.

1.11 Unitary Representations

At the beginning of Section 1.7 (compare Fig. 1.8) we made use of the following fact: If a representation ϑ consists of orthogonal transformations and if an invariant subspace of ϑ exists, then the subspace orthogonal to it is also invariant. In order to cover **complex** vector spaces likewise, we generalize the notion of real orthogonality. First let us recall some of the notions and statements from linear algebra.

The **inner product** in a complex finite- or infinite-dimensional vector space V is an operation that assigns to each ordered pair x, y of vectors of V a unique complex number $\langle x, y \rangle$ such that the following properties are satisfied (the bar stands for the complex conjugate number):

$$\langle x + x', y \rangle = \langle x, y \rangle + \langle x', y \rangle \tag{1.67}$$

$$\langle \alpha x, y \rangle = \alpha \langle x, y \rangle \qquad \alpha \in \mathbf{C} \tag{1.68}$$

$$\langle x, y \rangle = \overline{\langle y, x \rangle} \tag{1.69}$$

$$x \neq 0 \Rightarrow \langle x, x \rangle > 0 \tag{1.70}$$

The last equation makes sense, because (1.69) implies that $\langle x, x \rangle$ is real.

The non-negative real number

$$|x| = \sqrt{\langle x, x \rangle} \tag{1.71}$$

is called the **length** or the **norm** of the vector x.

The metric induced by the inner product is called **the hermitian metric**[7]. It follows immediately from the properties of the inner product that

$$\langle x, \alpha y + \beta y' \rangle = \overline{\alpha} \langle x, y \rangle + \overline{\beta} \langle x, y' \rangle \tag{1.72}$$

Furthermore, the zero vector has norm 0.

Two vectors x, y with the property

$$\langle x, y \rangle = 0 \tag{1.73}$$

are said to be **orthogonal** to each other (with respect to the chosen inner product $\langle x, y \rangle$).

Definition. A representation $\vartheta : s \to D(s)$ of a group G in V is said to be **unitary** with respect to a given inner product $\langle x, y \rangle$ if

$$\langle D(s)x, D(s)y \rangle = \langle x, y \rangle \qquad \forall x, y \in V, \ \forall s \in G \tag{1.74}$$

Interpretation. *The linear transformations $D(s)$ of a unitary representation in V all preserve lengths and carry orthogonal vectors into orthogonal vectors.*

---

[7]Hermite, Charles (1822 -1901). 1870 professor at Sorbonne. Solved the general quintic equation by elliptic functions. Proved that the number e was transcendental.

Theorem 1.5 *Every unitary representation ϑ is completely reducible.*

Proof. Let V_1 be a proper subspace of the representation space V invariant under $\vartheta : s \to D(s)$. Then the orthogonal subspace

$$V_2 = \{y | \langle y, x \rangle = 0 \ \ \forall x \in V_1\} \tag{1.75}$$

of V_1 is also invariant under ϑ, since for linear transformations $D(s)$

$$\langle D(s)y, x \rangle = \langle y, D(s^{-1})x \rangle = 0 \tag{1.76}$$

because ϑ is unitary and $D(s^{-1})x \in V_1$. Therefore $D(s)y \in V_2$. This process can be iterated so long as there is a proper invariant subspace in V_1 or V_2. Finally, after a finite number of steps, we obtain a decomposition of ϑ into nothing but irreducible constituents. $\qquad\qquad\square$

Another consequence is

Theorem 1.6 *Every representation ϑ of a finite group G is unitary with respect to a suitably chosen inner product and hence completely reducible.*

Proof. Let $D(s)$ be the representing matrices of ϑ relative to some given basis of the representation space. Furthermore, let $|G|$ be the order of the group G. The inner product[8] (T stands for transpose)

$$\langle x, y \rangle = \frac{1}{|G|} \sum_{s \in G} [D(s)x]^T \cdot \overline{[D(s)y]} \tag{1.77}$$

is invariant under the representation. The reason is that for an arbitrary fixed representing matrix $D(a)$

$$
\begin{aligned}
\langle D(a)x, D(a)y \rangle &= \frac{1}{|G|} \sum_{s \in G} [D(s)D(a)x]^T \cdot \overline{[D(s)D(a)y]} \\
&= \frac{1}{|G|} \sum_{s \in G} [D(sa)x]^T \cdot \overline{[D(sa)y]} = \langle x, y \rangle \quad (1.78)
\end{aligned}
$$

because if s varies through the group G, then so does sa (see Theorem 1.1). Finally, complete reducibility follows from Theorem 1.5. $\qquad\qquad\square$

Remark. The fundamental idea of summing over the group originates from Hurwitz[9].

---

[8] We leave it to the reader to verify that the right side of (1.77) conforms to the conditions (1.67) through (1.70) (the factor $1/|G|$ is inessential).

[9] Hurwitz, Adolf (1859-1919). Studies in Munich and Leipzig; associate professor in Königsberg from 1884, full professor at ETH, Zürich from 1892 on.

1.11.1 Unitary Representations and Unitary Matrices

Definition. A **basis** $e_1, e_2, \ldots$ of a vector space is said to be **orthonormal** if the basis vectors are mutually orthogonal and each of length 1, that is, if

$$\langle e_i, e_j \rangle = \delta_{ij} = \begin{cases} 1 & \text{if } i = j \\ 0 & \text{otherwise} \end{cases} \tag{1.79}$$

Now we consider a finite-dimensional representation (with representation space V) described by matrices relative to an orthonormal basis $e_1, e_2, \ldots, e_n$.

Then for any two vectors x and y with the respective components x_μ and y_ν the given inner product $\langle x, y \rangle$ is of the simple form

$$\langle x, y \rangle = \langle \sum_{\mu=1}^{n} x_\mu e_\mu, \sum_{\nu=1}^{n} y_\nu e_\nu \rangle = \sum_{\mu,\nu=1}^{n} x_\mu \overline{y_\nu} \langle e_\mu, e_\nu \rangle = \sum_{\mu=1}^{n} x_\mu \overline{y_\mu} = x^T \cdot \overline{y} \tag{1.80}$$

Here we used (1.67), (1.68), (1.72), and (1.79).

In the language of matrices, a unitary representation according to (1.74) means that the n columns of any representing matrix $D = (d_{ij})$ describe an orthonormal basis of V. This is because the kth column is the image of the kth basis vector e_k. Written as an equation in the matrix entries, this means that by (1.80)

$$\sum_{\rho=1}^{n} d_{\rho k} \overline{d_{\rho \ell}} = \delta_{k\ell} \qquad k, \ell = 1, 2, \ldots, n \tag{1.81}$$

These n^2 equations are equivalent to the matrix equation (where again T stands for transpose)

$$D^T \overline{D} = E = \text{identity matrix} \tag{1.82}$$

Definition. A complex non-singular square **matrix** D with property (1.82) is said to be **unitary**.

By $D^{-1}D = DD^{-1} = E$, the following equivalences hold:

$$D \text{ is unitary} \Leftrightarrow D^{-1} = \overline{D^T} \Leftrightarrow \overline{D} D^T = E \tag{1.83}$$

The last equation $\overline{D} D^T = E$ says that along with the columns, the rows of a unitary matrix also form an orthonormal basis.

We summarize the foregoing as follows:

Every finite-dimensional unitary representation is described by unitary matrices relative to some orthonormal basis. Conversely, if a representation is described by unitary matrices relative to an orthonormal basis $e_1, e_2, \ldots, e_n$, then this representation is unitary with respect to the inner product

$$\langle u, v \rangle = \sum_{i=1}^{n} u_i \overline{v_i} = u^T \overline{v} \tag{1.84}$$

(u$_i$ and v$_i$ are the components of the vectors u and v of the left hand side of Equation (1.84) relative to the e-basis). The reason is that any given basis may be considered orthonormal on introducing an inner product by (1.84).

Consequently, we mean by a unitary representation one that is described by unitary matrices, tacitly assuming that while interpreting matrices as linear transformations of the representation space we consider the underlying basis to be orthonormal.

PROBLEMS

Problem 1.1 Write the group table for the Klein four group $(V \cong D_2)$ denoting its elements by e, a, b, c. Is this group abelian? The group table furnishes a realization by permutations and hence a natural 4-dimensional representation. Find the irreducible invariant subspaces and the corresponding representations.

Problem 1.2 (The complete representation theory of the symmetric group S_4) The unit representation ϑ_1^1 and the alternating representation ϑ_2^1 of dimension 1 are already at hand.

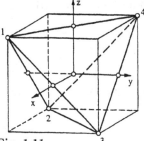

Fig. 1.11

In order to find further representations, regard the four objects of the symmetric group S_4 as vertices of a regular tetrahedron shown in Fig. 1.11 (the coordinate axes join the midpoints of two edges each). In this way we obtain a 3-dimensional representation ϑ_1^3 in the coordinate system x, y, z, namely, all proper and improper isometries taking the regular tetrahedron onto itself.

Combining each of the improper isometries of ϑ_1^3 that take the regular tetrahedron onto itself with a reflection about the origin produces another 3-dimensional representation ϑ_2^3 (all proper isometries taking the cube onto itself).

According to the figure, each operation of S_4 permutes the three (non-oriented) coordinate axes, whence it can be regarded as an operation of S_3. The well known representation theory of S_3 thereby produces a representation ϑ_1^2 of dimension 2 for the group S_4.

a) Show that the representations found above all are irreducible.

b) Show that ϑ_1^3 and ϑ_2^3 are inequivalent representations.

c) Show that, up to equivalence, the five representations make up the complete list of irreducible representations.

d) Find the representing matrices of all of the irreducible representations for the two elements

$$\begin{vmatrix} 1 & 2 & 3 & 4 \\ 2 & 1 & 3 & 4 \end{vmatrix} \text{ and } \begin{vmatrix} 1 & 2 & 3 & 4 \\ 2 & 3 & 4 & 1 \end{vmatrix}$$

Problem 1.3 Exhibit the equivalence transformation T between the representation ϑ_2 of Example 1.8 and the representation ϑ_2' given at the beginning of Section 1.5.

Problem 1.4 Show that the Kronecker product of two unitary representations is again unitary.

Problem 1.5 Determine all irreducible invariant subspaces of the Kronecker product $\vartheta_2 \otimes \vartheta_2$ of S_3, with ϑ_2 taken from Example 1.8. Use the fact that two 1-dimensional invariant subspaces exist. (This is an easy consequence from the theory in Section 5.5). What is the completely reduced form of the representing matrices of the elements a and c (Example 1.6)? Which of the irreducible representations of S_3 occurs? (see Example 1.13).

Problem 1.6 For the cyclic group $C_3 = \{e, a, a^2\}$ compute the inner product (1.77) which is invariant under the representation

$$a \rightarrow \begin{bmatrix} -1 & -1 \\ 1 & 0 \end{bmatrix}$$

Verify its invariance.

Problem 1.7 Let G be an arbitrary group. Show that a so-called **left translation** of G defined by

$$s \rightarrow as \qquad \forall s \in G \qquad (a \text{ fixed})$$

is a bijection of G onto G (i.e., one-to-one and onto). In addition, the set of all left translations forms a transformation group (with range G) which is isomorphic with G. This statement is a generalization of Theorem 1.3.

Chapter 2

Linear Operators with Symmetries

We now proceed to considerations that are of great importance for the applications. If a problem has a certain **symmetry**, that is, if it is invariant under a certain set of **symmetry operations**[1], then it can be considerably simplified and is often solvable only by exploiting this symmetry. The study of these symmetries may lead to new theoretic findings. As an example we may mention quantum mechanics.

A source for such considerations is Schur's[2] lemma.

2.1 Schur's Lemma

Assumptions.

1. *Let two irreducible representations $\vartheta_1 : s \to D_1(s)$ and $\vartheta_2 : s \to D_2(s)$ of a group G be given. The matrices $D_i(s)$ act in the representation space V_i of dimension n_i $(i = 1, 2)$. They describe linear transformations relative to some basis in V_i.*

2. *Let P be a rectangular matrix of length n_1 and height n_2 such that*
$$P \quad \cdot \quad D_1(s) \quad = \quad D_2(s) \quad \cdot \quad P \quad \forall s \in G$$

Claim. Either

---

1.   *P is the null matrix*

or (in the exclusive sense)

2. $n_1 = n_2$, *and P is non-singular; i.e.,* $D_1(s) = P^{-1}D_2(s)P$
$\forall s \in G$. *In other words, the representations are equivalent to each other, and P is the associating coordinate transformation.*

Proof. P can be regarded as a linear map $V_1 \to V_2$: To each n_1-dimensional vector of V_1 it assigns an n_2-dimensional vector of V_2. The commutation relations in Assumption 2 say the following:

First transforming an arbitrary vector $x \in V_1$ by $D_1(s)$ and then mapping the transformed vector by P into some vector in V_2 has the same effect as first sending x by P into a vector of V_2 and then transforming its image by $D_2(s)$.

We denote by $P(V_1)$ the image of V_1 under P. Thus

$$P(V_1) = \{y \mid y = Px, x \in V_1\} \subset V_2$$

$P(V_1)$ *is a subspace of* V_2 *which is invariant under the representation* ϑ_2. For, if y is an arbitrary vector in $P(V_1)$, that is, $Px = y$, then Assumption 2 implies

$$D_2(s)y = D_2(s)Px = PD_1(s)x \in P(V_1)$$

According to Assumption 1, the representation ϑ_2 is irreducible. This means that there are no proper invariant subspaces of V_2. Hence the following holds:

Either the **1st alternative** : $P(V_1) = $ null space
i.e., $Px = 0 \ \forall x \in V_1$
i.e., $P = $ null matrix
or else the **2nd alternative** : $P(V_1) = V_2 \Rightarrow n_1 \geq n_2$

At this stage, let us introduce the **kernel** K of the map P. By definition it consists of all vectors $z \in V_1$ that are mapped into the zero vector: $K = \{z \mid Pz = 0\}$.

K *is a subspace of* V_1 *invariant under the representation* ϑ_1. This is true, because obviously $D_1(s)z \in K$ by

$$PD_1(s)z = D_2(s)Pz = 0 \qquad \forall z \in K$$

Since V_1 also is transformed irreducibly, we have that $K = V_1$ or $K = $ null space. The first case implies that $P = $ null matrix, which is impossible due to $P(V_1) = V_2$. The second case implies that P maps only the zero vector into the zero vector. Thus $n_2 \geq n_1$. But above we showed that $n_2 \leq n_1$, hence $n_1 = n_2$. And so P is a square matrix and the corresponding linear map has kernel 0. From this it follows that P is non-singular. □

Remark. The proof never used the representation property. *The lemma (whence also the fundamental theorem stated in Section 2.3) holds even if s varies through any arbitrary set, as long as the two sets of matrices $D_1(s)$ and $D_2(s)$ are irreducible.* Even singular matrices are permitted here.

2.2 Symmetry of a Matrix

Let M be a given linear operator mapping an n-dimensional complex vector space V into itself. Let G be an arbitrary group, ϑ a representation defined on G and acting on V. Furthermore, let $D(s), s \in G$, be the representing matrices describing ϑ relative to some given basis. For simplicity, we denote again by M the matrix describing the linear operator M relative to this basis.

Throughout this section we assume that the, in general, complex matrix M commutes with every representing matrix; that is,

$$MD(s) = D(s)M \qquad \forall s \in G \qquad (2.1)$$

If (2.1) holds, we say, "M *has the symmetry* of the representation ϑ", or somewhat less precisely, "M *has the symmetry of the group* G". $D(s)$ are called the **symmetry operations**. For applications we are interested in, among other things, the eigenvalue problem of M:

$$Mx = \lambda x \qquad \lambda \in C,\, x \neq 0 \qquad (2.2)$$

Theorem 2.1 *If M has the symmetry of the representation ϑ, then every eigenspace E_λ of M corresponding to the eigenvalue λ is an invariant subspace of ϑ.*

Proof. From (2.1) and (2.2) we have

$$MD(s)x = D(s)Mx = \lambda D(s)x, \qquad \text{i.e.,} \qquad D(s)x \in E_\lambda$$

$\square$

Remark. It is not true that in the case of a completely reducible representation such an eigenspace must necessarily be transformed irreducibly by ϑ. (For this purpose see, e.g., Example 2.5).

By the group theory method for solving problems with symmetries, however, we mean the use of the fundamental theorem presented in the next section. It is some kind of a "converse" of Theorem 2.1. Based on the knowledge of irreducible invariant subspaces of the representation, we make statements about invariant subspaces of M.

The following serves as a preparatory theorem:

Theorem 2.2 *Let $\vartheta : s \to D(s)$ be an irreducible representation of the group G in the representation space V, and let Q be a linear operator $V \to V$ having the symmetry of ϑ; i.e., $QD(s) = D(s)Q$ for all $s \in G$. Then $Q = \lambda \cdot$ identity.*

Proof. Let λ be an eigenvalue of Q. This always exists (by the fundamental theorem of algebra: The characteristic polynomial has at least one complex-valued root). By hypothesis we have $QD(s) = D(s)Q$. Hence $Q' = Q - \lambda E$ also commutes with $D(s)$. We employ Schur's lemma with $\vartheta = \vartheta_1 = \vartheta_2$. Since $\det Q' = 0$, the first alternative applies. Hence $Q' =$ null matrix. Therefore, $Q = \lambda E$. $\square$

2.2.1   Representations of Abelian Groups

Theorem 2.3 *Every irreducible representation ϑ of an abelian group G is 1-dimensional.*

Proof. Let $D(b)$ be any fixed matrix of the irreducible representation. Since the group is abelian, we have

$$D(s)D(b) = D(b)D(s) \qquad \forall s \in G$$

Hence by Theorem 2.2: $D(b) = \lambda(b) \cdot E$. This holds for all $b \in G$, hence the representing matrices are all diagonal. But then the representation ϑ must be 1-dimensional, because by hypothesis it is irreducible.   □

Theorem 1.6 implies

Corollary 2.4   *Every completely reducible representation of an abelian group (whence every representation of a finite abelian group) is equivalent to a representation by diagonal matrices.*

2.3   The Fundamental Theorem

Let ϑ be a given n-dimensional completely reducible representation of an arbitrary group G. Let the complete reduction of ϑ be given by

$$\vartheta = c_1\vartheta_1 \oplus c_2\vartheta_2 \oplus \ldots \oplus c_N\vartheta_N \tag{2.3}$$

where the ϑ_j are irreducible and mutually inequivalent representations of G. The numbers $c_j \in \{1, 2, 3, \ldots\}$ indicate the **multiplicity** and n_j the dimension of ϑ_j. Accordingly, the representation space V of ϑ decomposes into

$$V = V_1 \oplus V_2 \oplus \ldots \oplus V_N \tag{2.4}$$

Here V_j consists of c_j invariant subspaces $V_j^1, V_j^2, \ldots, V_j^{c_j}$ each of which has dimension n_j and transforms after the manner of ϑ_j. More precisely, for each ρ the representation obtained by restricting ϑ to the space V_j^ρ is equivalent to ϑ_j. The $(c_j n_j)$-dimensional subspaces V_j of V are called the **isotypic components** (or the **conglomerates**) of type ϑ_j for ϑ.

Remark. *By construction, the isotypic decomposition (2.4) is unique (exercise).* However, *the decomposition of an isotypic component V_j into the n_j-dimensional irreducible subspaces is in general not unique* as the following Examples 2.1 through 2.4 shall show.

Example 2.1 Let ϑ_1 be the unit representation; in ϑ it may appear repeatedly. Then V_1 consists of all vectors that remain fixed under ϑ. Every subspace V_1^ρ is then spanned by a basis vector of V_1. But since the choice of a basis is not unique, this explains the statement given in the remark above.

Here and in the following, we shall arrange the $n_j c_j$ basis vectors of an isotypic component V_j in an **array**:

$$\left.\begin{array}{ccccccc} V_j^1: & b_1^1 & b_2^1 & b_3^1 & \ldots & b_{n_j}^1 \\ V_j^2: & b_1^2 & b_2^2 & b_3^2 & \ldots & b_{n_j}^2 \\ \vdots & \vdots & \vdots & \vdots & & \vdots \\ V_j^{c_j}: & b_1^{c_j} & b_2^{c_j} & b_3^{c_j} & \ldots & b_{n_j}^{c_j} \end{array}\right\} \tag{2.5}$$

The ρth row contains the basis vectors of the subspace V_j^ρ. All these subspaces are transformed equivalently. That is, if $D_j^\alpha(s)$ and $D_j^\beta(s)$, $s \in G$, are the representing matrices in V_j^α and V_j^β, then there exists a non-singular $n_j \times n_j$ matrix T with

$$D_j^\alpha(s) = T^{-1} D_j^\beta(s) T \qquad \forall s \in G \tag{2.6}$$

As will be seen later, in the applications results are often obtained only if *the subspaces* $V_j^1, V_j^2, \ldots, V_j^{c_j}$ *can be transformed not only equivalently but even in exactly the same way; that is, if the corresponding matrices satisfy*

$$D_j^1(s) = D_j^2(s) = \ldots = D_j^{c_j}(s) = D_j(s) \qquad \forall s \in G \tag{2.7}$$

Furthermore, we define:
    If the c_j rows of the array (2.5) transform in exactly the same way under the representation ϑ, then we say that the $c_j n_j$ basis vectors or the array are **symmetry adapted** with respect to ϑ.
    If the array is not symmetry adapted, then it can be made symmetry adapted by choice of a suitable new basis in the various subspaces.

Example 2.2 We consider Example 1.12 and recall that a vector in the representation space is given by values at the nine points of the square. The array of the 4-dimensional isotypic component with $n_j = c_j = 2$ in (1.48) on the right-most part (dots stand for zeros) is:

$$V_4 \left\{ \begin{array}{c} \begin{array}{l} V_4^1: \quad u_1 = \begin{array}{ccc} \cdot & \cdot & \cdot \\ -1 & \cdot & 1 \\ \cdot & \cdot & \cdot \end{array} \quad v_1 = \begin{array}{ccc} \cdot & 1 & \cdot \\ \cdot & \cdot & \cdot \\ \cdot & -1 & \cdot \end{array} \end{array} \\ \rule{9cm}{0.4pt} \\ \begin{array}{l} V_4^2: \quad u_2 = \begin{array}{ccc} -1 & \cdot & 1 \\ \cdot & \cdot & \cdot \\ -1 & \cdot & 1 \end{array} \quad v_2 = \begin{array}{ccc} 1 & \cdot & 1 \\ \cdot & \cdot & \cdot \\ -1 & \cdot & -1 \end{array} \end{array} \end{array} \right. \tag{2.8}$$

The vertical reflection σ and the 45° reflection μ (see Fig. 1.6) transform the two row spaces V_4^1 and V_4^2 in exactly the same way, as is expressed by (1.49). The array (2.8) is symmetry adapted, because the two elements σ and μ form a set of generators.

Example 2.3 On the other hand, the same 4-dimensional isotypic component can be spanned by

$$
V_4' \begin{cases}
V_4^{1'}:\quad u_1' = \begin{pmatrix} \cdot & \cdot & \cdot \\ -1 & \cdot & 1 \\ \cdot & \cdot & \cdot \end{pmatrix} \quad v_1' = \begin{pmatrix} \cdot & 1 & \cdot \\ \cdot & \cdot & \cdot \\ \cdot & -1 & \cdot \end{pmatrix} \\[2ex]
\hline \\[-1ex]
V_4^{2'}:\quad u_2' = \begin{pmatrix} \cdot & \cdot & 1 \\ \cdot & \cdot & \cdot \\ -1 & \cdot & \cdot \end{pmatrix} \quad v_2' = \begin{pmatrix} 1 & \cdot & \cdot \\ \cdot & \cdot & \cdot \\ \cdot & \cdot & -1 \end{pmatrix}
\end{cases}
\tag{2.9}
$$

Here, too, the row spaces are invariant under ϑ, but they are not transformed in the same way: The array (2.9) is not symmetry adapted. The subspaces are identical with those of Example 2.2 (that is, $V_4^1 = V_4^{1'}$ and $V_4^2 = V_4^{2'}$). In (2.9) we merely chose another basis.

Example 2.4 The sums of columns of Example 2.2 also span an invariant subspace, which together with the first row forms an array:

$$
V_4'' \begin{cases}
V_4^{1''}:\quad u_1'' = \begin{pmatrix} \cdot & \cdot & \cdot \\ -1 & \cdot & 1 \\ \cdot & \cdot & \cdot \end{pmatrix} \quad v_1'' = \begin{pmatrix} \cdot & 1 & \cdot \\ \cdot & \cdot & \cdot \\ \cdot & -1 & \cdot \end{pmatrix} \\[2ex]
\hline \\[-1ex]
V_4^{2''}:\quad u_2'' = \begin{pmatrix} -1 & \cdot & 1 \\ -1 & \cdot & 1 \\ -1 & \cdot & 1 \end{pmatrix} \quad v_2'' = \begin{pmatrix} 1 & 1 & 1 \\ \cdot & \cdot & \cdot \\ -1 & -1 & -1 \end{pmatrix}
\end{cases}
\tag{2.10}
$$

This array (2.10) is symmetry adapted. However, the decomposition of the isotypic component differs from that of Example 2.2; for, $V_4^{2''} \neq V_4^2$. This shows that the decomposition of an isotypic component into irreducible subspaces need not be unique.

Before we continue the theory, we first state the following: With respect to a symmetry adapted array of the isotypic component V_j, an element $s \in G$ is represented by a matrix $\Sigma_j(s)$ of the form

$$
\Sigma_j(s) = \begin{pmatrix} D_j(s) & & & 0 \\ & D_j(s) & & \\ & & \ddots & \\ 0 & & & D_j(s) \end{pmatrix} \qquad \begin{array}{l} c_j \text{ identical blocks } D_j \\ \text{along the major diagonal} \end{array}
\tag{2.11}
$$

In other words, $\Sigma_j(s)$ is a direct sum of c_j matrices $D_j(s)$.

We now return to the original situation (2.3). We pick two isotypic components V_j and V_k. In them, the representing matrices $\Sigma_i(s)$ and $\Sigma_k(s)$ relative to symmetry adapted arrays act according to (2.11). Again we study the commutation relations and thus obtain a **generalization of Schur's lemma** in the following form:

Theorem 2.5 *Let P be a rectangular matrix of $c_j n_j$ columns and $c_k n_k$ rows, and let*

$$P \cdot \Sigma_i(s) = \Sigma_k(s) \cdot P \qquad \forall s \in G \quad i, k \text{ fixed.} \tag{2.12}$$

1. If $i \neq k$ (i.e., V_i and V_k are inequivalently transformed under ϑ), then

$$P = null \ matrix \tag{2.13}$$

2. If $i = k$, then

$$P = \begin{bmatrix} \mu_{11}^i E & \mu_{12}^i E & \cdots & \mu_{1c_i}^i E \\ \mu_{21}^i E & \mu_{22}^i E & \cdots & \mu_{2c_i}^i E \\ \vdots & \vdots & & \vdots \\ \mu_{c_i 1}^i E & \mu_{c_i 2}^i E & \cdots & \mu_{c_i c_i}^i E \end{bmatrix} \tag{2.14}$$

where E is the $n_i \times n_i$ identity matrix and $\mu_{\alpha\beta}^i$ are complex numbers.

Proof. We subdivide P into blocks

$$P = \begin{bmatrix} P_{11} & P_{12} & \cdots & P_{1c_i} \\ P_{21} & P_{22} & \cdots & P_{2c_i} \\ \vdots & \vdots & & \vdots \\ P_{c_k 1} & P_{c_k 2} & \cdots & P_{c_k c_i} \end{bmatrix} \tag{2.15}$$

where all the $P_{\ell m}$ have the same length n_i and the same height n_k. By rewriting (2.12) blockwise and using (2.11), we find

$$P_{\ell m} D_i(s) = D_k(s) P_{\ell m} \qquad \forall \ell, m \tag{2.16}$$

(a) $i \neq k$. Schur's lemma and inequivalence of ϑ_i and ϑ_k imply the first alternative:
$$P_{\ell m} = null \ matrix \qquad \forall \ell, m$$
whence $P = null$ matrix.

(b) $i = k$. In this case, (2.16) says that the operator $P_{\ell m}$ has the symmetry of the irreducible representation ϑ_i. Hence by Theorem 2.2, the matrix $P_{\ell m}$ is a multiple $\mu_{\ell m}^i E$ of the $n_i \times n_i$ identity matrix. □

The Fundamental Theorem. *Let $\vartheta : s \to D(s)$ be an n-dimensional completely reducible representation of some group G. Let the complete reduction be given by*

$$\vartheta = c_1\vartheta_1 \oplus c_2\vartheta_2 \oplus \ldots \oplus c_N\vartheta_N \qquad (2.17)$$

where the ϑ_j are the irreducible and mutually inequivalent representations of G. The number $c_j \in \{1,2,3,\ldots\}$ indicates the multiplicity and n_j the dimension of ϑ_j. Hence for each j, there exist c_j irreducible subspaces $V_j^1, V_j^2, \ldots, V_j^{c_j}$ that are invariant under ϑ and that transform in a manner equivalent to ϑ_j. Together they form the isotypic component V_j of type ϑ_j. The $n_j c_j$ basis vectors of V_j are arranged in an array where the basis of V_j^ρ is written in the ρth row:

$$V_j \begin{cases} V_j^1: & b_1^1 & b_2^1 & b_3^1 & \ldots & b_{n_j}^1 \\ V_j^2: & b_1^2 & b_2^2 & b_3^2 & \ldots & b_{n_j}^2 \\ \vdots & \vdots & \vdots & \vdots & & \vdots \\ V_j^{c_j}: & b_1^{c_j} & b_2^{c_j} & b_3^{c_j} & \ldots & b_{n_j}^{c_j} \end{cases} \qquad (2.18)$$

Assumptions.

1. *Let M be a linear operator in the representation space of a completely reducible representation $\vartheta : s \to D(s)$ and have the symmetry of ϑ; i.e.,*

$$M \cdot D(s) = D(s) \cdot M \qquad \forall s \in G \qquad (2.19)$$

2. *The arrays of all of the various isotypic components are symmetry adapted; i.e., the following holds for all j: Each of the c_j rows of (2.18) transforms in exactly the same way under ϑ. Thus the representing matrices relative to the corresponding basis of the representation space have the form (2.21) with (2.11).*

Assertions.

1. *Each column of the various arrays (2.18) spans a c_j-dimensional subspace invariant under the operator M.*

2. *M transforms each column of the same array in exactly the same way.*

Assertions 1 and 2 may be reworded as follows:

If for all j the basis vectors of an isotypic component V_j are numbered by columns, that is, in this way:

then relative to this basis the operator will be a matrix M of the following form: M decomposes into a direct sum of

$$\left.\begin{array}{lll} n_1 & square\ matrices\ M_1\ of\ length & c_1 \\ n_2 & square\ matrices\ M_2\ of\ length & c_2 \\ \vdots & & \vdots \\ n_N & square\ matrices\ M_N\ of\ length & c_N \end{array}\right\} \tag{2.20}$$

This is also called the *block diagonalization* of M.

Proof. Relative to a complete list of symmetry adapted arrays, the representing matrices have the form:

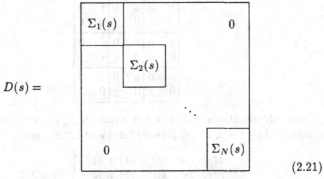

$$\tag{2.21}$$

We decompose the matrix M (which describes the linear operator relative to the basis mentioned above) into blocks M_{pq} according to (2.21):

$$M = \begin{bmatrix} M_{11} & M_{12} & \cdots & M_{1N} \\ M_{21} & M_{22} & \cdots & M_{2N} \\ \vdots & \vdots & & \vdots \\ M_{N1} & M_{N2} & \cdots & M_{NN} \end{bmatrix} \tag{2.22}$$

Hence M_{pq} is a block of height $n_p c_p$ and length $n_q c_q$. Blockwise multiplication of (2.19), together with (2.21) and (2.22), yields

$$M_{pq} \cdot \Sigma_q(s) = \Sigma_p(s) \cdot M_{pq} \qquad \forall p, q \tag{2.23}$$

Theorem 2.5 implies

(a) if $p \neq q$:

$$M_{pq} = 0 \tag{2.24}$$

In other words, each isotypic component is transformed into itself under M, hence is invariant under M.

(b) if $p = q$:

$$M_{pp} = (\mu^p_{\alpha\beta} E) \qquad \alpha, \beta = 1, 2, \ldots, c_p \tag{2.25}$$

This corresponds to the matrix P in (2.14) where E is the $n_p \times n_p$ identity matrix. □

For further illustration we consider a particular isotypic component V_ℓ, where $c_\ell = 3, n_\ell = 2$:

$$
\begin{array}{cc}
b_1^1 & b_2^1 \\
b_1^2 & b_2^2 \\
b_1^3 & b_2^3
\end{array}
\tag{2.26}
$$

Relative to the sequence of basis vectors

$$
b_1^1, b_2^1, b_1^2, b_2^2, b_1^3, b_2^3
\tag{2.27}
$$

the operator M restricted to the isotypic component V_ℓ above has the structure

$$
\left[
\begin{array}{cc|cc|cc}
r & 0 & s & 0 & t & 0 \\
0 & r & 0 & s & 0 & t \\
\hline
u & 0 & v & 0 & w & 0 \\
0 & u & 0 & v & 0 & w \\
\hline
x & 0 & y & 0 & z & 0 \\
0 & x & 0 & y & 0 & z
\end{array}
\right]
\tag{2.28}
$$

To simplify notations, the indexed numbers $\mu_{\alpha\beta}^p (p = \ell)$ in (2.25) were substituted by $r, s, t, \ldots, z$. Hence (2.27) and (2.28) read

$$
\left.
\begin{array}{rcl}
Mb_i^1 & = & rb_i^1 + ub_i^2 + xb_i^3 \\
Mb_i^2 & = & sb_i^1 + vb_i^2 + yb_i^3 \\
Mb_i^3 & = & tb_i^1 + wb_i^2 + zb_i^3
\end{array}
\right\} \quad i = 1, 2
\tag{2.29}
$$

Relative to the rearranged sequence of the same basis vectors

$$
b_1^1, b_1^2, b_1^3, b_2^1, b_2^2, b_2^3
\tag{2.30}
$$

the operator M, according to (2.29), has the structure

$$
\left[
\begin{array}{ccc|ccc}
r & s & t & & & \\
u & v & w & & 0 & \\
x & y & z & & & \\
\hline
 & & & r & s & t \\
 & 0 & & u & v & w \\
 & & & x & y & z
\end{array}
\right]
\tag{2.31}
$$

thus manifesting the invariance of the two 3-dimensional subspaces under M. They differ only in the order of the basis vectors. (2.27) labels the isotypic component in question along rows, while (2.30) labels it along columns.

An immediate consequence of the fundamental theorem (see (2.20)) is

Theorem 2.6 *If in the vector space V the linear operator M possesses the symmetry of the completely reducible representation*

$$\vartheta = c_1\vartheta_1 \oplus c_2\vartheta_2 \oplus \ldots \oplus c_N\vartheta_N, \qquad c_j \geq 1$$

(ϑ_j being mutually inequivalent irreducible representations) and if a symmetry adapted basis in V is known, then the eigenvalue problem of M (i.e., the problem of determining the eigenvalues and eigenvectors) reduces to the c_j-dimensional eigenvalue problems

$$|M_j - \lambda E| = 0 \qquad j = 1, 2, \ldots, N$$

Thus the computation of the eigenvalues (of algebraic multiplicity $\geq n_j$) requires solving a c_jth order equation each.

Statements.

1. **The fundamental theorem is the precise formulation of the group theoretic method for solving linear problems with symmetries.**

2. Obviously, the following problems have yet to be mastered:

   (a) Decomposing the representation space into the various isotypic components.

   (b) Decomposing the isotypic components into the various irreducible subspaces of ϑ.

   (c) Determining symmetry adapted basis vectors in the various isotypic components.

3. The fundamental theorem includes the following special cases:

   (a) G consists of the identity e alone. Then $D(e) = E$, $n_1 = 1$, $c_1 = 1$. Consequently, there is no group theoretic reduction of M.

   (b) ϑ is irreducible; that is, $n_1 = n$, $c_1 = 1$. Consequently, $M = \lambda \cdot$ identity.

   (c) **Theorem 2.7** *Let ϑ_ℓ be an irreducible representation occurring exactly once as a constituent of a completely reducible representation ϑ (thus $c_\ell = 1$). Then the invariant subspace associated with ϑ_ℓ is also a subspace[3] of an eigenspace of each of the linear operators M that commute with ϑ.*

Example 2.5 (Eigenvalue problem of a vibrating square membrane) (compare Examples 1.9 and 1.12). Let a membrane be stretched over a square region of Fig. 1.6 and clamped along the boundary. Then

---

[3]It is possible that the same eigenvalue also appears in other subspaces.

the displacement $u(x, y)$ perpendicular to the plane of the paper satisfies the partial differential equation

$$\Delta u + \lambda u = 0 \qquad \text{where } u = 0 \text{ on the boundary} \qquad (2.32)$$

$\Delta = \frac{\partial^2}{\partial x^2} + \frac{\partial^2}{\partial y^2}$ is the Laplacian operator and $\sqrt{\lambda}$ the frequency of the oscillation.

Discretization: u is approximated by a lattice function v on the nine points. The negative Δ-operator is approximated by the cross operator (2.33); see (Schwarz, 1989, Section 10.1.2).

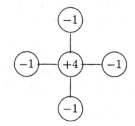

$$(2.33)$$

This defines the discretized 9-dimensional problem

$$Mv = \lambda v \qquad (2.34)$$

in the following way:

The ith equation of (2.34) is obtained by translating the cross operator into the point no. i. Hence, for example, the fifth and the first equations (see Fig. 1.6) read

$$\begin{aligned} 4v_5 - v_2 - v_4 - v_6 - v_8 &= \lambda v_5 \\ 4v_1 - v_2 - v_4 &= \lambda v_1 \end{aligned} \qquad (2.35)$$

Since the Δ-operator possesses the symmetry of isometries (see Section 8.6), it is a plausible consequence that the discretized cross operator has the symmetry of the dihedral group D_4.

In (1.48) we have tabulated all arrays for the isotypic components V_j of the 9-dimensional representation in the four framed boxes. There the following holds for the multiplicities c_j and the dimensions n_j of the corresponding irreducible subspaces V_j^i:

	Multiplicity c_j	Dimension n_j	
1st isotypic component	3	1	
2nd isotypic component	1	1	(2.36)
3rd isotypic component	1	1	
4th isotypic component	2	2	

All arrays are symmetry adapted: This is clear in the 1-dimensional cases. For the fourth isotypic component this follows from the equations (1.49).

Hence the assumptions of the fundamental theorem all are satisfied. This theorem says that each column of each symmetry adapted array spans a subspace invariant under M. In fact:

$$
Mx_1 = \begin{array}{rrr} 0 & -1 & 0 \\ -1 & 4 & -1 \\ 0 & -1 & 0 \end{array}
$$
$$= +4x_1 - x_2$$

$$My = 4y \qquad Mu_1 = \qquad Mv_1 =$$
$$4u_1 - u_2 \qquad 4v_1 - v_2$$

$$
Mx_2 = \begin{array}{rrr} -2 & 4 & -2 \\ 4 & -4 & 4 \\ -2 & 4 & -2 \end{array}
$$
$$= -4x_1 + 4x_2 - 2x_3$$

$$Mz = 4z \qquad Mu_2 = \qquad Mv_2 =$$
$$-2u_1 + 4u_2 \qquad -2v_1 + 4v_2$$

$$
Mx_3 = \begin{array}{rrr} 4 & -2 & 4 \\ -2 & 0 & -2 \\ 4 & -2 & 4 \end{array}
$$
$$= \qquad -2x_2 + 4x_3$$

$$(2.37)$$

The computation of (2.37) is preferably done by use of the cross operator (2.33) without using the corresponding matrix, because in this way we can avoid unnecessary multiplications by zero elements. This will be of particular interest when programming high-dimensional problems (see Section 3.3).

Relative to the basis vectors arranged along columns

$$x_1, x_2, x_3, y, z, u_1, u_2, v_1, v_2$$

we obtain by (2.37) the following matrix for M:

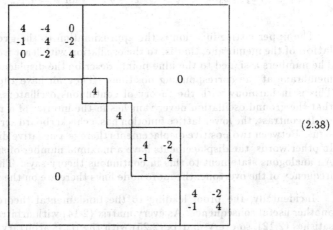

$$(2.38)$$

The characteristic polynomial of M factors into polynomials corresponding to the blocks in (2.38). Hence computing the eigenvalues reduces to solving a cubic and a quadratic equation:

$$(4 - \lambda)^3 - 8(4 - \lambda) = 0, \text{ eigenvalues} : 4, \ 4 \pm 2\sqrt{2} \qquad (2.39)$$

$$(4 - \lambda)^2 - 2 = 0, \text{ eigenvalues} : 4 \pm 2\sqrt{2} \qquad (2.40)$$

Accordingly, computing the corresponding eigenvectors reduces to solving homogeneous singular systems of equations in 3 or 2 unknowns, respectively. The following table gives a complete list of the eigenspaces of M, with E_λ denoting the eigenspace corresponding to the eigenvalue λ.

λ	Dimension of E_λ	Basis of E_λ
4	3	$-2x_1 + x_3; \quad y; \quad z$
$4 + 2\sqrt{2}$	1	$2x_1 - \sqrt{2}x_2 + x_3$
$4 - 2\sqrt{2}$	1	$2x_1 + \sqrt{2}x_2 + x_3$
$4 + \sqrt{2}$	2	$-\sqrt{2}u_1 + u_2; -\sqrt{2}v_1 + v_2$
$4 - \sqrt{2}$	2	$\sqrt{2}u_1 + u_2; \sqrt{2}v_1 + v_2$

$$(2.41)$$

Remark. This example shows that an eigenspace E_λ need not necessarily be an irreducible subspace of the representation, because E_λ is 3-dimensional for $\lambda = 4$, but the irreducible representations of the group D_4 have dimensions ≤ 2.

Hence the eigenvectors corresponding to the least and the greatest eigenvalue, interpreted as lattice functions, have the form:

$$2x_1 + \sqrt{2}x_2 + x_3 = \begin{array}{ccc} 1 & \sqrt{2} & 1 \\ \sqrt{2} & 2 & \sqrt{2} \\ 1 & \sqrt{2} & 1 \end{array} \quad \text{for } \lambda = 4 - 2\sqrt{2}$$

$$(2.42)$$

$$2x_1 - \sqrt{2}x_2 + x_3 = \begin{array}{ccc} 1 & -\sqrt{2} & 1 \\ -\sqrt{2} & 2 & -\sqrt{2} \\ 1 & -\sqrt{2} & 1 \end{array} \quad \text{for } \lambda = 4 + 2\sqrt{2}$$

The upper lattice function is the approximation to the **ground oscillation** of the membrane, that is, to the oscillation with least frequency $\sqrt{\lambda}$. The numbers assigned to the nine points describe the displacements of the membrane at the corresponding positions. They all have the same sign. This is in harmony with the theory of continuous oscillators, which says that the ground oscillation never vanishes in the interior of a membrane.

In contrast, the lower lattice function has a checkerboard arrangement of signs: Between two positive displacements there is a negative displacement. In other words, the displacements have a maximal number of sign changes. An analogous statement to this in continuous theory says: The larger the frequency of the overtone, the more node lines there are on the membrane.

Incidentally, the proof leading to the fundamental theorem provides another useful consequence. As every matrix (2.14) with arbitrary μ-values satisfies (2.12), so every matrix (2.25) with the $(c_p)^2$ arbitrary μ-values

$$\mu_{\alpha\beta}^p \qquad \alpha, \beta = 1, 2, \ldots, c_p; \ p = 1, 2, \ldots, N \qquad (2.43)$$

satisfies the relation (2.23) for $p = q$. Thus the set of all blocks M_{11}, M_{22}, ..., M_{NN} in (2.22) arranged along the major diagonal contains a total of

$$\sum_{j=1}^{N} c_j^2 \text{ complex numbers}$$

This proves

Theorem 2.8 *The set of all linear operators M commuting with the completely reducible representation $\vartheta = c_1\vartheta_1 \oplus c_2\vartheta_2 \oplus \ldots \oplus c_N\vartheta_N$ (ϑ_j pairwise inequivalent and irreducible) has exactly*

$$\sum_{j=1}^{N} c_j^2 \qquad free\ complex\ parameters \qquad (2.44)$$

By describing the linear operators M relative to a symmetry adapted basis numbered along columns in each of the isotypic components (2.18) (see text ahead of (2.20)) we obtain all matrices M that reduce into arbitrary matrices $M_p = (\mu_{\alpha\beta}^p)$ along the major diagonal as described in (2.20).

Obviously, Theorem 2.8 holds in any coordinate system. In passing to another coordinate system via the matrix S, we obtain a matrix $S^{-1}MS$, which has entries that are linear combinations of $\mu_{\alpha\beta}^p$.

PROBLEMS

Problem 2.1 (a) The permutation

$$t = \left|\begin{array}{ccccc} 1 & 2 & 3 & \ldots & (n-1) & n \\ 2 & 3 & 4 & \ldots & n & 1 \end{array}\right.$$

generates a cyclic group $C_n = \{t, t^2, \ldots, t^n = e\}$. The representing matrix T of the element t is

$$T = \begin{bmatrix} 0 & 0 & \ldots & 0 & 1 \\ 1 & 0 & & & 0 \\ 0 & 1 & & & \vdots \\ & 0 & & & \vdots \\ \vdots & \vdots & & 0 & 0 \\ 0 & 0 & \ldots & 1 & 0 \end{bmatrix} \qquad (2.45)$$

Show that

$$\lambda_k = e^{ik\frac{2\pi}{n}}, \qquad x_k = \begin{bmatrix} 1 \\ \lambda_k^{n-1} \\ \lambda_k^{n-2} \\ \vdots \\ \lambda_k \end{bmatrix}, \quad k = 0, 1, 2, \ldots, n-1 \qquad (2.46)$$

are the eigenvalues and the eigenvectors of the natural representation $t^\alpha \to T^\alpha$.

(b) Verify that the fundamental theorem in Section 2.3 implies the following: If an $n \times n$ square matrix M satisfies

$$MT = TM \tag{2.47}$$

then every eigenvector of T is also an eigenvector of M.

(c) In particular, show that the so-called **cyclic matrix**

$$M = \begin{bmatrix} a_1 & a_2 & a_3 & \cdots & a_n \\ a_n & a_1 & a_2 & \cdots & a_{n-1} \\ a_{n-1} & a_n & a_1 & \cdots & a_{n-2} \\ \vdots & & & & \vdots \\ a_2 & a_3 & a_4 & \cdots & a_1 \end{bmatrix} \qquad a_i \in \mathbb{C} \tag{2.48}$$

has the property (2.47), and with its aid compute the eigenvalues of M.

Remark. A more general version of this problem (c) is phrased as Problem Nr. 859 in the journal "Elemente der Mathematik", (volume 36, p. 45, 1981). A solution to it is given in volume 37, pp. 60-61, 1982.

Problem 2.2 (a) By comparing the coefficients of the polynomial on the left hand side of the quadratic equation $x^2 + px + q = 0$ with those of the characteristic polynomial of a cyclic 2×2 matrix and using Problem 2.1, deduce a formula for the solution of quadratic equations.

(b) Do the corresponding for the case $n = 3$ by choosing the special value $a_1 = 0$. The characteristic polynomial of M then is

$$\lambda^3 - 3a_2 a_3 \lambda - (a_2^3 + a_3^3) \tag{2.49}$$

Problem 2.1 furnishes the three roots of this cubic polynomial. By identifying this polynomial with the general reduced cubic polynomial

$$\lambda^3 + p\lambda + q \tag{2.50}$$

one obtains **Cardano's**[4] **formulas** for solving cubic equations. (We owe this problem to S. Pelloni.)

Problem 2.3 Let the unit circle be segmented by n equally spaced points P_k. Let $f(\varphi)$, where φ is the polar angle, be a function defined on the circle such that at P_k it takes on the values

$$f_k = f(\varphi_k) = f(k\frac{2\pi}{n}) \qquad k = 0, 1, 2, \ldots, n-1$$

As a discretization of the differential operator

$$\frac{d^2}{d\varphi^2}$$

---

[4]Cardano, Girolamo (1501-76) was a physician, philosopher, mathematician, natural scientist, astrologer, and an interpreter of dreams. He lived as a universal scholar in Milano.

we take

$$f_{k+1} - 2f_k + f_{k-1}$$

Here the indices are understood additively up to multiples of n. Determine by group theory methods the approximated eigenvalues of the differential operator; that is, solve the problem

$$f_{k+1} - 2f_k + f_{k-1} = \lambda f_k$$

Problem 2.4 Consider the vector space of all continuously differentiable functions $f(\varphi)$ defined on the unit circle with polar angle φ, also consider the differential operator

$$M = \frac{d}{d\varphi}$$

Determine the eigenfunctions of M and, using Theorem 2.1, also determine the irreducible representations of the continuous group G of all proper rotations of the circle.

Problem 2.5 To supplement (2.42), discuss the remaining forms of eigenoscillations. Also make comparisons with the analytic theory. (Courant, Hilbert, 1953, vol. 1, Ch. V, Sec. 5.4).

Chapter 3

Symmetry Adapted Basis Functions

3.1 Illustration by Dihedral Groups

The dihedral group D_m consists of m rotations and m reflections of the real plane carrying a regular m-sided polygon into itself (see Section 1.9.2).

Let $\mathcal{N}$ be a (finite) set of n points in the plane such that it is sent into itself by each of the transformations of D_m (permutations of the n points).

Thus the position of these n points is already determined by that of the points inside a closed angular region which itself is enclosed by two adjacent half axes of reflections forming an angle of π/m (Fig. 3.1).

Fig. 3.1

Such an unbounded angular region, including the two rays and the origin O, is called a **fundamental domain** of D_m. The shaded area in Fig. 3.2 shows an example of it for D_8. Evidently, the following holds:

$$\text{Number of fundamental domains of } D_m = |D_m| = 2 \cdot m \qquad (3.1)$$

For the general case of the group D_m, we pick in Fig. 3.2 a special (shaded) fundamental domain and define (see Fig. 3.1)

$$\left.\begin{aligned}
i \;&=\; \text{number of points of } \mathcal{N} \text{ in the interior of the} \\
&\quad\; \text{fundamental domain (i.e., without the points} \\
&\quad\; \text{on the boundary rays or at the origin O)} \\
h \;&=\; \text{number of points of } \mathcal{N} \text{ on the} \\
&\quad\; \text{vertical boundary ray, the origin O excluded} \\
k \;&=\; \text{number of points of } \mathcal{N} \text{ on the} \\
&\quad\; \text{other boundary ray, the origin O excluded} \\
\delta \;&=\; \begin{cases} 1 & \text{if } O \in \mathcal{N} \\ 0 & \text{if } O \notin \mathcal{N} \end{cases} \\
&\text{Here } i, h, k \in \{0, 1, 2, \dots\} \text{ and } i + h + k > 0.
\end{aligned}\right\} \quad (3.2)$$

The motivation for this can be found in the applications, where the elements of the set $\mathcal{N}$ may be atoms of a molecule, the discretization points in a numerical process, or other things.

3.1.1   The Representation ϑ_{per}

We now consider an arbitrary function on $\mathcal{N}$, that is, a correspondence assigning numbers to the n points of $\mathcal{N}$. Each transformation $s \in D_m$ permutes the points of the set $\mathcal{N}$, whence also the numbers associated with them. If the n points are numbered in an arbitrary way, then each $s \in D_m$ corresponds to a permutation $\pi(s)$ of the numbers $1, 2, \dots, n$.

Let us denote by $\Pi(s)$ the permutation matrix of the natural representation of $\pi(s)$. Then as in (1.30) we obtain a well-defined n-dimensional representation

$$s \to \Pi(s) \tag{3.3}$$

of D_m, in the following denoted by ϑ_{per}. The representation space consists of functions defined on the set $\mathcal{N}$, with the basis functions given as follows: They take on the value 1 at point no. i and the value 0 elsewhere.[1]

3.1.2   The Notion of the Orbit

Let us consider an arbitrary point P of the $(i + h + k + \delta)$ elements of $\mathcal{N}$ in a fundamental domain (see Fig. 3.1) of the plane $\mathbf{R}^2$.

The subset of $\mathcal{N}$ consisting of all images of P under the group D_m

$$\{s(P) \mid s \in D_m\} \tag{3.4}$$

is obviously invariant under D_m and is called the **orbit of P under D_m in the plane $\mathbf{R}^2$.** We distinguish the orbit types $(t_1), (t_2), (t_3)$, which shall be explained by aid of Fig. 3.2, employing the group D_8 as an example.

---

[1] The functions on $\mathcal{N}$ may equally well be denoted by vectors. The function value at point no. i then is the ith component of the given basis.

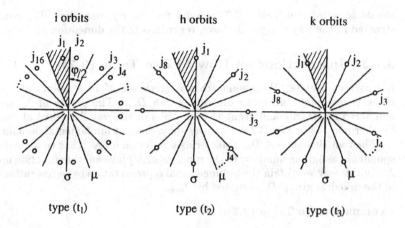

Fig. 3.2

The two reflections σ and μ about the corresponding lines form a generating set of D_m. The natural numbers j_k label the points of the orbit in question. For a fixed numbering in $\mathcal{N}$ they are uniquely specified. If the origin O belongs to $\mathcal{N}$, there will be an additional type (t_0) consisting of this point O alone.

In the general case of D_m, the orbits are classified according to Fig. 3.2. Namely,

Orbit type	Number of points per orbit	Number of orbits
(t_0)	1	δ
(t_1)	$2m$	i
(t_2)	m	h
(t_3)	m	k

$$(3.5)$$

Let us consider a fixed orbit. Then we find that each of the $2m$ fundamental domains contains exactly one point of that orbit.

Since distinct orbits have no points in common, Table (3.5) implies that the total number of points of $\mathcal{N}$ is

$$n = m(2i + h + k) + \delta \tag{3.6}$$

In the following, we describe an arbitrary orbit by a finite sequence of numbers

$$j_1, j_2, j_3, \ldots$$

in clockwise numbering (see Fig. 3.2) where the first number j_1 relates to some particular fixed fundamental domain (the shaded area in Fig. 3.2).

The n-dimensional representation space of ϑ_{per} now consists of all functions on $\mathcal{N}$.

When listing the symmetry adapted basis functions in Table (3.7), we made use of the concept of the orbit and the group structure of D_m. It *should be noted that Table (3.7) is valid for any representation* ϑ_{per} constructed in the way described above, regardless of the dimension n.

3.1.3   Instructions on How to Use Table (3.7)

Let us recapitulate: We assume that a given set $\mathcal{N}$ of n points of the plane is invariant under the dihedral group D_m. The points of $\mathcal{N}$ are numbered arbitrarily and grouped into orbits in the way described above. Furthermore, as in Fig. 3.2 we identify some distinct fundamental domain.

Then an element $s \in D_m$ transforms a function on $\mathcal{N}$ (that is, a correspondence assigning numbers to the n points of $\mathcal{N}$) into another function on $\mathcal{N}$. In this way we obtain the n-dimensional representation by permutations of the dihedral group D_m denoted by ϑ_{per}.

Explanations to Table (3.7):

1. The table describes a complete basis of n **orthogonal symmetry adapted functions** on $\mathcal{N}$. This means that these functions span irreducible subspaces of ϑ_{per} such that subspaces transformed equivalently are in fact transformed in exactly the same way. In other words, we obtain a **symmetry adapted complete reduction of** ϑ_{per}.

2. The head column contains the labels j_k of those points of $\mathcal{N}$ that belong to some arbitrarily chosen orbit. The numbering is clockwise (compare Fig. 3.2 for D_8) where the first number j_1 relates to the orbit point of the specific fundamental domain.

3. The head row contains the irreducible representations of the dihedral group D_m discussed in Section 1.9.2. Note that for odd m, the third and fourth columns are empty, since in this case the representations ϑ_3^1 and ϑ_4^1 do not exist.

4. For each orbit and each irreducible 1-dimensional representation ϑ_ℓ^1, the table gives the values of a symmetry adapted basis function on this orbit. This basis function takes on the value 0 at every point of $\mathcal{N}$ that fails to belong to the orbit. A void entry belonging to a certain orbit and representation means that there exists no corresponding basis function.

   For the 2-dimensional representations ϑ_ℓ^2 the table accordingly gives two basis functions in each of the pairs of columns 5a, 5b and 6a, 6b. Statements analogous to those given for ϑ_ℓ^1 hold for void entries and zero values.

5. If the origin O belongs to $\mathcal{N}$ (that is, if $\delta = 1$), then there is one more symmetry adapted basis function, namely the one corresponding to the unit representation ϑ_1^1. Its only nonzero value is taken on at O and equals 1.

Table 3.7:

Orbit points labeled in accordance with Fig. 3.2	ϑ_1^1	ϑ_2^1	(ϑ_3^1)	(ϑ_4^1)	ϑ_ℓ^2 $\ell = 1,2,\ldots$		$\begin{cases} \frac{m}{2}-1, \\ \text{if } m \text{ even} \\ \frac{m-1}{2}, \\ \text{otherwise} \end{cases}$	
i orbits of type (t_1)								
j_1	1	1	1	-1	1	1	-1	1
j_2	1	-1	1	1	1	1	1	-1
j_3	1	1	-1	1	ω_ℓ	$\overline{\omega}_\ell$	$-\omega_\ell$	$\overline{\omega}_\ell$
j_4	1	-1	-1	-1	ω_ℓ	$\overline{\omega}_\ell$	ω_ℓ	$-\overline{\omega}_\ell$
j_5	1	1	1	-1	ω_ℓ^2	$\overline{\omega}_\ell^2$	$-\omega_\ell^2$	$\overline{\omega}_\ell^2$
j_6		-1	1	1	ω_ℓ^2	$\overline{\omega}_\ell^2$	ω_ℓ^2	$-\overline{\omega}_\ell^2$
j_7	1	1	-1	1	ω_ℓ^3	$\overline{\omega}_\ell^3$	$-\omega_\ell^3$	$\overline{\omega}_\ell^3$
j_8	1	-1	-1	-1	ω_ℓ^3	$\overline{\omega}_\ell^3$	ω_ℓ^3	$-\overline{\omega}_\ell^3$
$\vdots$	$\vdots$	$\vdots$	$\vdots$	$\vdots$	$\vdots$	$\vdots$	$\vdots$	$\vdots$
j_{2m-1}	1	1	-1	1	ω_ℓ^{m-1}	$\overline{\omega}_\ell^{m-1}$	$-\omega_\ell^{m-1}$	$\overline{\omega}_\ell^{m-1}$
j_{2m}	1	-1	-1	-1	ω_ℓ^{m-1}	$\overline{\omega}_\ell^{m-1}$	ω_ℓ^{m-1}	$-\overline{\omega}_\ell^{m-1}$
h orbits of type (t_2)								
j_1	1		1		1	1		
j_2	1		-1		ω_ℓ	$\overline{\omega}_\ell$		
j_3	1		1		ω_ℓ^2	$\overline{\omega}_\ell^2$		
j_4	1		-1		ω_ℓ^3	$\overline{\omega}_\ell^3$		
$\vdots$	$\vdots$		$\vdots$		$\vdots$	$\vdots$		
j_{m-1}	1		1		ω_ℓ^{m-2}	$\overline{\omega}_\ell^{m-2}$		
j_m	1		-1		ω_ℓ^{m-1}	$\overline{\omega}_\ell^{m-1}$		
k orbits of type (t_3)								
j_1	1			-1	α_ℓ^{2m-1}	$\overline{\alpha}_\ell^{2m-1}$		
j_2	1			1	α_ℓ	$\overline{\alpha}_\ell$		
j_3	1			-1	α_ℓ^3	$\overline{\alpha}_\ell^3$		
j_4	1			1	α_ℓ^5	$\overline{\alpha}_\ell^5$		
$\vdots$	$\vdots$			$\vdots$	$\vdots$	$\vdots$		
j_{m-1}	1			-1	α_ℓ^{2m-5}	$\overline{\alpha}_\ell^{2m-5}$		
j_m	1			1	α_ℓ^{2m-3}	$\overline{\alpha}_\ell^{2m-3}$		
Basis functions denoted by	r	y	z	u	v	v'	w	w'
Column no.	1	2	3	4	5a	5b	6a	6b

Header spanning title: Symmetry adapted basis functions for the representations ϑ_{per} of the dihedral groups D_m

Here $\omega_\ell = e^{i\ell 2\pi/m}, \alpha_\ell = e^{i\ell\pi/m}$ ($i =$ imaginary unit), and the bar means complex conjugate.

As an exercise, the reader may construct Table (1.47) from Table (3.7). In that former example we considered the representation ϑ_{per} with $i = 0$, $h = k = \delta = 1$.

If the representing transformations of the two generating reflections σ and μ (see Fig. 3.2) are denoted by

$$R = \Pi(\sigma), \qquad Q = \Pi(\mu) \tag{3.8}$$

and the basis functions in Table (3.7) according to the second to last row[2], then one can verify the following transformation behavior by working with a diagram corresponding to Fig. 3.2 (see Example 1.12):

ϑ_1^1	$Rr =$	r	$Qr =$	r
ϑ_2^1	$Ry =$	$-y$	$Qy =$	$-y$
ϑ_3^1	$Rz =$	z	$Qz =$	$-z$
ϑ_4^1	$Ru =$	$-u$	$Qu =$	u

ϑ_ℓ^2	$Rv = v'$	$Qv = \overline{\omega}_\ell v'$
	$Rv' = v$	$Qv' = \omega_\ell v$
	$Rw = w'$	$Qw = \overline{\omega}_\ell w'$
	$Rw' = w$	$Qw' = \omega_\ell w$

$$\tag{3.9}$$

where

$$\ell = 1, 2, \ldots \begin{cases} (m/2) - 1, & \text{if } m \text{ is even} \\ (m-1)/2, & \text{otherwise} \end{cases} \tag{3.10}$$

The transformation behavior (3.9) corresponds exactly to that of the irreducible representations (1.59) and (1.60).

From Table (3.7) we find the multiplicities of the irreducible representations of D_m appearing in ϑ_{per}:

Representation	Multiplicity
ϑ_1^1	$i + h + k + \delta$
ϑ_2^1	i
ϑ_3^1	$(i + h)$
(ϑ_4^1)	$(i + k)$
ϑ_ℓ^2	$2i + h + k$

For odd m, the third and fourth rows are suppressed. ℓ is given by (3.10). $\tag{3.11}$

3.1.4 Real Orthonormal Symmetry Adapted Basis

Let us construct a real orthonormal symmetry adapted basis. Since all nonzero function values in Table (3.7) are roots of unity, the square of the norm of a basis function with respect to the usual inner product is equal to the number of points in the corresponding orbit. Hence, in accordance with (3.5):

A symmetry adapted basis function of type (t_1) has the norm $\sqrt{2m}$, a symmetry adapted basis function of type $(t_2), (t_3)$ has the norm $\sqrt{m}$. $\tag{3.12}$

---

[2] The transformations (3.9) are meant to be such that a basis function belonging to some orbit is carried into a basis function of the same orbit.

The basis functions of 1-dimensional irreducible subspaces are already real and can easily be normalized by virtue of (3.12).

From the pairs of basis functions v, v' and w, w' of the 2-dimensional irreducible subspaces, one obtains by linear combination real, symmetry adapted, orthogonal pairs of basis functions $\tilde{v}, \tilde{v}'$ and $\tilde{w}, \tilde{w}'$, namely

$$\left.\begin{array}{ll} \tilde{v} = (v + v')/2 & \tilde{v}' = (v - v')/(2i) \\ \tilde{w} = (w + w')/(2i) & \tilde{w}' = (w - w')/2 \end{array}\right\} \tag{3.13}$$

with the norms[3]

$$|\tilde{v}| = |\tilde{v}'| = |\tilde{w}| = |\tilde{w}'| = \sqrt{m} \tag{3.14}$$

Remarkable for computer storage is the fact that *all basis functions (3.13), regardless of the number n of points in $\mathcal{N}$ take on only the following nonzero values:*

$$\cos(k\varphi/2), \quad \sin(k\varphi/2) \qquad k = 0, 1, 2, \ldots, 2m - 1; \text{ where } \varphi = 2\pi/m \tag{3.15}$$

Remark. The paper (Fässler, Mäder, 1980) provides a table of symmetry adapted basis vectors for permutation representations of all finite point groups (e.g., the icosahedral group) of dimensions 2 and 3.

3.2 Application in Quantum Physics

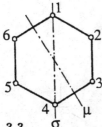

Let us consider the following simple problem related to the **benzol ring** in quantum physics of molecules. The carbon atoms labeled 1, 2, ..., 6 are arranged in the vertices of a regular hexagon, as shown in Fig. 3.3.

Fig. 3.3

Calculation of the so-called molecular orbitals and their energy levels leads by a theory due to E. Hückel (see (Mathiak, Stingl, 1969, Ch. 5)) to an eigenvalue problem asssociated with the following operator H (which is an approximation to the Hamiltonian operator):

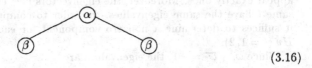

$$\tag{3.16}$$

---

[3] We recommend that the reader verify these properties by a short calculation.

Here f denotes a function defined on the 6 vertices with values f_i ($i =$ number of the carbon atom). Find the eigenfunctions f in the sense that, for example, the first eigenvalue equation reads

$$\alpha f_1 + \beta(f_2 + f_6) = \lambda f_1 \qquad \text{where } \lambda \text{ is an eigenvalue} \qquad (3.17)$$

The ith eigenvalue equation is obtained by moving the "angle" (3.16) until the weight α in the middle coincides with atom no. i and the weights β coincide with the two atoms that are connected with atom no. i. The second eigenvalue equation, for example, is

$$\alpha f_2 + \beta(f_1 + f_3) = \lambda f_2 \qquad (3.18)$$

Here α denotes the Coulomb integral and β the resonance integral.

The operator H has the symmetry of the 6-dimensional representation ϑ_{per} of the dihedral group D_6, where $\mathcal{N}$ is the set of vertices of the hexagon. The reason is that the benzol ring is carried into itself by, say, the reflections σ and μ. Hence we may apply the theory discussed in Section 3.1. In accordance with (3.2),

$$h = 1, \qquad i = k = \delta = 0 \qquad (3.19)$$

We can read the multiplicities from (3.11):

$$\vartheta_{\text{per}} = \vartheta_1^1 \oplus \vartheta_3^1 \oplus \vartheta_1^2 \oplus \vartheta_2^2 \qquad (3.20)$$

By Table (3.7) and Fig. 3.2, the six symmetry adapted vectors are

where $\omega_\ell = e^{i\ell\pi/3}$, $\qquad \ell = 1, 2$.

Theorem 2.7 says that all of the six vectors listed above are eigenvectors of H, because by virtue of (3.20) all irreducible representations of ϑ_{per} appear exactly once. Moreover, the eigenvectors v, v' corresponding to the same ℓ have the same eigenvalues. In order to compute the eigenvalues, it suffices to determine a nonzero component for each of Hr, Hz, and Hv ($\ell = 1, 2$).

Since $\omega_\ell + \overline{\omega_\ell} = \pm 1$, the eigenvalues are

$$\alpha + 2\beta, \quad \alpha - 2\beta, \quad \alpha + \beta, \quad \alpha + \beta, \quad \alpha - \beta, \quad \alpha - \beta \qquad (3.21)$$

3.3 Application to Finite Element Method

From the ample choice of available material we pick the following examples: elliptic equations of second order (e.g., Laplace and Poisson equations), equations of fourth order of type $\Delta\Delta$ (arising in elasticity theory) and the associated eigenvalue problems (determination of oscillation frequencies), parabolic and hyperbolic equations. Here we mainly work in 2- or 3-dimensional domains on which the problems are to be solved. To be concrete, let us consider the elliptic problem

$$\Delta u(x_1, x_2) + \rho(x_1, x_2) \cdot u(x_1, x_2) = f(x_1, x_2) \qquad (3.22)$$

on a domain D of the plane. Here both D and the function ρ admit the dihedral group D_m; that is,

$$\left.\begin{array}{rcl} s(D) & = & D \\ \rho(sx) & = & \rho(x), \qquad x \in D \end{array}\right\} \qquad \forall s \in D_m \qquad (3.23)$$

Moreover, the values of u are prescribed on some part of the boundary ∂D, while on the remaining part of it the values of $\partial u/\partial n$ (exterior normal derivative) are prescribed. *However, neither the boundary value functions nor the load function f in (3.22) need to obey any symmetry conditions.*

3.3.1 Discretization and Symmetry of the Problem

In dealing with the **finite element method** (see (Schwarz, 1989, Ch. 10), (Schwarz, 1988)), we subdivide the original domain D into polygons (3-dimensional domains are subdivided into polyhedra), called the elements. One frequently uses triangles (or tetrahedra); that is, one triangulizes the domain D. The solution to the partial differential equation is approximated on each element by a simple function v in a way such that each of these individual functions, when glued together, extends to a continuous function on the whole domain D. For simplicity, we describe the procedure from now on for a plane triangulized domain. We make the assumption (justifiable by its success) that on each of the triangles the solution functions are linear of the form

$$c_0 + c_1 x_1 + c_2 x_2 \qquad (3.24)$$

The constants c_i are determined by the (still unknown) function values of v taken on at each of the three vertices of an element. Therefore, the variation of the approximation to the energy integral reduces the elliptic boundary value problem (3.22) to a system of linear equations

$$Mv = b \qquad (3.25)$$

with given b. Here $M = (m_{ij})$ is a real positive-definite $n \times n$ matrix with the property $m_{ij} = m_{ji}$, and the components v_i of the vector v are the function values at arbitrarily labeled triangulization points.

If a configuration of the elements is so chosen that the triangulization mesh is carried into itself under the transformations of the dihedral group D_m (see Section 3.3.2 below), then the n triangulization points (making up the set $\mathcal{N}$) are also carried into themselves; in other words, they are permuted. Thus again, a representation ϑ_{per} by permutations

$$s \to D_{\text{per}}(s) \tag{3.26}$$

of the dihedral group D_m in the representation space of the functions v is obtained. The following is true:

$$M \cdot D_{\text{per}}(s) = D_{\text{per}}(s) \cdot M \qquad \forall s \in D_m \tag{3.27}$$

Hence the assumptions of the fundamental theorem in Section 2.3 are satisfied, and Table (3.7) provides a group theoretic transformation of M in block form, leading to a simpler version of problem (3.25).

Remark. Commutativity in (3.27) was anticipated, because the Laplacian commutes with all isometries of the plane. In the following elliptic boundary value problem we show how to verify (3.27) in an application.

3.3.2   Elliptic Boundary Value Problem

Let us consider the problem (3.22). The given function ρ is assumed to have rotational symmetry, while f need not be symmetric. Furthermore, let u be given on the boundary. The triangulization mesh is chosen as in Fig. 3.4.

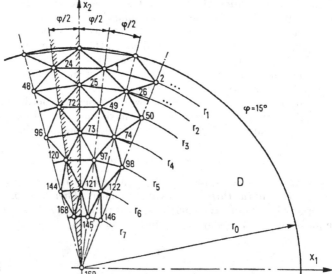

Fig. 3.4

It possesses the symmetry of the dihedral group D_{24}. As an approximation, we assume on each of the triangles that the solutions be linear functions

(3.24). Hence the set $\mathcal{N}$ consists of all triangulization points (that are not on the boundary). In Fig. 3.4 they are numbered from 1 through 169. By definition (3.2) we have

$$i = 0, \quad h = 3, \quad k = 4, \quad \delta = 1 \qquad (3.28)$$

and using (3.6) one verifies that $n = 169$.

Structure of M

The shaded fundamental domain of Fig. 3.4 contains 8 points. They are classified by the following configuration: The 25th equation of (3.25) has the form

$$a \cdot v_{25} + r(v_1 + v_{24}) + s(v_{26} + v_{48}) + t(v_{49} + v_{72}) = b_{25} \qquad (3.29)$$

and is associated with the star shaped figure with six "arms" in Fig. 3.5. Similar stars are associated with the points labeled 25, 26, 27, ..., 48 of the same orbit.

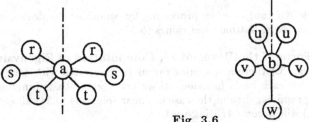

Fig. 3.5   **Fig. 3.6**

There are other stars with six arms, but in general with different directions and different weights, namely for the points labeled 24, 72, 73, 120, and 121.

The star belonging to point no. 168 has five arms (Fig. 3.6). Similar stars are associated with other points of the same orbit.

Lastly, the center no. 169 is associated with a star with 24 arms, carrying equal weights except at the center.

For the star in Fig. 3.5 we need to store only the four values a, r, s, t. The corresponding holds for the other stars, so that by the symmetric structure of the matrix M, we need to store 30 values only.

As was announced earlier, M commutes with the permutation representation of the dihedral group D_{24}. The reason is that the configuration of all 169 stars along with the associated weights are carried into themselves by the operations of D_{24}.

The Transformed Problem

Replacing a basis of the operation space by a new one always requires converting the data to the new basis in the beginning and converting the results back to the old basis at the end.

Let S be the transformation that takes the original basis functions into the orthonormal, real, symmetry adapted basis functions given in Table

(3.7) and Section 3.1.4. The columns of S consist of these tabulated symmetry adapted functions. Thus the matrix S is real and unitary, hence orthogonal.

In the current example, we shall find the solution to (3.25) in three steps:

$$
\left.
\begin{array}{ll}
1. & \text{Compute } \widetilde{M} = S^T M S \text{ and } \tilde{b} = S^T b \\
2. & \text{Solve the problem } \widetilde{M}\tilde{v} = \tilde{b} \\
3. & \text{Compute } v = S\tilde{v}
\end{array}
\right\}
\qquad (3.30)
$$

On substituting (3.28) into (3.11), and by the fundamental theorem of Section 2.3, we obtain the following statement:

$$
\widetilde{M} \text{ decomposes into }
\left\{
\begin{array}{ll}
1 & 8 \times 8 \text{ matrix} \\
1 & 4 \times 4 \text{ matrix} \\
1 & 3 \times 3 \text{ matrix} \\
22 & 7 \times 7 \text{ matrices,} \\
  & \text{where 2 of each are equal}
\end{array}
\right.
\qquad (3.31)
$$

Remark. Of course, when processing by computer one does not store all of the matrix S, but only the values (3.15).

Band Form of the Boxes of $\widetilde{M}$, Computation of Eigenvalues

In Fig. 3.4 there is a unique radius r_k associated with each orbit and hence with each basis function. If we arrange the basis functions by decreasing radii r_k, then in the case of linear solutions (3.24) all submatrices in (3.31) will become **tridiagonal**.

This follows immediately from the structure of the matrix: Action of M on a basis vector corresponding to some orbit with radius r_k gives a linear combination of at most three basis vectors. This is because the operator M modifies only the orbits of adjacent rays in Fig. 3.4 (see Figures 3.5 and 3.6).

At this stage, let us emphasize the fact that the *tridiagonal form has the added merit of providing the Sturm's chains directly, enabling us to compute the eigenvalues and hence the frequencies.* Furthermore, it should be noted *that the band structure of the boxes is still valid for polynomial solution functions of higher degrees* (additional points on the edges of the elements). The band width increases with the degree.

Computational Complexity and Condition Number

If we compute the afore-mentioned boundary value problem without utilizing symmetry, then, with an appropriate numbering as in Fig. 3.4, we obtain a band matrix of 169 rows with band width $b = 24$. Generally speaking, the computational complexity for solving a problem with n unknowns and band width b by a direct method involves $\approx n^2 \cdot b/2$ multiplications or divisions. Solving our example by group theory methods (that is, taking advantage of the symmetry), however, reduces the computational complexity by a factor $1/m = 1/24$, all computations of (3.30) included, see (Fässler, 1976).

Moreover, let us point out that splitting a system of linear equations into subsystems **improves numerical conditioning** of the problem.

3.3.3 Heat conduction

We consider the following parabolic equation on the disk D with boundary ∂D:

$$\frac{\partial u(x,y,t)}{\partial t} = \Delta u(x,y,t) - f(x,y,t) - \rho(x,y,t) \cdot u(x,y,t) \qquad (3.32)$$

The initial and boundary conditions are

$$u(x,y,0) = \alpha(x,y) \qquad (3.33)$$
$$u(x,y,t)|_{\partial D} = \varphi(s,t) \qquad (3.34)$$

Here the given functions f, α, φ on D or its boundary ∂D (s is the arc length on the circle) need not obey any symmetry conditions. However, $\rho(x,y,t)$ should have rotational symmetry.

Now we use the same triangulization for D as in Fig. 3.4, and we denote the approximation of the solution $u(x,y,t)$ in the $n = 169$ points by the vector

$$v(t) = [v_1(t), v_2(t), \ldots, v_{169}(t)]^T \qquad (3.35)$$

Thus we need to determine 169 functions of time $v_i(t)$. The approximated Laplacian operator Δ is described by (3.29), so that (3.32) is approximated by

$$\frac{dv(t)}{dt} = Mv(t) - b(t) \qquad (3.36)$$

with initial conditions resulting from (3.33).

In order to solve the above ordinary differential equations with constant coefficients simultaneously, we now discretize also the time variable with increment h. Among the various methods, the trapezoidal rule (Henrici, 1982), (Schwarz, 1989), being numerically stable, is chosen here.

In explicit numerical processes[4], to ensure stability, the increment h must not exceed certain bounds.

Integrating both sides of (3.36) over the time interval h ($\int_t^{t+h} \ldots$) approximately yields

$$v(t+h) - v(t) = \frac{h}{2}M[v(t+h) + v(t)] - \frac{h}{2}[b(t+h) + b(t)] \qquad (3.37)$$

If $v(t)$ is known, then $v(t+h)$ can be calculated from (3.37):

$$(\frac{h}{2}M - E)v(t+h) = (-\frac{h}{2}M - E)v(t) + \frac{h}{2}[b(t+h) + b(t)] \qquad (3.38)$$

E is the identity matrix.

---

[4] For example, processes of type Runge-Kutta.

The matrix M and thus also $(\frac{h}{2}M - E)$ have the symmetry of the dihedral group D_{24}. Therefore, to split this system of linear equations according to (3.31) we can employ the real, orthonormal, symmetry adapted basis (see Section 3.1.4)

As (3.38) has to be solved for each time interval (starting with $t = 0$, h, $2h$, $3h$, ...), the *group theory method considerably simplifies calculations.*

If the solution function is not required after each time interval, but rather after several time intervals, then the computational complexity reduces more still, since converting the results back and forth via the matrix S accounts for most of the computational complexity.

Remark. The group theory method may be of interest also for explicit integration methods. It relaxes the stability bounds.

For further applications we refer to the works (Qui, Zhong, 1981) and (Guy, Mangeol, 1981), the latter containing nonlinear examples.

3.4   Perturbed Problems with Symmetry

We consider the linear operator

$$M = A + B$$

with the property that A shall commute with a completely reducible representation $\vartheta : s \to D(s)$ of an arbitrary group G; i.e.,

$$AD(s) = D(s)A \qquad \forall s \in G \tag{3.39}$$

and that B shall be "small" relative to A in some not yet specified sense. The perturbation B of M need not obey any symmetry conditions. In perturbation theory it is customary to use the following method:

We first solve the unperturbed problem by use of symmetry and then, proceeding from this first approximation, we solve the perturbed problem. This is a useful method as will be shown formally through the example of a system of linear inhomogeneous equations

$$Mx = (A + B)x = b \tag{3.40}$$

Let ξ be the solution of the unperturbed problem

$$A\xi = b$$

(A non-singular), which we may have found by use of symmetry. On subtracting the latter equation from (3.40), we obtain

$$A(x - \xi) + Bx = A(x - \xi) + B(x - \xi) + B\xi = 0$$

In a first approximation, the summand in the middle can be neglected, because both B and the difference vector $d = x - \xi$ are "small". Thus in a second approximation, the simultaneous equations

$$Ad = \eta$$

with the given inhomogeneity $\eta = -B\xi$ must be solved. But to do this, we can exploit the symmetry of A once again and so we finally obtain as a second approximation to the perturbed problem (3.40), that $x = \xi + d$.

3.5. Fast Fourier Transform on Finite Groups

Here we introduce an algorithm that makes considerable use of Chapter 2 (symmetry adapted bases). We shall also mention some of the possible applications of it.

This section is mainly based on (Diaconis, Rockmore, 1988), which covers more detailed design ideas for a fast Fourier transform on finite groups, in particular for "small" symmetric groups S_n. New architectures (parallel processor) are also considered and questions still open are raised.

3.5.1 Definitions and Properties

Let $f : G \to \mathbf{C}$ be a complex-valued function on a finite group G and $\vartheta : s \to D(s)$, $s \in G$, be an n-dimensional representation of G. The group $G' = \{D(s) \mid s \in G\}$ of $n \times n$ matrices is homomorphic to G.

Definition. The **Fourier transform** of f at the representation ϑ of G is defined by

$$\hat{f}(\vartheta) = \sum_{s \in G} f(s) \cdot D(s) \tag{3.41}$$

Thus $\hat{f}(\vartheta)$ is a complex $n \times n$ matrix.

Let $\vartheta_j : s \to D_j(s)$, $s \in G$, be the *entire collection* of irreducible inequivalent representations of the group G with $\dim(\vartheta_j) = n_j$.

Definition. The set of Fourier transforms $\hat{f}_j(\vartheta)$ at *all* irreducible representations $\vartheta_j : s \to D_j(s)$ *of the group* G is called the **spectrum** of f.

The spectrum determines f by means of

Theorem 3.1 (Fourier inversion formula)

$$f(s) = \frac{1}{|G|} \sum_j n_j \cdot \operatorname{tr}[D_j(s^{-1}) \cdot \hat{f}(\vartheta_j)] \tag{3.42}$$

Proof. From definition (3.41) we obtain

$$\hat{f}(\vartheta_j) \cdot D_j(s^{-1}) = f(s) \cdot E + \sum_{t \neq s} f(t) \cdot D_j(t \cdot s^{-1})$$

where E is the $n_j \times n_j$ identity matrix and $t \in G$.

On passing to traces on both sides, we obtain

$$\text{tr}[\hat{f}(\vartheta_j) \cdot D_j(s^{-1})] = n_j \cdot f(s) + \sum_{t \neq s} f(t) \cdot \chi_j(t \cdot s^{-1})$$

where χ_j is the character of ϑ_j.

Now we multiply both sides by n_j and sum over all j. This together with (5.165) yields

$$\sum_j n_j \cdot \text{tr}[\hat{f}(\vartheta_j) \cdot D_j(s^{-1})] = |G| \cdot f(s) + \sum_{t \neq s} f(t) \sum_j n_j \chi_j(t \cdot s^{-1})$$

Property (5.175) of characters, with the first factor being the character of the neutral element e and $n_j = $ dimension of ϑ_j, yields

$$\sum_j n_j \chi_j(p) = 0 \qquad \forall p \neq e$$

Thus the inner sum on the right hand side becomes zero and

$$\sum_j n_j \cdot \text{tr}[\hat{f}(\vartheta_j) \cdot D_j(s^{-1})] = |G| \cdot f(s)$$

$\square$

Definition. The function

$$(f \star g)(s) = \sum_{t \in G} f(st^{-1}) \cdot g(t) \qquad s \in G \tag{3.43}$$

is called the **convolution** of f and g.

The following is true:

$$\widehat{f \star g}(s) = \hat{f}(s) \cdot \hat{g}(s) \tag{3.44}$$

The details may be left as an exercise to the reader.

From the Fourier inversion formula (3.42) we obtain the convolution

$$(f \star g)(s) = \frac{1}{|G|} \sum_j n_j \cdot \text{tr}[D_j(s^{-1}) \cdot \hat{f}(\vartheta_j) \cdot \hat{g}(\vartheta_j)] \qquad s \in G \tag{3.45}$$

3.5.2   Direct and Fast Algorithm

In various applications, some of which will be collected at the end, a fast algorithm for calculating spectra and convolutions is of great importance, as such an algorithm has to be carried out frequently. It will prove that a fast algorithm for calculating spectra must be used in (3.45) in order to obtain a good algorithm for calculating the convolution.

We assume that all ireducible and inequivalent representations ϑ_j of the group G are stored in a computer memory for every $s \in G$.

Direct computation of the spectrum of f in (3.41) and the convolution $f \star g$ in (3.45) requires order $|G|^2$ multiplications and additions, because $|G| \cdot \sum_j n_j^2 = |G|^2$ by (1.50).

The main idea in designing a fast Fourier transform on G is outlined in the following:

Let us take a subgroup $H \subset G$ with $q = |G|/|H|$.

With the coset representatives chosen to be $e = s_0, s_1, s_2, \ldots, s_{q-1}$, the group G can be written as

$$\bigcup_i s_i H = G$$

so that the Fourier transform (3.41) becomes

$$\hat{f}(\vartheta) = \sum_{k=0}^{q-1} \sum_{h \in H} f(s_k \cdot h) \cdot D(s_k \cdot h) = \sum_{k=0}^{q-1} D(s_k) \sum_{h \in H} f_k(h) \cdot D(h) \quad (3.46)$$

with $f_k(h) = f(s_k \cdot h)$.

In general, the restriction of an irreducible representation ϑ on a subgroup is *not* irreducible; for, the irreducible representation ϑ of G *restricted* on the subgroup H decomposes into

$$\vartheta|_H = \bigoplus_k c_k \vartheta_k^H \qquad c_k \in \{0, 1, 2, \ldots\} \qquad (3.47)$$

where the ϑ_k^H denote the inequivalent irreducible representations of the subgroup H.

We may assume that the representing matrices $D(h)$ in (3.46) relate to a symmetry adapted basis. This assumption is often satisfied for abelian groups, but also for symmetric groups $G = S_n$ if the representations are constructed by means of the Young tableaux; (see (Stanton, 1988), which gives an elementary introduction to Young tableaux and the representations of symmetric groups).

In this case, the $D(h)$ in (3.46), being related to a symmetry adapted basis, are block diagonal. The idea is now the following:

In order to evaluate the inner sum in (3.46), it is enough to just calculate *one* block for each of the irreducible inequivalent representations of H appearing in (3.47). For them, $c_k \geq 1$.

By iterating this technique for a tower of subgroups

$$G \supset G_1 \supset G_2 \supset \ldots \supset \{e\}$$

we can again reduce the computational complexity. This qualitatively describes the fast Fourier transform on some (in general non-commutative) group G.

Let us denote its computational complexity by $\text{Fast}(G)$. If we use just one subgroup H and work with a symmetry adapted basis, then, owing to (3.46), we obtain an upper bound

$$\text{Fast}(G) \le q \cdot |H|^2 + (q-1) \cdot |G|$$

for the number of multiplications and additions. Here the first term comes from the inner sum, while the second term comes from the outer sum in (3.46), assuming that the computational complexity for the product of two $n \times n$ matrices is n^2 (recall that $D(s_k)$ is, in general, not block diagonal). Here it needs to be said that this assumption is rather optimistic. For in most cases, one uses algorithms with complexity n^3. A theoretical lower bound has exponent 2. The currently best algorithm, though, has exponent 2.38; compare (Coppersmith and Winograd, 1987). But it requires a larger overhead, so that its use only pays for very large matrices.

Dividing both sides by $|G|$, and bearing in mind that $|G| = q \cdot |H|$, gives

$$\frac{\text{Fast}(G)}{|G|} \le |H| - (k-1)$$

Equality holds only if all irreducible representations of $|H|$ occur. The upper bound is minimal for $|H| \approx \sqrt{|G|}$. Then $k \approx |H|$, and hence

$$\text{Fast}(G) \le 2 \cdot |G|^{3/2}$$

On especially favorable occasions one may obtain, by choosing an optimal tower, the order of

$$\text{Fast}(G) = \text{const} \cdot |G| \cdot \log |G|$$

We shall show this for the case $G = C_n$ (cyclic group) in the following subsection.

3.5.3   The Classical Fast Fourier Transform (FFT)

This special case is treated in (Henrici, 1982), where also an implementation of the algorithm is given; also see (Weaver, 1988).

The group C_N may be regarded as the multiplicative group of the complex Nth roots of unity $\{1, \omega, \omega^2, \ldots, \omega^{N-1}\}$, where $\omega = e^{i\frac{2\pi}{N}}$. It may also be regarded as the additive group of the integers modulo N, since $C_N \cong \mathbf{Z}_N$.

For $N = p \cdot q$,

$$C_p = \{1, \omega^q, \omega^{2q}, \ldots, \omega^{(p-1)q}\}$$

is a subgroup of $C_{p \cdot q}$.

Our coset representatives are $s_k = e^{ik\frac{2\pi}{N}}$, $(k = 0, 1, 2, \ldots, N\text{-}1)$. With the short notation $\hat{f}_j = \hat{f}(\vartheta_j)$ for the Fourier transform of f at frequency j and with the 1-dimensional irreducible representations of (1.52), Equation (3.46) becomes

$$\hat{f}_j = \sum_{k=0}^{q-1} \omega^{kj} \cdot \sum_{m=0}^{p-1} f_k(m) \cdot \omega^{qjm} \qquad j = 0, 1, 2, \ldots, N\text{-}1 \qquad (3.48)$$

where $\omega = e^{i\frac{2\pi}{N}}$ and $f_k(m) = f(mq + k)$.

Note that the arguments are to be taken modulo N.

The *direct computation* of the spectrum $\hat{f}_0, \hat{f}_1, \ldots, \hat{f}_{N-1}$ involves N^2 *(complex) multiplications and additions.*

Let us now consider the special case where N is a power of 2: $N = 2^r$. If we choose the subgroup $C_{N/2} \subset C_N$, then $q = 2$. Equation (3.48), with the additive representatives 0 and 1, implies the **duplication property**

$$\hat{f}_j = \sum_{m=0}^{N/2-1} f(2m)\omega^{2jm} + \omega^j \cdot \sum_{m=0}^{N/2-1} f(2m+1)\omega^{2jm} \qquad (3.49)$$

This property is fundamental to the FFT in the original literature (Cooley, Tukey, 1965). However, (Goldstine, 1977) shows that its idea is rooted in a paper by Carl Friedrich Gauss (1777-1855).

By recursively applying the duplication property to the tower of subgroups $C_{2^r} \supset C_{2^{r-1}} \supset C_{2^{r-2}} \supset \ldots \supset C_1 = \{1\}$, we obtain the computational complexity

$$\text{Fast}(C_N) = 2 \cdot \text{Fast}(C_{N/2}) + 1$$

The initial condition is $\text{Fast}(C_1) = 0$, as there are no operations required for the trivial group. The solution to this recursion is

$$\text{Fast}(C_{2^r}) = \frac{1}{2}2^r \cdot r$$

Thus in our case of $G = C_{2^r}$, the *fast Fourier transform* indeed uses *only* $\frac{1}{2}|G| \cdot \lg|G|$ *complex multiplications and additions* (lg is the logarithm to the base 2).

This computational complexity is definitely smaller than that of the direct method, particularly for large $|G| = 2^r$. The transformation taking a discrete function, the **data** $f = (f_0, f_1, f_2, \ldots, f_{N-1})^T$, into its spectrum $\hat{f} = (\hat{f}_0, \hat{f}_1, \hat{f}_2, \ldots, \hat{f}_{N-1})^T$ for the special case $G = C_N$ is given by

$$\hat{f} = F \cdot f$$

where F (the so-called **Fourier operator**) is an $n \times n$ matrix with entries $f_{ij} = e^{ij\frac{2\pi}{N}}$.

For our special case the Fourier inversion formula (3.42) yields

$$f = \frac{1}{N}\overline{F} \cdot \hat{f} \tag{3.50}$$

Thus the matrix $\frac{1}{\sqrt{N}}F$ is unitary. This is also due to the orthogonality (5.141) of the characters of the group C_N, which are the rows of F.

For the convolution vector $f \star g$, Equation (3.45) yields

$$F \cdot (f \star g) = (\hat{f}_0 \cdot \hat{g}_0, \hat{f}_1 \cdot \hat{g}_1, \ldots, \hat{f}_{N-1} \cdot \hat{g}_{N-1})^T$$

In applying the operator $F^{-1} = \frac{1}{N}\overline{F}$ to both sides of this equation, we obtain

$$f \star g = \frac{1}{N}\overline{F}(\hat{f}_0 \cdot \hat{g}_0, \hat{f}_1 \cdot \hat{g}_1, \ldots, \hat{f}_{N-1} \cdot \hat{g}_{N-1})^T \tag{3.51}$$

Evidently, the actions of both $\overline{F}$ and F (by use of the FFT) can be carried out with $\frac{1}{2}|G| \cdot \lg|G|$ operations. Therefore, the latter equation gives rise to

Theorem 3.2 *The convolution of the two sequences f and g of length $|G| = 2^r$ for a cyclic group G can be performed in no more than (approximately)*

$$\frac{3}{2}|G| \cdot \lg|G|$$

complex multiplications and additions. (lg is the logarithm to the base 2).

Remark. Directly computing the convolution by use of its definition would require $|G|^2$ complex multiplications and additions, a number which increases considerably with large $|G|$. (Henrici, 1982) provides an explicit implementation of the FFT.

3.5.4   Applications and Remarks

A **time series** is a finite sequence of observations of a physical quantity x at equidistant times: $t_k = t_0 + \Delta t \cdot k$ ($\Delta t = $ const, $k = 0, 1, 2, \ldots, N\text{-}1$).

The quantity x may be a **brain current** as recorded in an encephalogram or the displacement of a seismograph during an **earthquake**. The length of a time series requires values as large as 2^{12}.

Frequently, important conclusions on the nature of a physical process can be inferred from analyzing its time series. For instance, in electroencephalography, epilepsy can be discovered in this way.

Digital image processing also deals with lengths of sizes such as 2^{12}, 2^{16} (Lenz, 1990). The direct methods become very difficult in the applications mentioned above, because the same transform has to be carried out a large number of times.

Classical FFT is used to **multiply and evaluate polynomials** and is the basis for the **modern fast way of multiplying integers**. We shall take a closer look at this:

Let
$$P(x) = p_0 + p_1 x + p_2 x^2 + \ldots + p_{N-1} x^{N-1}$$
$$Q(x) = q_0 + q_1 x + q_2 x^2 + \ldots + q_{N-1} x^{N-1}$$

be polynomials of degree $< N = 2^r$.

Their product

$$R(x) = P(x) \cdot Q(x) = r_0 + r_1 x + r_2 x^2 + \ldots + r_{2N-2} x^{2N-2}$$

is a polynomial of degree $\leq 2N - 2$ with coefficients

$$r_i = p_0 q_i + p_1 q_{i-1} + p_2 q_{i-2} + \ldots + p_i q_0$$

where for $i \geq N$, the coefficients with subscripts $\geq N$ on the right side are defined to be zero.

The r_i are identical with the *convolutions* of the two subsequences from the left with lengths $(i + 1)$ of

$$p_0, p_1, p_2, \ldots, p_{N-1}, 0, 0, \ldots, 0$$

$$q_0, q_1, q_2, \ldots, q_{N-1}, \underbrace{0, 0, \ldots, 0}_{N \text{ zeros}}$$

Remark. If the variable x takes the place of the base b (e.g. 10) and the coefficients those of the digits from 0 through $b - 1$ of the numbers in question, then this algorithm of multiplying polynomials provides a **fast technique for multiplying large numbers**. Here the case where r_i is greater than b still needs to be treated.

(Good, 1958) gives an interesting application of the classical FFT to **statistics**: Non-abelian transforms are used in theoretical considerations, such as the **rate of convergence of Markov chains** towards their stationary distribution (see Section 6.2).

More applications of non-abelian transforms may be found in (Rothaus, Thompson, 1966) or in (Chihara, 1987), the latter dealing with invertibility of Radon transforms.

Chapter 4

Continuous Groups With Representations

4.1 Continuous Matrix Groups

We resume Example 1.5 where the group G of all proper rotations of the space $\mathbf{R}^3$ was introduced. After choosing a cartesian coordinate system (i.e., an orthonormal basis) in $\mathbf{R}^3$, we can describe such a rotation by the following matrix A: the kth basis vector is rotated into a vector a_k with coordinates a_{1k}, a_{2k}, a_{3k}. Thus these coordinates must be written in the kth column of A, thus

$$A = \begin{bmatrix} a_{11} & a_{12} & a_{13} \\ a_{21} & a_{22} & a_{23} \\ a_{31} & a_{32} & a_{33} \end{bmatrix}$$

The three vectors a_k are orthogonal unit vectors. Therefore, the inner product of two distinct columns $= 0$, and the inner product of a column with itself $= 1$. In short,

$$A^T A = E \qquad \text{or equivalently} \qquad A^{-1} = A^T \tag{4.1}$$

Here A^T is the transposed matrix and E the 3×3 identity matrix. Real matrices with the property (4.1) are called **orthogonal**. Consequently, the inverse of an orthogonal matrix is obtained by interchanging rows and columns.

The orthogonal matrices (4.1) form the so-called **orthogonal group** **O(3)**. Obviously, O(3) is a faithful representation of the group of all proper and improper rotations ($\det A = \pm 1$). In most considerations we deal with the proper rotations; that is, besides (4.1) the corresponding matrices must have the property

$$\det A = +1 \tag{4.2}$$

This subgroup of O(3) satisfying (4.2) is called the **special orthogonal group SO(3)** and is a faithful representation of the group of all proper rotations. It follows that G and SO(3) are isomorphic. The matrices of O(3) and SO(3) depend differentiably upon three free parameters. The number 3 comes from the fact that by (4.1) the 9 unknown matrix entries of A must satisfy the 6 equations

$$\sum_{\lambda=1}^{3} a_{\lambda i} \cdot a_{\lambda k} = \delta_{ik} = \begin{cases} 1, & \text{if } i = k \\ 0, & \text{otherwise} \end{cases} \qquad i, k = 1, 2, 3 \text{ and } i \le k$$

Groups of linear transformations are described by matrices relative to a fixed coordinate system. Important for applications are those matrix groups whose matrix entries can be described differentiably (at least in a neighborhood of E) by certain parameters. All the examples listed in Table (4.3) have this property. They belong to the so-called **continuous matrix groups**. For our purposes it is not necessary to generally and precisely define this topological notion.

Continuous linear groups of $n \times n$ matrices:

Symbol	Name of group	Properties of matrix A		f
GL(n,C)	general linear	complex,	$\det A \ne 0$	$2n^2$
GL(n,R)	general linear	real,	$\det A \ne 0$	n^2
SL(n,C)	special linear (unimodular)	complex,	$\det A = 1$	$2n^2 - 2$
SL(n,R)	special linear (real unimodular)	real,	$\det A = 1$	$n^2 - 1$
U(n)	unitary	complex,	$A^T \bar{A} = E$	n^2
SU(n)	special unitary	complex,	$A^T \bar{A} = E,$ $\det A = 1$	$n^2 - 1$
O(n)	orthogonal	real,	$A^T A = E$	$\frac{1}{2}n(n-1)$
SO(n)	special orthogonal (proper orthogonal)	real,	$A^T A = E,$ $\det A = 1$	$\frac{1}{2}n(n-1)$

(4.3)

Here f is the number of free real parameters, $\bar{A}$ the complex conjugate of A, and A^T the transpose of A.

Clearly, all matrix groups are subgroups of the general linear group. Furthermore,

$$\text{GL}(n, \mathbf{R}) \supset \text{SL}(n, \mathbf{R}) \supset \text{SO}(n), \qquad \text{U}(n) \supset \text{O}(n) \supset \text{SO}(n), \text{ etc.}$$

4.1.1 Comments on U(n)

The elements of U(n) were already discussed in Section 1.11.1: By definition, an $n \times n$ matrix $A = (a_{\alpha\beta})$ is unitary if

$$A^T \bar{A} = E \tag{4.4}$$

This condition is equivalent to the fact that the matrix entries must obey the following equations

$$\sum_{\lambda=1}^{n} a_{\lambda i} \cdot \bar{a}_{\lambda k} = \delta_{ik} = \begin{cases} 1, & \text{if } i = k \\ 0, & \text{if } i \neq k \end{cases} \quad i, k = 1, 2, \ldots n \text{ and } i \leq k \quad (4.5)$$

The number f is obtained as follows: For $i \neq k$, Equation (4.5) describes $1 + 2 + \ldots + (n - 1) = n(n - 1)/2$ complex or $n(n - 1)$ real equations (considering the real and imaginary parts separately). For $i = k$, Equation (4.5) describes n real equations. But the general linear group $GL(n, \mathbf{C})$ possesses $2n^2$ real parameters. Hence for $U(n)$, we have

$$f = 2n^2 - n(n - 1) - n = n^2 \quad (4.6)$$

4.1.2 Comments on SU(n)

We use the result in (4.6) to determine f. The additional condition $\det A = 1$ implies that there is one more real equation than in the case of unitary groups, because the absolute value of the determinant equals 1 for unitary matrices. Hence

$$f = n^2 - 1 \quad (4.7)$$

We shall investigate SU(2) in more detail. Let the matrix A of this group be of the form

$$A = \begin{bmatrix} a & b \\ x & y \end{bmatrix}$$

where $a\bar{a} + b\bar{b} = 1$ (norm of the first row). The orthogonality condition $a\bar{x} + b\bar{y} = 0$ for the two row vectors implies

$$\bar{x} = -\lambda b, \qquad \bar{y} = \lambda a$$

for some scalar factor $\lambda \in \mathbf{C}$. Thus the determinant condition gives

$$\det A = ay - bx = \bar{\lambda}(a\bar{a} + b\bar{b}) = 1 \quad \text{whence} \quad \lambda = 1$$

Therefore, the matrices of SU(2) must be of the form

$$A = \begin{bmatrix} a & b \\ -\bar{b} & \bar{a} \end{bmatrix} \quad \text{with} \quad a\bar{a} + b\bar{b} = 1 \quad (4.8)$$

Remark. We leave it to the reader to check the accuracy of the remaining values of f in Table (4.3).

4.1.3 Connectedness of Continuous Groups

Definition. A continuous $n \times n$ matrix group G is said to be **connected** if every matrix $A_0 \in G$ is associated with a continuous 1-parameter subset $A(t)$, $(0 \leq t \leq 1)$, of G with

$$A(0) = \text{identity matrix}, \qquad A(1) = A_0 \quad (4.9)$$

Example 4.1 The orthogonal group $O(n)$ is not connected. It decomposes into the two portions with either $\det = +1$ or $\det = -1$. **Reason**: $\det A(t)$ is continuous on a continuous 1-parameter subset $A(t)$.

Example 4.2 The unitary group $U(n)$ is connected. **Reason**: A well-known theorem in linear algebra says that for any unitary matrix A_0 there exists a unitary $n \times n$ matrix S such that $\tilde{A}_0 = S^{-1}A_0S$ is diagonal. Its diagonal entries are complex numbers of absolute value 1, hence of the form $e^{i\varphi_k}$ ($\varphi_k \in \mathbf{R}$, $k = 1, 2, \ldots, n$). By successively transforming the φ_k into 0 one obtains a continuous subset $\tilde{A}(t)$ with $\tilde{A}(1) = \tilde{A}_0$ and $\tilde{A}(0) = E$. Hence $A(t) = S\tilde{A}(t)S^{-1}$ (S fixed) carries $A_0 = A(1)$ continuously into E.

Example 4.3 The special linear group $SL(2,\mathbf{C})$ is connected. **Reason**: For any given matrix $A_0 \in SL(2,\mathbf{C})$ there is an $S \in SL(2,\mathbf{C})$ such that $\tilde{A}_0 = S^{-1}A_0S$ has triangular form:

$$\tilde{A}_0 = \begin{bmatrix} a & b \\ 0 & 1/a \end{bmatrix}, \quad a = a_1 + ia_2 \neq 0, \quad b = b_1 + ib_2$$

A_0 is carried continuously into the identity matrix by $b_1 \to 0$, $b_2 \to 0$, $a_1 \to 1$, $a_2 \to 0$.

Remark. The continuous matrix groups in Table (4.3), except $O(n)$ and $GL(n, \mathbf{R})$, are connected.

4.2 Relationship Between Some Groups

We start from the special linear group $SL(2,\mathbf{C})$ of all complex 2×2 matrices A with $\det A = 1$. Let X be a singular or non-singular 2×2 matrix. Its four complex entries may be thought of as coordinates of a 4-dimensional vector in $\mathbf{C}^4$. Then

$$Y = AX\bar{A}^T \tag{4.10}$$

defines a linear transformation in $\mathbf{C}^4$ carrying an X-vector into a Y-vector, because

$$A(\alpha X_1 + \beta X_2)\bar{A}^T = \alpha AX_1\bar{A}^T + \beta AX_2\bar{A}^T \qquad \alpha, \beta \in \mathbf{C} \tag{4.11}$$

This is a 4-dimensional representation of $SL(2,\mathbf{C})$; for, if we combine (4.10) with

$$Z = BY\bar{B}^T \qquad B \in SL(2,\mathbf{C}) \tag{4.12}$$

then we find that

$$Z = BAX\bar{A}^T\bar{B}^T = (BA)X\overline{(BA)}^T \tag{4.13}$$

In view of (4.10), $A = E$ (identity matrix) transforms the vectors X identically. Which other members A of $SL(2,\mathbf{C})$ are associated with the identity

transformation of $\mathbf{C}^4$?
For such an A,

$$X = AX\bar{A}^T \qquad \forall X \tag{4.14}$$

On substituting $X = E$, we obtain

$$A\bar{A}^T = E \qquad \text{whence} \qquad A^{-1} = \bar{A}^T \tag{4.15}$$

Thus (4.14) implies

$$X = AXA^{-1} \qquad \text{or equivalently} \qquad AX = XA \qquad \forall X \tag{4.16}$$

The set of 2×2 matrices X forms an irreducible system of matrices. Hence Theorem 2.2 implies that A is a multiple of the identity matrix. Furthermore, by definition (4.10) of our 4-dimensional representation and as $\det A = 1$, it follows that the two matrices

$$E = \begin{bmatrix} 1 & 0 \\ 0 & 1 \end{bmatrix}, \qquad -E = \begin{bmatrix} -1 & 0 \\ 0 & -1 \end{bmatrix} \tag{4.17}$$

and only these two are associated with the identity transformation of $\mathbf{C}^4$.

4.2.1 Relationship between SL(2, C) and the Lorentz group

Now we restrict our attention to the matrices X of the form

$$X^T = \overline{X} \tag{4.18}$$

They are said to be **hermitian**. In (4.10), the matrix Y is also hermitian, because

$$Y^T = \overline{A}X^T A^T = \overline{A}\,\overline{X}A^T = \overline{Y} \tag{4.19}$$

According to (4.18), a general hermitian matrix is characterized by real entries along the main diagonal and by complex conjugate pairs of entries off the main diagonal ($x_{ik} = \bar{x}_{ki}$).

We suggest to set the 2×2 hermitian matrices X as follows:

$$X = \begin{bmatrix} -x_1 + x_4 & x_2 + ix_3 \\ x_2 - ix_3 & x_1 + x_4 \end{bmatrix} \qquad x_i \in \mathbf{R} \tag{4.20}$$

Y shall be described likewise by using the four real numbers y_1, y_2, y_3, y_4. Thus (4.10), together with (4.13) and (4.19), provides a real linear transformation τ_A of the 4-dimensional vectors $(x_1, x_2, x_3, x_4)^T$ into the vectors $(y_1, y_2, y_3, y_4)^T$ with the property that

$$\tau_A \cdot \tau_B = \tau_{AB} \qquad \forall A, B \in \mathrm{SL}(2,\mathbf{C}) \tag{4.21}$$

Hence the correspondence

$$A \rightarrow \tau_A \qquad A \in \mathrm{SL}(2,\mathbf{C}) \tag{4.22}$$

is in fact a **homomorphism** (see (1.22)) which, in addition, is continuous[1]. From (4.10) it follows that

$$\det Y = |\det A|^2 \cdot \det X = \det X$$

that is, $y_4^2 - y_1^2 - y_2^2 - y_3^2 = x_4^2 - x_1^2 - x_2^2 - x_3^2$

Consequently, our linear transformations τ_A in $\mathbf{R}^4$ leave the quadratic form

$$x_1^2 + x_2^2 + x_3^2 - x_4^2 \tag{4.23}$$

invariant and are thus known as the **Lorentz transformations**.

Claim I. *The set of all image matrices $\{\tau_A \mid A \in SL(2, \mathbf{C})\}$ in (4.22) is connected, and $\det \tau_A = 1$.*
Proof. Let $x \in \mathbf{R}^4$ and F be a real symmetric matrix with the quadratic form $x^T F x$. Let further $x = Ty$ be a real linear transformation leaving a non-degenerate quadratic form (i.e., $\det F \neq 0$) invariant. Under these assumptions

$$y^T F y = x^T F x = (Ty)^T F (Ty) = y^T (T^T F T) y \qquad \forall y \in \mathbf{R}^4$$

Hence

$$F = T^T F T$$

holds and by passing to the determinants, we obtain

$$\det F = |\det T|^2 \cdot \det F$$

Moreover, since T is real,

$$|\det T|^2 = 1 \qquad \text{thus} \qquad \det T = \pm 1$$

Thus with $T = \tau_A$ and F a diagonal matrix with diagonal entries 1, 1, 1, -1, we obtain

$$\det \tau_A = \pm 1 \qquad \forall A \in SL(2,\mathbf{C}) \tag{4.24}$$

Continuity of the homomorphism (4.22) and the fact that the group $SL(2,\mathbf{C})$ is connected imply the first part of Claim I. Furthermore, (4.24) gives

$$\det \tau_A = +1 \tag{4.25}$$

$\square$

---

[1] i.e., the matrix entries of τ_A vary continuously with the matrix entries of A.

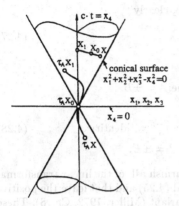

Fig. 4.1

Thus in particular, invariance of the quadratic form (4.23) implies that the so-called **Minkowski cone**

$$x_1^2 + x_2^2 + x_3^2 - x_4^2 \leq 0 \qquad (4.26)$$

is transformed invariantly under the τ_A (Fig. 4.1).

Claim II. *The upper half of the Minkowski cone is transformed into itself under the transformations τ_A (that is, $x_4 > 0$ implies $y_4 > 0$). In the language of relativity theory, τ_A carries an event in the future into another such event. ($x_4 = c \cdot t = $ time axis, $c = $ speed of light in vacuum).*

Proof. For the special case of $x_1 = x_2 = x_3 = 0$, $x_4 = 1$, the matrix $X = E$ in (4.20). Hence in (4.10)

$$Y = A\bar{A}^T$$

By taking the traces[2], it follows for $A = (a_{ik})$ that

$$2y_4 = \sum_{i,k=1}^{2} |a_{ik}|^2 > 0$$

Thus $y_4 > 0$.

It remains to show that for an **arbitrary** vector X of the cone (4.26) with positive fourth component $x_4 > 0$ the transformed vector $\tau_A X$ has the property $y_4 > 0$.

In Fig. 4.1 let the vector with $x_1 = x_2 = x_3 = 0$, $x_4 = 1$ be denoted by X_1. Suppose that the fourth component $\tau_A X \leq 0$. Then by continuity[3] there is a vector $X_0 \neq 0$ on the line segment joining X_1 and X whose transformed vector $\tau_A X_0$ has the component $y_4 = 0$. Hence X_0 is the zero vector, contradicting (4.25). □

In (4.17) we already stated that the **kernel**[4] of the homomorphism $A \to \tau_A$ consists of the two 2×2 matrices E and $-E$. Hence *in this way, every matrix τ_A is associated with two distinct matrices, namely A and $-A$.*

---

[2]The trace of a matrix is defined as the sum of its diagonal elements.

[3]Every linear transformation is continuous.

[4]The kernel of a homomorphism consists of all elements that are mapped into the neutral element.

The reason is that from definition (4.10), clearly

$$\tau_A = \tau_{-A} \tag{4.27}$$

If on the other hand

$$\tau_A = \tau_B \text{ for some } A \neq \pm B$$

then

$$\tau_{AB^{-1}} = \tau_A \cdot \tau_B^{-1} = \tau_A \cdot \tau_A^{-1} = \text{ identity} \tag{4.28}$$

contradicting property (4.17), since $AB^{-1} \neq \pm E$.

Remark. Equations (4.10) and (4.20) furnish **all** of the linear transformations τ_A that satisfy properties (4.23) and (4.25) and that leave the positive half of the Minkowski cone ($x_4 \geq 0$) invariant (Miller, 1972, Ch. 8). These form the so-called **group of all proper positive Lorentz transformations** $L = \{\tau_A \mid A \in SL(2, C)\}$ or the **proper positive Lorentz group** L.

We have thus proved the following theorem:

Theorem 4.1 *The continuous homomorphism $A \to \tau_A$ defined by (4.10) and (4.20) between the groups $SL(2, C)$ and L has the property that each proper positive Lorentz transformation $\tau_A \in L$ is associated with exactly 2 unimodular matrices, namely A and $-A$ (see Fig. 4.2).*

In other words, the two groups are isomorphic provided we identify any two matrices of $SL(2, C)$ that differ only in their signs. This is denoted by

$$L \cong SL(2, C)/(E, -E) \tag{4.29}$$

Moreover, L is a connected continuous group, since $SL(2, C)$ is.

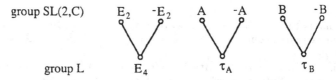

group SL(2,C) E_2 $-E_2$ A $-A$ B $-B$

group L E_4 τ_A τ_B

Fig. 4.2   where E_k = k×k identity matrix

In $SL(2, C)$ we have the 1-parameter continuous subgroup

$$A(\omega) = \begin{bmatrix} e^\omega & 0 \\ 0 & e^{-\omega} \end{bmatrix}, \quad \omega \in \mathbf{R} \tag{4.30}$$

An explicit calculation by means of (4.10) and (4.20) gives

$$\left. \begin{array}{lll} y_1 = & (\cosh 2\omega) \cdot x_1 & - & (\sinh 2\omega)x_4, & y_2 = x_2 \\ y_4 = & -(\sinh 2\omega) \cdot x_1 & + & (\cosh 2\omega)x_4, & y_3 = x_3 \end{array} \right\} \tag{4.31}$$

This together with $x_1 = x$, $y_1 = y$, $x_4 = ct$, $y_4 = ct'$ and

$$\cosh 2\omega = \frac{1}{\sqrt{1 - \frac{v^2}{c^2}}}, \qquad \sinh 2\omega = \frac{\frac{v}{c}}{\sqrt{1 - \frac{v^2}{c^2}}}$$

yields the transformation laws

$$y = \frac{1}{\sqrt{1 - \frac{v^2}{c^2}}}(x - vt), \qquad t' = \frac{1}{\sqrt{1 - \frac{v^2}{c^2}}}(t - \frac{v}{c^2}x) \qquad (4.32)$$

for the relativistic translation in direction of the x-axis.

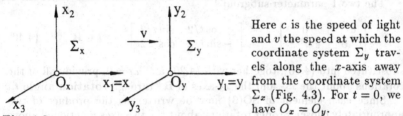

Fig. 4.3

Here c is the speed of light and v the speed at which the coordinate system Σ_y travels along the x-axis away from the coordinate system Σ_x (Fig. 4.3). For $t = 0$, we have $O_x = O_y$.

For $v \ll c$, the transformation laws (4.32) convert into the classical transformation laws of kinematics

$$y = x - vt, \qquad t' = t \qquad (4.33)$$

4.2.2 Relationship Between SU(2) and SO(3)

Our group SL(2, **C**) contains the subgroup of the special unitary matrices of the form (4.8). Thus the considerations of the preceding Section 4.2.1 apply to this group. For $A \in$ SU(2), Equation (1.83) implies

$$\bar{A}^T = A^{-1} \qquad (4.34)$$

and our main equation (4.10) specializes to

$$Y = AXA^{-1} \qquad (4.35)$$

By passing to the traces[5] we obtain

$$\text{tr}(Y) = \text{tr}((AX)A^{-1}) = \text{tr}(A^{-1}(AX)) = \text{tr}((A^{-1}A)X) = \text{tr}(X) \qquad (4.36)$$

Observing (4.20), we obtain

$$y_4 = x_4 \qquad (4.37)$$

Thus the linear subspace with $x_4 = 0$ is invariant under the linear transformations τ_A with $A \in$ SU(2). In the following, we denote them by τ'_A. Hence in the remaining part $\mathbf{R}^3$, Equation (4.10) applied to the hermitian

---

[5]$\text{tr}(VW) = \text{tr}(WV)$ for $n \times n$ matrices V and W (see (5.128)).

matrices (4.20) with $x_4 = 0$ is a group of real transformations of x_1, x_2, x_3. By (4.23) they leave the quadratic form

$$x_1^2 + x_2^2 + x_3^2 \qquad (4.38)$$

invariant.

In analogy to the preceding Section 4.2.1 we may conclude that the set of all image matrices $\{\tau_A' \mid A \in SU(2)\}$ is connected and that, furthermore,

$$\det \tau_A' = 1 \qquad (4.39)$$

Thus we are dealing with proper rotations of $\mathbf{R}^3$.

The two 1-parameter subgroups

$$\begin{bmatrix} e^{it/2} & 0 \\ 0 & e^{-it/2} \end{bmatrix}, \quad \begin{bmatrix} \cos t/2 & \sin t/2 \\ -\sin t/2 & \cos t/2 \end{bmatrix} \quad t \in \mathbf{R} \qquad (4.40)$$

substituted into (4.10), together with (4.20) and $x_4 = 0$, provide all of the rotations about the x_1- and the x_3-axes of $\mathbf{R}^3$ through **rotation angle** t.

Since each element of $SO(3)$ may be written as the product of two appropriately chosen proper rotations about the two axes mentioned above (see Fig. 1.3), this proves that (4.35) describes **all** proper rotations of $\mathbf{R}^3$. In analogy to Theorem 4.1, we obtain

Theorem 4.2 *The continuous homomorphism $A \to \tau_A'$ defined by (4.10) and (4.20) between the groups $SU(2)$ and $SO(3)$ has the property that each proper rotation $\tau_A' \in SO(3)$ is associated with exactly 2 special unitary matrices, namely A and $-A$ (see Fig. 4.4).*

In other words, the two groups are isomorphic provided we identify two matrices of $SU(2)$ that differ only in their signs; that is,

$$SO(3) \cong SU(2)/(E, -E) \qquad (4.41)$$

Moreover, $SO(3)$ is a connected continuous group.

Remark. The equation

$$\begin{bmatrix} -y_1 & y_2 + iy_3 \\ y_2 - iy_3 & y_1 \end{bmatrix} = \begin{bmatrix} a & b \\ -\bar{b} & \bar{a} \end{bmatrix} \begin{bmatrix} -x_1 & x_2 + ix_3 \\ x_2 - ix_3 & x_1 \end{bmatrix} \begin{bmatrix} \bar{a} & -b \\ \bar{b} & a \end{bmatrix} \qquad (4.42)$$

hence describes for any choice of two complex numbers

$$\begin{aligned} a &= \alpha_1 + i\alpha_2 \\ b &= \beta_1 + i\beta_2 \end{aligned} \quad \text{with } \alpha_i, \beta_i \in \mathbf{R} \text{ and } \alpha_1^2 + \alpha_2^2 + \beta_1^2 + \beta_2^2 = 1 \quad (4.43)$$

(i.e., three real free parameters) a proper rotation, namely

$$\begin{bmatrix} y_1 \\ y_2 \\ y_3 \end{bmatrix} = \begin{bmatrix} (\alpha_1^2 + \alpha_2^2 - \beta_1^2 - \beta_2^2) & 2(-\alpha_1\beta_1 - \alpha_2\beta_2) & 2(-\alpha_1\beta_2 + \alpha_2\beta_1) \\ 2(\alpha_1\beta_1 - \alpha_2\beta_2) & (\alpha_1^2 - \alpha_2^2 - \beta_1^2 + \beta_2^2) & 2(-\alpha_1\alpha_2 - \beta_1\beta_2) \\ 2(\alpha_1\beta_2 + \alpha_2\beta_1) & 2(\alpha_1\alpha_2 - \beta_1\beta_2) & (\alpha_1^2 - \alpha_2^2 + \beta_1^2 - \beta_2^2) \end{bmatrix} \begin{bmatrix} x_1 \\ x_2 \\ x_3 \end{bmatrix}$$
$$(4.44)$$

Furthermore, **every** orthogonal 3×3 matrix with det $= +1$ can be described by (4.44). This is called the **Cayley**[6] **parametrization** of the group SO(3). It specifies a two-to-one relation between the two 3-parameter groups SU(2) and SO(3): Every matrix of SU(2) is associated with a unique matrix of SO(3). Conversely, every matrix of SO(3) is associated with exactly two matrices of SU(2) (see Fig. 4.4).

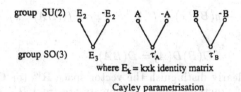

where $E_k = k \times k$ identity matrix

Cayley parametrisation

Fig. 4.4

4.3 Constructing Representations

First let G be an arbitrary group of real or complex $m \times m$ matrices A. Furthermore, let F be a given complex, not necessarily finite-dimensional vector space of functions

$$f(x_1, x_2, \ldots, x_m) = f(x) \qquad x \in \mathbf{R}^m \text{ or } \mathbf{C}^m \qquad (4.45)$$

with the usual laws for addition and scalar multiplication of functions:

$$(f + g)(x) = f(x) + g(x), \qquad (\alpha f)(x) = \alpha f(x) \quad \text{with } \alpha \in \mathbf{C} \qquad (4.46)$$

Now we construct a representation $A \to D(A)$ of the group G with the representation space F as follows:

$$D(A) : f(x) \to g(x) = f(A^{-1}x)$$

or more explicitly,

$$\left.\begin{array}{c} D(A) : f(x_1, x_2, \ldots, x_m) \to g(x_1, x_2, \ldots, x_m) = f(y_1, y_2, \ldots, y_m) \\[2mm] \text{where} \quad \begin{bmatrix} y_1 \\ y_2 \\ \vdots \\ y_m \end{bmatrix} = \underbrace{\begin{bmatrix} \alpha_{11} & \alpha_{12} & \cdots & \alpha_{1m} \\ \alpha_{21} & \alpha_{22} & \cdots & \alpha_{2m} \\ \cdots & \cdots & & \cdots \\ \alpha_{m1} & \alpha_{m2} & \cdots & \alpha_{mm} \end{bmatrix}}_{A^{-1}} \begin{bmatrix} x_1 \\ x_2 \\ \vdots \\ x_m \end{bmatrix} \end{array}\right\}$$

$$(4.47)$$

The α_{ik} are the matrix entries of the inverse A^{-1}. The construction above is known as the **transplantation of the function** $f(x)$ or the **action of the matrix** A **on** $f(x)$.

---

[6]Cayley, Arthur (1821-1895). Professor in Cambridge, England. Worked on the theory of invariants as well as on analytic and projective geometry.

$D(A)$ indeed describes a **linear** transformation, because

$$D(A) : (\alpha f + \beta g)(x) \to (\alpha f + \beta g)(A^{-1}x) = (\alpha f)(A^{-1}x) + (\beta g)(A^{-1}x)$$

and furthermore, it **satisfies the representation property**; for, if first A and then B act on the elements of the function space according to (4.47), then we obtain

$$D(B) : g(X) \to g(B^{-1}x) = f(A^{-1}B^{-1}x) = f((BA)^{-1}x),$$

hence

$$D(B)D(A) = D(BA)$$

Note that we must clearly distinguish the vector space $\mathbf{R}^m$ (or $\mathbf{C}^m$) and the vector space F of functions whose arguments are from $\mathbf{R}^m$ (or $\mathbf{C}^m$, respectively).

We leave it to the reader to verify that the representations ϑ_{nat} in Section 3.1 can also be explained by way of transplantation if in (4.45) only functions defined on a finite set $\mathcal{N}$ of points x are allowed.

Remark. If in (4.47) we had replaced A^{-1} by A, then this would have resulted in a "reversed" homomorphism $D(A)D(B) = D(BA)$. But this entails no substantial consequences.

4.3.1   Irreducible Representations of SU(2)

Now we consider the special function space F of homogeneous polynomials of nth degree $P_n(x_1, x_2)$ in the two variables x_1 and x_2; it is spanned by the monomials

$$x_1^n, \quad x_1^{n-1}x_2, \quad x_1^{n-2}x_2^2, \ldots, x_1 x_2^{n-1}, \quad x_2^n \tag{4.48}$$

which we choose as a basis in F. By (4.8), an arbitrary matrix from SU(2) has the form

$$A = \begin{bmatrix} a & b \\ -\bar{b} & \bar{a} \end{bmatrix} \qquad \text{where} \qquad a\bar{a} + b\bar{b} = 1$$

According to (4.4), its inverse A^{-1} is given by first transposing A and then passing to the complex conjugate. Hence (4.47) reads

$$P_n(x_1, x_2) \to P_n(y_1, y_2)$$
$$\text{where } y_1 = \bar{a}x_1 - bx_2, \quad y_2 = \bar{b}x_1 + ax_2 \tag{4.49}$$

Thus (4.49) describes an $(n+1)$-dimensional representation. In the case of $n = 2$, Equation (4.49) implies for the basis functions (4.48) that

$$
\begin{array}{rcll}
x_1^2 & \to & y_1^2 & = & \bar{a}^2 x_1^2 - 2\bar{a}b x_1 x_2 + b^2 x_2^2 \\
x_1 x_2 & \to & y_1 y_2 & = & \bar{a}\bar{b} x_1^2 + (a\bar{a} - b\bar{b}) x_1 x_2 - ab x_2^2 \\
x_2^2 & \to & y_2^2 & = & \bar{b}^2 x_1^2 + 2a\bar{b} x_1 x_2 + a^2 x_2^2
\end{array}
\tag{4.50}
$$

Hence the representing matrices relative to this basis have the form

$$D(A) = \begin{bmatrix} \bar{a}^2 & \overline{ab} & \bar{b}^2 \\ -2\bar{a}b & (a\bar{a} - b\bar{b}) & 2a\bar{b} \\ b^2 & -ab & a^2 \end{bmatrix} \tag{4.51}$$

In physics, one puts[7]

$$\ell = \frac{n}{2} \qquad (n = \text{degree of the polynomial}) \tag{4.52}$$

The corresponding representations of dimension

$$n + 1 = 2\ell + 1 \tag{4.53}$$

are denoted by

$$\vartheta_\ell \qquad \text{where} \qquad \ell = 0, \frac{1}{2}, 1, \frac{3}{2}, 2, \ldots \tag{4.54}$$

We remark here that $\vartheta_{1/2}$ is not the identity representation, but

$$\begin{bmatrix} a & b \\ -\bar{b} & \bar{a} \end{bmatrix} \xrightarrow{\vartheta_{1/2}} \begin{bmatrix} \bar{a} & \bar{b} \\ -b & a \end{bmatrix}$$

the complex conjugate. The reason is that in the special case of $n = 1$, the variables x_1, x_2 are not only coordinates but also the basis functions of F. The transformations behave differently for these two ways of interpretation.

In Example 5.11 we shall prove the following:

  (I) The **representations** ϑ_ℓ $(\ell = 0, \frac{1}{2}, 1, \frac{3}{2}, 2, \ldots)$ are **irreducible and**, since they differ in their dimensions, **mutually inequivalent**.

 (II) The list of representations ϑ_ℓ contains **all** continuous irreducible representations of SU(2) (**completeness**).

(III) Every continuous representation of SU(2) is **completely reducible**.

4.3.2 Irreducible Representations of SO(3)

From the ϑ_ℓ one obtains — possibly double-valued — irreducible representations of the group of proper rotations by describing its elements by the Cayley parametrization (4.44) and passing to SU(2). For simplicity, we denote these representations by ϑ_ℓ again.

By virtue of the isomorphism (4.41) described by this parametrization, not only (I) but also the statements (II) and (III) hold for the thus constructed representations of SO(3).

---

[7] see Section 7.5.1.

Which of the representations ϑ_ℓ of SO(3) are one-to-one, which are double-valued? This question can be answered by just considering the representing matrices of

$$ - E_2 = \begin{bmatrix} -1 & 0 \\ 0 & -1 \end{bmatrix} \tag{4.55} $$

Definition (4.47) together with (4.48) yields

$$ x_1^k x_2^{n-k} \rightarrow (-1)^k x_1^k (-1)^{n-k} x_2^{n-k} = (-1)^n x_1^k x_2^{n-k} \qquad k = 0, 1, \ldots, n \tag{4.56} $$

Hence with $n = 2\ell$,

$$ - E_2 \rightarrow (-1)^{2\ell} E_{2\ell+1} \tag{4.57} $$

Here $E_{2\ell+1}$ denotes the $(2\ell + 1) \times (2\ell + 1)$ identity matrix. Consequently,

$$ - E_2 \rightarrow \begin{cases} E_{2\ell+1}, & \text{for integral } \ell \\ -E_{2\ell+1}, & \text{for non-integral } \ell \end{cases} \tag{4.58} $$

These results are summarized in

Theorem 4.3 *For each number $\ell = 0, \frac{1}{2}, 1, \frac{3}{2}, 2, \ldots$ there exists a continuous irreducible $(2\ell + 1)$-dimensional representation ϑ_ℓ of the two groups SO(3) and SU(2).*

(a) *For SU(2) one obtains ϑ_ℓ by applying the elements of SU(2) to the monomials (4.48) with $n = 2\ell$.*

(b) *The corresponding irreducible representations of SO(3) are constructed by parametrizing each element of SO(3) after Cayley and representing the corresponding elements of SU(2) as in (a).*

The representations with even dimensions (i.e., ℓ not integral) are double-valued. The representations with odd dimensions (i.e., ℓ integral) are one-to-one.

The list given above is complete; that is, there are no other continuous irreducible representations of SU(2) and SO(3). Moreover, every representation of the two groups is completely reducible.

4.3.3 Complete Reduction of Representations of SU(2) and SO(3)

Since we know that a continuous representation ϑ of these groups is completely reducible, our next problem will be to reduce it into the ϑ_ℓ. When studying ϑ, we shall restrict our attention to certain abelian subgroups as is the standard technique in representation theory of continuous groups. A representation of such subgroups often characterizes a representation of the whole group, as will be seen later.

We demonstrate this technique through the example of a continuous N-dimensional representation ϑ of SO(3)[8]. As this group consists of the proper rotations of the (x_1, x_2, x_3)-space $\mathbf{R}^3$, it contains an abelian subgroup consisting of the rotations about the x_1-axis

$$\begin{bmatrix} y_1 \\ y_2 \\ y_3 \end{bmatrix} = \begin{bmatrix} 1 & 0 & 0 \\ 0 & \cos t & -\sin t \\ 0 & \sin t & \cos t \end{bmatrix} \begin{bmatrix} x_1 \\ x_2 \\ x_3 \end{bmatrix} \qquad (4.59)$$

This subgroup is also called a **torus** T and is isomorphic with SO(2). Its elements are characterized by the rotation angle t, whence denoted by $[t]$. The group multiplication corresponds to the addition of the t-values modulo 2π.

The representation mentioned above,

$$\vartheta : s \to D(s), \qquad s \in \text{SO}(3)$$

restricted to the torus T is equivalent to a representation by diagonal matrices due to complete reducibility (proof in Section 5.7) and Corollary 2.4. Instead of writing down the matrices, we just write the entries of the main diagonal

$$f_1([t]), f_2([t]), \ldots, f_N([t]) \qquad (4.60)$$

We shall call this the **representation by diagonals**. From

$$\{D([t'])\}^q = \pm E \quad \text{where} \quad t' = \frac{p}{q} \cdot 2\pi, p \text{ and } q \in \{0, 1, 2, 3, \ldots\}, \quad q \neq 0$$

it follows for rational multiples of 2π that

$$\{f_k([t'])\}^q = \pm 1 \quad \text{whence} \quad |f_k([t'])| = 1, \quad k = 1, 2, \ldots, N$$

Finally, continuity of ϑ implies

$$|f_k([t])| = 1 \qquad \forall t \in \mathbf{R}, \quad k = 1, 2, \ldots, N$$

Hence the matrix entries are complex numbers of absolute value 1:

$$f_k([t]) = e^{i\lambda_k t} \qquad (4.61)$$

Because of

$$f_k([2\pi]) = \pm 1$$

it follows that

$$\lambda_k \in \{\ldots, -2, -\frac{3}{2}, -1, -\frac{1}{2}, 0, \frac{1}{2}, 1, \frac{3}{2}, 2, \ldots\}$$

is **half integral**.

---

[8]Via the Cayley parametrization this also settles the case of representations of SU(2).

The numbers λ_k are called the **weights** of the representation ϑ of SO(3). This notation is justifiable by the fact that the $f_k([t])$ describe **all** eigenvalues of the matrices of ϑ, because for any rotational matrix $s \in$ SO(3) there exists a $u \in$ SO(3) such that

$$s = u^{-1}[t]u \tag{4.62}$$

In other words, in an appropriate coordinate system every proper rotation has the form (4.59). For the representing matrices, Equation (4.62) implies

$$D(s) = D(u^{-1})D([t])D(u) \tag{4.63}$$

Hence the eigenvalues of $D(s)$ are identical with those of $D([t])$.

Next we determine the weights of the irreducible representations ϑ_ℓ described in Theorem 4.3. Equation (4.40) implies that (4.59) is associated with

$$\begin{bmatrix} e^{it/2} & 0 \\ 0 & e^{-it/2} \end{bmatrix} \tag{4.64}$$

These diagonal matrices applied to the monomials (4.48) in the sense of transplantations (4.47) yield with $\ell = n/2$

$$x_1^k x_2^{n-k} \rightarrow (e^{-it/2}x_1)^k(e^{it/2}x_2)^{2\ell-k} = e^{i(\ell-k)t}x_1^k x_2^{n-k} \quad k = 0,1,\ldots,n \tag{4.65}$$

Hence the elements of the torus are represented by the diagonal matrices

$$[t] \rightarrow e^{i\ell t}, e^{i(\ell-1)t}, \ldots, e^{-i\ell t} \tag{4.66}$$

We may record this as follows:

Theorem 4.4 *The weights of the irreducible representation ϑ_ℓ of Theorem 4.3 are $\ell, \ell - 1, \ell - 2, \ldots, -\ell$.*

Now let ϑ be a continuous N-dimensional representation of SO(3) or SU(2) with the weights

$$\ell_1, \ell_2, \ldots, \ell_N$$

Some of the ℓ-values may occur repeatedly. Let the numbering be such that ℓ_1 is a dominant weight meaning that

$$\ell_k \leq \ell_1 \qquad \forall k = 2, 3, \ldots, N \tag{4.67}$$

Here ϑ_ℓ denotes the irreducible representation with dominant weight ℓ.

By complete reducibility the representation ϑ decomposes into certain irreducible representations. Of course, ϑ_ℓ, where $\ell = \ell_1$, must be a constituent of ϑ; for, by Theorem 4.4 none of the ϑ_ℓ with $\ell < \ell_1$ furnishes the dominant weight ℓ_1. Hence there is a subspace V_1 of the representation space V which is transformed invariantly under ϑ after the manner of ϑ_{ℓ_1}. Theorem 4.4 also says that this subspace "consumes" the weights $\ell_1, \ell_1 - 1, \ldots, -\ell_1$ of ϑ. If after deleting the values $\ell_1, \ell_1 - 1, \ldots, -\ell_1$ from these weights of ϑ there are still some left over, then one repeats the procedure for the remaining subspace of V. In other words, we again determine the dominant weight among the remaining values, etc.

Remarks.

1. Some of the ϑ_ℓ may appear repeatedly (i.e., multiplicity > 1), namely if and only if a dominant weight occurs repeatedly.

2. This process produces just the subscripts of the irreducible representations with their multiplicities of appearance in ϑ but by no means the different irreducible subspaces. The latter problem will be discussed in Sections 7.3 and 7.4, and (for certain cases) also in the next Section 4.4.

These results are summarized in the following

Algorithm (Complete reduction of the representations of SO(3) and SU(2))

Let ϑ be a given continuous representation of SO(3) or SU(2). In the case of SO(3), consider the representing matrices $D([t])$ of all rotations of $\mathbf{R}^3$ about an arbitrary fixed axis (e.g., one of the cartesian coordinate axes x_1, x_2, x_3) through the angle of rotation t. In the case of SU(2), consider the representing matrices of the corresponding abelian subgroups of SU(2), (e.g., the matrices (4.64)).

If the $D([t])$ are already in diagonal form with entries $e^{i\ell_k t}$ along the main diagonal, then the half integers ℓ_k are the weights of the representation ϑ.

If not, then the exponential terms $e^{i\ell_k t}$ are determined as the eigenvalues of $D([t])$.

Let ℓ_1 be a dominant weight; that is, $\ell_1 \geq \ell_k \ \forall k$. Then the irreducible representation ϑ_{ℓ_1} (described in Theorem 4.3) is a constituent of ϑ. Further reduction consists in deleting the weights $\ell_1, \ell_1 - 1, \ldots, -\ell_1$ from the entire set of all weights and repeating the procedure if necessary.

This **method of weights** was introduced into representation theory by E. Cartan[9].

4.4 Clebsch-Gordan Coefficients, Spherical Functions

Here we shall discuss some important applications of the preceding Section 4.3. Also we shall determine the invariant irreducible subspaces of Kronecker products of representations of SO(3).

---

[9] Cartan, Elie (1869-1951). Worked mainly on continuous groups and, in particular, accomplished the classification of the semi-simple groups.

As an introductory example, let us consider the group SU(2) consisting of the matrices

$$\begin{bmatrix} a & b \\ -\bar{b} & \bar{a} \end{bmatrix} \quad \text{with} \quad \begin{cases} a = \alpha_1 + i\alpha_2 \\ b = \beta_1 + i\beta_2 \end{cases} \quad \text{and} \quad a\bar{a} + b\bar{b} = 1$$

The Cayley parametrization (4.44) provides a 3-dimensional representation of this group. In particular, the abelian subgroup

$$\begin{bmatrix} e^{it/2} & 0 \\ 0 & e^{-it/2} \end{bmatrix}$$

is represented by the torus T described in (4.59), because

$$\begin{array}{ll} \alpha_1 = \cos(t/2), & \alpha_2 = \sin(t/2) \\ \beta_1 = 0, & \beta_2 = 0 \end{array}$$

Its eigenvalues are $1, e^{it}, e^{-it}$, hence its weights $0, 1, -1$. Consequently, this representation is equivalent to ϑ_1 and, therefore, irreducible.

4.4.1   Spherical Functions

Spherical functions (or spherical harmonics) u_ℓ of ℓth degree are homogeneous and harmonic polynomials of ℓth degree in the three cartesian coordinates x_1, x_2, x_3; that is, they are linear combinations of the monomials

$$x_1^\alpha x_2^\beta x_3^\gamma \quad \text{where} \quad \alpha + \beta + \gamma = \ell, \quad \ell = 0, 1, 2, \ldots \quad (4.68)$$

that satisfy the Laplace differential equation

$$\Delta u_\ell = 0 \quad (4.69)$$

The vector space K_ℓ of spherical functions u_ℓ is $(2\ell + 1)$-dimensional, see (Weyl, 1950). For $\ell = 2$, we can choose the following basis

$$x_1^2 - x_2^2, \quad x_2^2 - x_3^2, \quad x_1 x_2, \quad x_1 x_3, \quad x_2 x_3 \quad (4.70)$$

Since (in the sense of (4.47)) a spherical rotation carries a harmonic function into another harmonic function, it follows that the vector space K_ℓ (ℓ arbitrary fixed) is invariant under spherical rotations.

We denote the action of a rotation $s \in SO(3)$ (described in (4.47)) by $D(s)$. In the following, we shall discuss the $(2\ell + 1)$-dimensional representations

$$s \to D(s) \quad s \in SO(3) \quad (4.71)$$

for different ℓ.

First we consider the case $\ell = 1$ with the coordinates x_1, x_2, x_3 as the spherical functions. The torus (4.59) is carried into itself ($t \to -t$) by the transformations $D(s)$. Hence the eigenvalues of $D(s)$ are $1, e^{it}, e^{-it}$, thus the weights $0, 1, -1$. Consequently, this representation is equivalent to ϑ_1 of SO(3) and this again confirms Cartan's method.

Next we look at the general case of spherical functions u_ℓ of ℓth degree. We introduce the spherical polar coordinates

$$
\begin{aligned}
r &= \text{radius} \\
\varphi &= \text{longitude (see Fig. 1.3)} \\
\Theta &= \text{colatitude (angle with } x_3\text{-axis)}
\end{aligned}
\tag{4.72}
$$

Then the spherical functions obviously take the form

$$u_\ell = r^\ell Y_\ell(\Theta, \varphi) \tag{4.73}$$

The $Y_\ell(\Theta, \varphi)$ are called **the spherical surface functions** or the **surface harmonics** of ℓth degree. In the space of these functions (ℓ fixed) there exists a basis of the form

$$Y_\ell^m(\Theta, \varphi) = e^{im\varphi} F_\ell^m(\Theta), \qquad m = \ell, \ell - 1, \ell - 2, \ldots, -\ell \tag{4.74}$$

What do the representing matrices $D([t])$ of the rotations about the x_3-axis look like?
If $[t]$ acts on Y_ℓ^m, then by (4.47)

$$r, \Theta = \text{const} \qquad \varphi \to \varphi - t \qquad \text{where } t = \text{angle of rotation} \tag{4.75}$$

hence by (4.74)

$$D([t])Y_\ell^m = e^{-imt}Y_\ell^m \qquad [t] = \text{rotation about the } x_3\text{-axis} \tag{4.76}$$

Hence the representation by the diagonal is given by

$$e^{-i\ell t}, e^{-i(\ell-1)t}, \ldots, e^{+i\ell t} \tag{4.77}$$

with the weights

$$-\ell, -(\ell - 1), \ldots, \ell$$

Thus the following holds:

Theorem 4.5 *The representations of* SO(3) *by spherical functions of ℓth degree ($\ell = 0, 1, 2, \ldots$) are equivalent to the irreducible representations ϑ_ℓ.*

4.4.2 The Kronecker Product $\vartheta_\ell \otimes \vartheta_m$

We wish to completely reduce the Kronecker product $\vartheta_\ell \otimes \vartheta_m$[10] of SO(3) (or SU(2)). Here ϑ_ℓ and ϑ_m are irreducible representations according to Theorem 4.3. In ϑ_ℓ the element $[t]$ of the torus T of SO(3) described in (4.59) is represented by the diagonal $e^{i\ell_j t}$ because of (4.64) and (4.66).

---

[10] Compare Section 1.10.

In the notation of basis functions of Section 1.10, the corresponding linear transformation then is

$$\varphi'_j = e^{i\ell_j t}\varphi_j \qquad j = 1, 2, \ldots, (2\ell + 1) \tag{4.78}$$

The basis functions φ_j coincide with the monomials (4.48).

The representing matrix of $[t]$ transforms in like manner the vectors of the representation space of ϑ_m:

$$\psi'_k = e^{im_k t}\psi_k \qquad k = 1, 2, \ldots, (2m + 1) \tag{4.79}$$

The representation $\vartheta_\ell \otimes \vartheta_m$ restricted to the elements of the torus T thus transforms the basis functions $(\varphi_j \psi_k)$ as follows:

$$(\varphi_j \psi_k)' = e^{i(\ell_j + m_k)t}(\varphi_j \psi_k) \tag{4.80}$$

In other words:

Theorem 4.6 *The $(2\ell + 1)(2m + 1)$ weights of the Kronecker product $\vartheta_\ell \otimes \vartheta_m$ are of the form $(\ell_j + m_k)$; that is, they are the sums of the weights $\ell_j = -\ell, -(\ell - 1), \ldots, \ell$ and $m_k = -m, -(m - 1), \ldots, m$ of the respective representations ϑ_ℓ and ϑ_m.*

In the following, we shall employ the diagram shown for $\ell = 3$, $m = 2$ in Fig. 4.5.

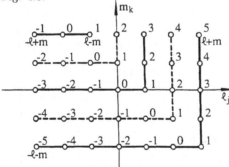

The ℓ_j- and m_k-axes are labeled by the weights of ϑ_j and ϑ_m, respectively. Furthermore, each point with coordinates (ℓ_j, m_k) is labeled by the weight $(\ell_j + m_k)$. There is a total of $(2\ell + 1)\cdot(2m + 1)$ points. Some of the weights of $\vartheta_j \otimes \vartheta_m$ occur repeatedly.

Fig. 4.5

The dominant weight of the Kronecker product is located at the upper right hand corner and has the value $5 = 3 + 2 = \ell + m$. Hence ϑ_5 is an irreducible constituent and consumes the weights 5, 4, $\ldots$, -5. This is indicated by the bold path in the diagram. Among the remaining weights, 4 is the dominant weight and provides the constituent ϑ_4, etc. Thus we finally obtain

$$\vartheta_3 \otimes \vartheta_2 = \vartheta_5 \oplus \vartheta_4 \oplus \vartheta_3 \oplus \vartheta_2 \oplus \vartheta_1$$

in accordance with the indicated paths. Dimension check gives: $7 \cdot 5 = 11 + 9 + 7 + 5 + 3$. The diagram for the general case implies

Theorem 4.7 *The complete reduction of the Kronecker product $\vartheta_\ell \otimes \vartheta_m$ where $\ell, m \in \{0, \frac{1}{2}, 1, \frac{3}{2}, 2, \ldots\}$ is described by the so-called Clebsch-Gordan series*

$$\vartheta_\ell \otimes \vartheta_m = \vartheta_{\ell+m} \oplus \vartheta_{\ell+m-1} \oplus \ldots \oplus \vartheta_{|\ell-m|} \tag{4.81}$$

Remark. By Theorem 4.3 the Kronecker product of SO(3) is one-to-one (double-valued) if and only if $(\ell + m)$ is an integer (not an integer).

Clebsch-Gordan Coefficients

Finally, we wish to explicitly find the complete reduction of $\vartheta_\ell \otimes \vartheta_m$ in terms of (4.81), that is, to find all its irreducible invariant subspaces. In Section 7.4 we shall derive an algorithm to compute these subspaces. This algorithm will be valid even for arbitrary representations of SO(3) (or SU(2)). Meanwhile, we content ourselves with just giving the recipe.

Recall that the irreducible representation ϑ_ℓ was constructed via monomials (4.48), where $n = 2\ell$. We may also write these monomials in the form

$$x_1^{\ell+\lambda} x_2^{\ell-\lambda} \quad \text{where} \quad \lambda = \ell, \ell - 1, \ldots, -\ell \tag{4.82}$$

We multiply (4.82) by suitable scalar factors and obtain a new basis[11]

$$X_\lambda = \frac{x_1^{\ell+\lambda} x_2^{\ell-\lambda}}{\sqrt{(\ell + \lambda)!(\ell - \lambda)!}} \quad \lambda = \ell, \ell - 1, \ldots, -\ell \tag{4.83}$$

If the matrix group SU(2) acts on the new basis as described in (4.47), then the matrices of the thus modified representation ϑ_ℓ become **unitary** (see Section 7.3).

Remark. One can readily verify that, in general, the matrices (4.51) are not unitary (Example: $a = b = \frac{1}{\sqrt{2}}$).

Similarly, for the second representation ϑ_m in the Kronecker product one takes

$$Y_\mu = \frac{y_1^{m+\mu} y_2^{m-\mu}}{\sqrt{(m + \mu)!(m - \mu)!}} \quad \mu = m, m - 1, \ldots, -m \tag{4.84}$$

Thus the representing matrices of both ϑ_ℓ and ϑ_m are unitary. This implies that **the Kronecker product is unitary** (see Problem 1.4).

According to (4.81), the Kronecker product $\vartheta_\ell \otimes \vartheta_m$ decomposes into the representations

$$\vartheta_\rho \text{ where } \rho = \ell + m, \ell + m - 1, \ldots, |\ell - m|$$

Next we need to determine the $(2\rho + 1)$ basis vectors that span the irreducible subspace of ϑ_ρ.

---

[11] Note that $0! = 1$.

We denote these basis vectors by Z_i^ρ $(i = \rho, \rho - 1, \ldots, -\rho)$. They are linear combinations of the functions $X_\lambda Y_\mu$. We set

$$\left. \begin{array}{l} Z_i^\rho = \sum_{\lambda+\mu=i} c_{\lambda\mu}^\rho X_\lambda Y_\mu, \qquad i = \rho, \rho - 1, \ldots, -\rho \\ \text{where } \lambda \in \{-\ell, -\ell+1, \ldots, \ell\}, \quad \mu \in \{-m, -m+1, \ldots, m\} \end{array} \right\} \quad (4.85)$$

The numbers $c_{\lambda\mu}^\rho$ are called the **Clebsch-Gordan coefficients** and are adequately tabulated in (Rotenberg, Bivins, Metropolis, Wooten, 1959). However, we recommend using the powerful MATHEMATICA computer system (Wolfram, 1988) with the package clebsch.m for explicit calculations. The following small table is reproduced from (van der Waerden, 1974):

(I) Representations $\vartheta_\ell \otimes \vartheta_{1/2}$

ℓ arbitrary, $m = \frac{1}{2}$ implies $\rho = \ell + \frac{1}{2}, \ell - \frac{1}{2}$ and $\mu = \frac{1}{2}, -\frac{1}{2}$.

	$\mu = 1/2$	$\mu = -1/2$
$\rho = \ell + \frac{1}{2}$	$\sqrt{\ell + \lambda + 1}$	$\sqrt{\ell - \lambda + 1}$
$\rho = \ell - \frac{1}{2}$	$-\sqrt{\ell - \lambda}$	$+\sqrt{\ell + \lambda}$

$\lambda = \ell, \ell - 1, \ldots, -\ell$

$\qquad (4.86)$

(II) Representations $\vartheta_\ell \otimes \vartheta_1$

ℓ arbitrary, $m = 1$ implies $\rho = \ell + 1, \ell, \ell - 1$ and $\mu = 1, 0, -1$.

	$\mu = 1$	$\mu = 0$	$\mu = -1$
$\rho = \ell + 1$	$\sqrt{\frac{1}{2}(\ell+\lambda+2)(\ell+\lambda+1)}$	$\sqrt{(\ell+\lambda+1)(\ell-\lambda+1)}$	$\sqrt{\frac{1}{2}(\ell-\lambda+2)(\ell-\lambda+1)}$
$\rho = \ell$	$-\sqrt{2(\ell+\lambda+1)(\ell-\lambda)}$	2λ	$\sqrt{2(\ell+\lambda)(\ell-\lambda+1)}$
$\rho = \ell - 1$	$\sqrt{\frac{1}{2}(\ell-\lambda)(\ell-\lambda-1)}$	$-\sqrt{(\ell+\lambda)(\ell-\lambda)}$	$\sqrt{\frac{1}{2}(\ell+\lambda)(\ell+\lambda-1)}$

$\lambda = \ell, \ell - 1, \ldots, -\ell$

$\qquad (4.87)$

We demonstrate the practical application for the case $\ell = 1, m = \frac{1}{2}$. Here

$$X_1 = \frac{x_1^2}{\sqrt{2}}, \quad X_0 = x_1 x_2, \quad X_{-1} = \frac{x_2^2}{\sqrt{2}}; \quad Y_{1/2} = y_1, \quad Y_{-1/2} = y_2$$

The transformation (4.50) modifies into

$$\begin{array}{rcl} X_1' & = & \bar{a}^2 X_1 - \sqrt{2}\,\bar{a}b X_0 + b^2 X_{-1} \\ X_0' & = & \sqrt{2}\,\bar{a}\bar{b} X_1 + (a\bar{a} - b\bar{b}) X_0 - \sqrt{2}\,ab X_{-1} \\ X_{-1}' & = & \bar{b}^2 X_1 + \sqrt{2}\,a\bar{b} X_0 + a^2 X_{-1} \end{array}$$

and (4.49) implies

$$Y_{1/2}' = \bar{a} Y_{1/2} - b Y_{-1/2} \qquad Y_{-1/2}' = \bar{b} Y_{1/2} + a Y_{-1/2}$$

The representation $\vartheta_1 \otimes \vartheta_{1/2}$ is obtained by writing down the products $X'_j Y'_k$; for example,

$$
\begin{aligned}
X'_1 Y'_{1/2} = {} & \bar{a}^3 X_1 Y_{1/2} - \sqrt{2}\,\bar{a}^2 b X_0 Y_{1/2} + \bar{a}b^2 X_{-1}Y_{1/2} \\
& -\bar{a}^2 b X_1 Y_{-1/2} + \sqrt{2}\,\bar{a}b^2 X_0 Y_{-1/2} - b^3 X_{-1}Y_{-1/2}
\end{aligned}
$$

and five similar equations. Application of the Clebsch-Gordan coefficients gives for the superscript $\rho = 1/2$ (which is omitted in the following):

$$
Z_{1/2} = \sqrt{2}X_1 Y_{-1/2} - X_0 Y_{1/2}, \qquad Z_{-1/2} = X_0 Y_{-1/2} - \sqrt{2}X_{-1}Y_{1/2} \quad (4.88)
$$

After some computation one finds

$$
\begin{aligned}
Z'_{1/2} &= \sqrt{2}\,X'_1 Y'_{-1/2} - X'_0 Y'_{1/2} \\
&= -\bar{a}X_0 Y_{1/2} + \sqrt{2}\,bX_{-1}Y_{1/2} + \sqrt{2}\,\bar{a}X_1 Y_{-1/2} - bX_0 Y_{-1/2} \\
Z'_{1/2} &= \bar{a}Z_{1/2} - bZ_{-1/2}, \qquad Z'_{-1/2} = \bar{b}Z_{1/2} + aZ_{-1/2}
\end{aligned}
$$

Thus we have found an invariant 2-dimensional subspace. The Kronecker product $\vartheta_\ell \otimes \vartheta_m$ transforms this subspace by way of the matrices

$$
\begin{bmatrix} \bar{a} & \bar{b} \\ -b & a \end{bmatrix}
$$

because $Z_{1/2}, Z_{-1/2}$ are the two basis vectors. This is precisely the representation $\vartheta_{1/2}$ (see equation after (4.54)).

4.4.3 Spherical Functions and Laplace Operator

Let the proper rotation group SO(3) act upon the 3-dimensional real vector space with the cartesian coordinates x_1, x_2, x_3. The diagonal matrices

$$
\begin{bmatrix} 1 & 0 & 0 \\ 0 & e^{it} & 0 \\ 0 & 0 & e^{-it} \end{bmatrix} \tag{4.89}
$$

are equivalent to the matrices (4.59). They act upon the homogeneous polynomials in x_1, x_2, x_3 of ℓth degree ($\ell = 0, 1, 2, \dots$) according to (4.47) and are linear combinations of the monomials

$$
x_1^\alpha x_2^\beta x_3^\gamma \text{ with } \alpha + \beta + \gamma = \ell \text{ and } \alpha, \beta, \gamma \in \{0, 1, 2, \dots, \ell\} \tag{4.90}
$$

For the special case $\ell = 2$, the latter are the monomials

$$
x_1^2, \quad x_2^2, \quad x_3^2, \quad x_1 x_2, \quad x_1 x_3, \quad x_2 x_3
$$

Since linear transformations carry a homogeneous polynomial of ℓth degree into another such polynomial, this obviously is a representation of SO(3). We denote it by $\vartheta_{\text{hom}}^{(\ell)}$.

Now we shall work with the operator

$$M = (x_1^2 + x_2^2 + x_3^2) \cdot \Delta \qquad (4.91)$$

where Δ is the Laplace operator $\frac{\partial^2}{\partial x_1^2} + \frac{\partial^2}{\partial x_2^2} + \frac{\partial^2}{\partial x_3^2}$. The operator M carries each homogeneous polynomial of ℓth degree into another such polynomial. Moreover, it has the **symmetry of the representation** $\vartheta_{\text{hom}}^{(\ell)}$. This is a consequence of the rotary invariance of the Laplacian, or equivalently, of the fact that the same results are obtained, irrespective of whether a function undergoes a rotation followed by the Laplacian or the reverse. Therefore, the hypotheses of the fundamental theorem in Section 2.3 are satisfied.

Determining the Weights of $\vartheta_{\text{hom}}^{(\ell)}$. For the representing matrices of the torus T we obtain from (4.47), (4.89), (4.90)

$$x_1^\alpha x_2^\beta x_3^\gamma \rightarrow x_1^\alpha \cdot (x_2 e^{-it})^\beta (x_3 e^{it})^\gamma = e^{i(\gamma - \beta)t} x_1^\alpha x_2^\beta x_3^\gamma$$

The pairs (β, γ) associated with the monomials (4.90) with **fixed** exponent α are

$$(\ell - \alpha, 0); \quad (\ell - \alpha - 1, 1); \quad (\ell - \alpha - 2, 2); \ldots; (0, \ell - \alpha)$$

Hence the corresponding weights $\gamma - \beta$ are

$$\ell - \alpha; \quad \ell - \alpha - 2; \quad \ell - \alpha - 4; \ldots; -(\ell - \alpha) \qquad (4.92)$$

We now tabulate (4.92) for different α-values:

α	Occurring weights $\gamma - \beta$								Rep.
0	ℓ	$(\ell - 2)$	$(\ell - 4)$	$(\ell - 6)$ $\ldots$	$\ldots$	$\ldots$	$\ldots$	$-\ell$	
1	$(\ell - 1)$	$(\ell - 3)$	$(\ell - 5)$		$\ldots$	$\ldots$	$\ldots$	$-(\ell - 1)$	ϑ_ℓ
2		$(\ell - 2)$	$(\ell - 4)$	$(\ell - 6)$ $\ldots$		$\ldots$	$-(\ell - 2)$		
3			$(\ell - 3)$	$(\ell - 5)$		$\ldots$	$-(\ell - 3)$		$\vartheta_{\ell-2}$
$\vdots$				$\vdots$					$\vdots$
ℓ				0					

$$(4.93)$$

By connecting two adjacent rows by a zigzag line, one finds

$$\vartheta_{\text{hom}}^{(\ell)} = \vartheta_\ell \oplus \vartheta_{\ell-2} \oplus \vartheta_{\ell-4} \oplus \ldots \oplus \begin{cases} \vartheta_0, & \text{if } \ell \text{ is even} \\ \vartheta_1, & \text{if } \ell \text{ is odd} \end{cases} \qquad (4.94)$$

Accordingly, the representation space V decomposes into

$$V = V_\ell \oplus V_{\ell-2} \oplus V_{\ell-4} \oplus \ldots \oplus \begin{cases} V_0, & \text{if } \ell \text{ is even} \\ V_1, & \text{if } \ell \text{ is odd} \end{cases} \qquad (4.95)$$

The irreducible subspaces V_j are, as we know, $(2j + 1)$-dimensional.

Since in (4.94) each of the irreducible representations occurs with multiplicity 1, Theorem 2.7 says that *each irreducible subspace in (4.95) is associated with an eigenvalue λ of M; that is, each nonzero vector of the subspace is an eigenvector of M corresponding to λ.*

Let us first discuss the subspace V_ℓ. Since ϑ_ℓ occurs just once in $\vartheta_{\text{hom}}^{(\ell)}$, each homogeneous polynomial transformed after the manner of ϑ_ℓ must lie in V_ℓ. By Theorem 4.5 one subspace of such polynomials is already known, namely the space K_ℓ of spherical functions u_ℓ of ℓth degree. Thus $K_\ell \subset V_\ell$. Equality of dimensions implies $K_\ell = V_\ell$. Hence V_ℓ is the space of all spherical functions u_ℓ of ℓth degree.

Therefore, to explicitly determine the invariant subspace V_ℓ of the representation $\vartheta_{\text{hom}}^{(\ell)}$ is equivalent to *constructing the spherical functions of ℓth degree by group theory methods.*

Next, let us consider $V_{\ell-2}$. It consists of the homogeneous polynomials of ℓth degree that are transformed after the manner of $\vartheta_{\ell-2}$. From Theorem 4.5 we know one subspace of such polynomials, namely that of the functions $r^2 u_{\ell-2}$, where $r^2 = x_1^2 + x_2^2 + x_3^2$. This follows from the rotary invariance of the function r^2. By an argument similar to the above, we find that $V_{\ell-2}$ consists of the functions $r^2 u_{\ell-2}$. Correspondingly, for all subspaces of (4.95) we have that

$$
\begin{array}{lll}
V_\ell & \text{consists of all functions} & u_\ell \\
V_{\ell-2} & \text{consists of all functions} & r^2 u_{\ell-2} \\
V_{\ell-4} & \text{consists of all functions} & r^4 u_{\ell-4} \\
\vdots & & \vdots
\end{array}
\tag{4.96}
$$

$$
\begin{array}{ll}
\text{where} & r^2 = x_1^2 + x_2^2 + x_3^2 \\
\text{and} & u_j = u_j(x_1, x_2, x_3) \text{ is an arbitrary spherical function} \\
& \text{of } j\text{th degree.}
\end{array}
$$

Eigenvalues of the Operator M. $\Delta u_\ell = 0$ implies $M u_\ell = 0$, whence the eigenvalue associated with the eigenspace V_ℓ is zero. To calculate the remaining eigenvalues it is favorable to employ the spherical coordinates r, Θ, φ as given in (4.72). Relative to these coordinates the operator M is

$$
M = \frac{\partial}{\partial r} \left(r^2 \frac{\partial}{\partial r} \right) + \Lambda
\tag{4.97}
$$

Here Λ does not vary with r.

According to (4.73), the spherical functions of jth degree in spherical coordinates have the form

$$
u_j = r^j Y_j(\Theta, \varphi)
$$

Equation (4.97) and $\Delta u_j = 0$ imply

$$
M u_j = j(j+1) u_j + \Lambda u_j = 0 \quad \text{whence} \quad \Lambda u_j = -j(j+1) u_j
\tag{4.98}
$$

Therefore, the spherical surface functions satisfy

$$\Lambda Y_j = -j(j+1)Y_j \qquad j = 0, 1, 2, \ldots \qquad (4.99)$$

In the following, we denote by $v_{\ell-k}$ the functions of the irreducible subspace $V_{\ell-k}$ in (4.95). By (4.96) they are of the form

$$v_{\ell-k} = r^k \cdot r^{\ell-k} Y_{\ell-k} = r^\ell Y_{\ell-k} \qquad k = 0, 2, 4, \ldots \text{ and } k \leq \ell$$

By (4.97), (4.98) and with $j = \ell - k$

$$M v_{\ell-k} = [\ell(\ell+1) - (\ell-k)(\ell-k+1)]v_{\ell-k} = k(2\ell-k+1)v_{\ell-k} \quad (4.100)$$

Hence the eigenvalue of M asssociated with the eigenspace $V_{\ell-k}$ is

$$k(2\ell - k + 1) \qquad k = 0, 2, 4, \ldots \text{ and } k \leq \ell \qquad (4.101)$$

For the special case of $\ell = 2$, Equation (4.95) implies $V = V_2 \oplus V_0$. By (4.96), the space V_0 is spanned by the function $x_1^2 + x_2^2 + x_3^2$. The eigenvalue for V_0 is given by (4.101) where $\ell = k = 2$. It equals 6, which may also be confirmed by direct computation in cartesian coordinates.

Completeness of spherical functions. We now consider the values of all functions in question on the surface of the unit sphere. Thereby the elements in the list (4.96) become **spherical surface functions** $Y_\ell, Y_{\ell-2}, \ldots$ The decomposition (4.95) has an interesting analytic conclusion:

Every homogeneous polynomial of lth degree in x_1, x_2, x_3 restricted by the constraint $x_1^2 + x_2^2 + x_3^2 = 1$ is the sum of spherical surface functions $Y_\ell, Y_{\ell-2}, \ldots$

This result extends to arbitrary polynomials, as they can be regarded as sums of homogeneous polynomials. Finally, by the Weierstrass approximation theorem, every continuous function of x_1, x_2, x_3 restricted to the spherical surface can be **uniformly approximated** by polynomials. Hence the following holds:

Theorem 4.8 *Every continuous function defined on the spherical surface can be uniformly approximated by a sum of spherical surface functions* Y_ℓ *($\ell = 0, 1, 2, \ldots$).*

4.5   The Lorentz group and SL(2,C)

In this section, we shall largely follow the pattern of study of Section 4.3. In contrast to the preceding, we shall state some of the theorems without supplying the proofs.

4.5.1   Irreducible Representations

As in Section 4.3 we construct a representation ϑ_ℓ for $\ell = 0, \frac{1}{2}, 1, \frac{3}{2}, 2, \ldots$ by applying an arbitrary complex 2×2 matrix $A \in \mathrm{SL}(2,\mathbf{C})$ to the $(2\ell+1)$ monomials

$$x_1^n, \; x_1^{n-1}x_2, \; x_1^{n-2}x_2^2, \ldots, x_2^n \qquad (4.102)$$

Here again $n = 2\ell$.

Irreducibility of the representations ϑ_ℓ of $\mathrm{SL}(2,\mathbf{C})$ follows immediately from $\mathrm{SU}(2)$ being a subgroup of $\mathrm{SL}(2,\mathbf{C})$. Unlike before, however, these are **not all** of the continuous irreducible representations. We obtain additional ones by conjugating each of the representing matrices $D(A)$ of ϑ_ℓ.

We denote the latter representations by $\overline{\vartheta_\ell}$. To prove inequivalence of ϑ_ℓ and $\overline{\vartheta_\ell}$ ($\ell > 0$) it suffices to show that there exist representing matrices with non-real trace, for the trace is invariant under coordinate transformations. Already in the simple case of the identity representation of $\mathrm{SL}(2,\mathbf{C})$, there are representing matrices with non-real trace. In contrast, the traces of the representations of $\mathrm{SU}(2)$ are all real, as will be shown in (5.204).

The following holds (Miller, 1972, Ch. 9):

Theorem 4.9 *All continuous irreducible representations of* $\mathrm{SL}(2,\mathbf{C})$ *are of the form*

$$\vartheta_{\ell,m} = \vartheta_\ell \otimes \overline{\vartheta_m} \qquad (4.103)$$

Here ℓ and m are independent of each other and range through the half integers $0, \frac{1}{2}, 1, \frac{3}{2}, 2, \ldots$ The representation ϑ_ℓ is constructed by action of the matrices $A \in \mathrm{SL}(2,\mathbf{C})$ upon the monomials (4.102). The representation $\overline{\vartheta_m}$ results from ϑ_m by conjugating all its representing matrices.

Moreover, every continuous representation of $\mathrm{SL}(2,\mathbf{C})$ *is completely reducible.*

Remark. The representations ϑ_ℓ and $\overline{\vartheta_m}$ are included by the special cases of $m = 0$ and $\ell = 0$, respectively.

Taking into consideration the isomorphism (4.29), one obtains — possibly double-valued — irreducible representations of the proper positive Lorentz group L. Hence Theorem 4.9 is also valid for the group L, due to the isomorphism (4.29) described by the parametrization between the two groups.

4.5.2   Complete Reduction

As mentioned in Section 4.3.3, the technique of complete reduction deals with the representing matrices of an appropriate abelian subgroup. The set of all diagonal matrices of $\mathrm{SL}(2,\mathbf{C})$ is such a subgroup. It can be described by

$$\begin{bmatrix} e^{(\tau+it)/2} & 0 \\ 0 & e^{-(\tau+it)/2} \end{bmatrix} \qquad (4.104)$$

using the two parameters τ and t.

The 2-dimensional parameter range corresponds to a cylindrical surface of infinite length. Each point on this surface (τ = longitudinal coordinate, t = angle) corresponds to a unique diagonal matrix of SL(2,C), and vice versa.

The matrices (4.104) applied to the monomials (4.102) according to (4.47) give, with $n = 2\ell$ and $\lambda = (\tau + it)/2$,

$$x_1^k x_2^{n-k} \overset{\vartheta_\ell}{\rightarrow} (e^{-\lambda}x_1)^k (e^\lambda x_2)^{n-k} = e^{(n-2k)\lambda} x_1^k x_2^{n-k}$$

that is,

$$x_1^k x_2^{2\ell-k} \overset{\vartheta_\ell}{\rightarrow} e^{(\ell-k)(\tau+it)} x_1^k x_2^{2\ell-k} \qquad k = 0,1,2,\ldots,2\ell \tag{4.105}$$

Equation (4.105) implies

$$x_1^k x_2^{2m-k} \overset{\overline{\vartheta_m}}{\rightarrow} e^{(m-k)(\tau-it)} x_1^k x_2^{2m-k} \qquad k = 0,1,2,\ldots,2m \tag{4.106}$$

The exponential terms in (4.105) and (4.106) are the entries in the main diagonal of the representing diagonal matrices of (4.104). They are of the form

$$e^{\alpha\tau+i\beta t} \tag{4.107}$$

hence are specified by the two real numbers α, β. Therefore, the ordered number-pair (α, β) is said to be the **weight** of the corresponding representation.

According to (4.105), the representation ϑ_ℓ has the weights

$$(\ell,\ell); (\ell-1,\ell-1); (\ell-2,\ell-2); \ldots; (-\ell,-\ell) \tag{4.108}$$

According to (4.106), the representation $\overline{\vartheta_m}$ has the weights

$$(m,-m); (m-1,1-m); (m-2,2-m); \ldots; (-m,m) \tag{4.109}$$

The considerations of Section 4.4.2 for Kronecker products apply to the representation $\vartheta_{\ell,m} = \vartheta_\ell \otimes \overline{\vartheta_m}$. Hence the weights of $\vartheta_{\ell,m}$ are

$$(i+j, i-j) \text{ where } \begin{cases} i &= \ell, \ell-1, \ldots, -\ell \\ j &= m, m-1, \ldots, -m \end{cases} \tag{4.110}$$

These weights may be thought of as 2-dimensional position vectors in the plane (α, β). The endpoints of all weights form a **weight diagram**. This is illustrated in Fig. 4.6 for the examples $\vartheta_{3/2,1}$ and $\vartheta_{1/2,1/2}$.

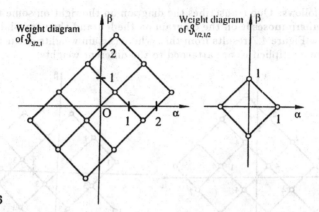

Fig. 4.6

For any irreducible representation $\vartheta_{\ell,m}$ there is a rectangle[12], shown in Fig. 4.6, which is symmetric with respect to the origin O, in 45° position, and made up of small squares of sides $\sqrt{2}$. In addition, the coordinates of all weight vectors are half integral.

This is readily shown: Starting from the weights in (4.108), all lying on the 45°-line, one adds the weights $(j, -j)$ of (4.109). This amounts to a parallel shift of the existing weights. The translation vector $(j, -j)$ is perpendicular to the 45°-line, etc.

Definition. A weight is said to be **dominant** if its first component is maximal.

For a given irreducible representation $\vartheta_{\ell,m}$ the dominant weight has the coordinates

$$p = \ell + m, \qquad q = \ell - m \qquad (4.111)$$

and is uniquely determined.

Conversely, if the dominant weight (p, q) is known, one can uniquely determine the corresponding irreducible representation $\vartheta_{\ell,m}$ by taking

$$\ell = \frac{p+q}{2}, \qquad m = \frac{p-q}{2} \qquad (4.112)$$

In other words, distinct irreducible representations are associated with distinct rectangles, and vice versa.

We have thus worked out a technique to completely reduce given representations of the positive proper Lorentz group L by use of Cartan's method of weights.

Example 4.4 (The Kronecker product $\vartheta_{3/2, 1} \otimes \vartheta_{1/2, 1/2}$, completely reduced) We form the sum vectors of the two diagrams in Fig. 4.6 (i.e., vector of one diagram + vector of the other diagram). In practice this is

---

[12] We also include the degenerate cases of a line segment and of a point ($\ell = 0$ or $m = 0$).

done as follows: One transcribes the diagram on the right on some tracing tissue, superimposes it on the diagram on the left and then translates it.

A new Figure 4.7 results from this, where certain weights occur repeatedly. The multiplicities are attached to the different weights.

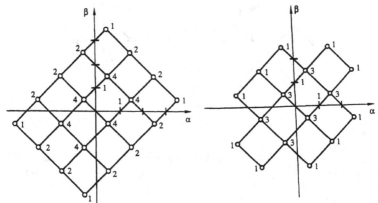

Fig. 4.7   **Fig. 4.8**

The dominant weight is $(7/2, 1/2)$, whence by (4.112) the irreducible representation $\vartheta_{2,3/2}$ is a constituent of $\vartheta_{3/2,1} \otimes \vartheta_{1/2,1/2}$. The diagram of $\vartheta_{2,3/2}$ itself has the shape of the rectangle in Fig. 4.7. But the weight vectors all occur with multiplicity 1. They are subtracted from Fig. 4.7 and thus furnish the weight diagram of Fig. 4.8 for the remaining invariant subspace.

We repeat the procedure with Fig. 4.8. The dominant weight $(5/2, 3/2)$ yields $\vartheta_{2,1/2}$ as another irreducible constituent, etc.

Finally, we obtain

$$\vartheta_{3/2,1} \otimes \vartheta_{1/2,1/2} = \vartheta_{2,3/2} \oplus \vartheta_{2,1/2} \oplus \vartheta_{1,3/2} \oplus \vartheta_{1,1/2} \qquad (4.113)$$

We recommend that the reader prove the **general** formula

$$\vartheta_{a,b} \otimes \vartheta_{c,d} = \bigoplus_{i=|a-c|}^{a+c} \bigoplus_{k=|b-d|}^{b+d} \vartheta_{i,k} \qquad (4.114)$$

by proceeding analogously.

Remark. The reader may be musing why in this chapter we investigate only a small selection of continuous groups. One reason is that for many continuous groups, for example $GL(n, \mathbf{R})$, there exist **non**-completely reducible representations. An example was shown in (1.46).

PROBLEMS

Problem 4.1 Show that the rotation (4.44) parametrized after Cayley
(a) has a rotation axis with components $(-\alpha_2, \beta_2, -\beta_1)$;
(b) describes a rotation through an angle φ given by $\cos \varphi/2 = \alpha_1$.
(Hint: Use the trace).

Problem 4.2 Let us confine our attention in (4.10) to the real 2×2 matrices A with $\det A = 1$. Equation (4.20) defines a representation of the group SL(2,R). Show that the representing transformations leave the subspace $x_3 = 0$ unchanged. Hence in this 3-dimensional vector space with the coordinates x_1, x_2, x_4, we obtain a group G of linear transformations leaving the quadratic form $(x_1^2 + x_2^2 - x_4^2)$ invariant. G is the group of **Laguerre circular geometry** and is a representation of SL(2,R).

Problem 4.3 The representation of SU(2) by matrices in (4.44) and the representation ϑ_1 of SU(2) in (4.51) both have the weights 1, 0, −1, thus are equivalent to each other.
(a) Verify that the representing matrices indeed have the same trace.
(b) In order to produce an equivalence transformation, let

$$A = \begin{bmatrix} \alpha_1 + i\beta_1 & \beta_2 + i\alpha_2 \\ -\beta_2 + i\alpha_2 & \alpha_1 - i\beta_1 \end{bmatrix} \quad \text{where } \alpha_1^2 + \alpha_2^2 + \beta_1^2 + \beta_2^2 = 1$$

act in the sense of transplantation (4.47) not on the monomials $x_1^2, x_1 x_2, x_2^2$, but on the new basis

$$\frac{1}{2}(x_1^2 - x_2^2), \qquad \frac{1}{2i}(x_1^2 + x_2^2), \qquad -x_1 x_2$$

(known as the approach by spinors). This corresponds to a representation equivalent to ϑ_1. Its matrices have precisely the form (4.44).

Problem 4.4 Show that for the spherical functions $u_\ell(x_1, x_2, x_3)$, recursion formulas of the form

$$x_k \cdot u_\ell = \alpha u_{\ell+1} + \beta r^2 u_{\ell-1} \quad \text{where } r^2 = x_1^2 + x_2^2 + x_3^2, \quad k = 1, 2, 3, \quad \ell \neq 0$$

hold. They are of importance for a selection rule in quantum mechanics (see Problem 7.10).
   Hint: Use the results (4.95), (4.96) on the decomposition of the space of homogeneous polynomials as well as orthogonality and completeness of the spherical functions.

Problem 4.5 The representation $\vartheta_{1/2} \otimes \vartheta_{1/2}$ of the special unitary group SU(2) contains the unit representation, as follows from the Clebsch-Gordan series. This means that in the 4-dimensional representation space, there is a nonzero vector v which is transformed identically under all representing matrices; i.e., $D(s)v = v \; \forall s \in SU(2)$.
   Determine this vector
(a) without using the tables of the Clebsch-Gordan coefficients,
(b) by using the tables of the Clebsch-Gordan coefficients.

Problem 4.6 Determine the 4-dimensional invariant irreducible subspace for the Kronecker product $\vartheta_1 \otimes \vartheta_{1/2}$ of SU(2) utilizing Table (4.86) of the Clebsch-Gordan coefficients.

Problem 4.7 Verify that the 3-dimensional representation (4.51) of SU(2) modified by (4.83) is unitary.

Problem 4.8 Show that
(a) The groups GL(n, **R**) are disconnected;
(b) The groups SU(n), SO(n), GL(n, **C**) are connected.

Problem 4.9 Let x_1, x_2 be the cartesian coordinates of the plane. Furthermore, let $\vartheta : s \to D(s)$ be the representation of the group SO(2) of all proper rotations about the origin, obtained by transplanting the following monomials according to (4.47)
(a) $\alpha x_1^2 + \beta x_1 x_2 + \gamma x_2^2$
(b) $\alpha x_1^3 + \beta x_1^2 x_2 + \gamma x_1 x_2^2 + \delta x_2^3$
where $\alpha, \beta, \gamma, \delta \in$ **C**.
    Reduce ϑ completely by aid of the operator $M = (x_1^2 + x_2^2) \cdot \Delta$, where Δ is the Laplacian.
    Do the same for the polynomials of third degree, i.e., for the linear combinations of
$$x_1^3, \qquad x_1^2 x_2, \qquad x_1 x_2^2, \qquad x_2^3$$

Problem 4.10 Show that the irreducible representations ϑ_ℓ and $\overline{\vartheta_\ell}$, where $\ell > 0$, of the group SL(2,**C**) are inequivalent to each other.

Chapter 5

Symmetry Adapted Vectors, Characters

Even though all considerations made in the subsequent sections are preliminarily meant for finite groups, most statements also extend to a larger class, namely to the class of all compact Lie groups. The latter will not be defined until later in Section 5.7, so as not to burden those readers who are primarily interested in finite groups.

5.1 Orthogonality of Representations

Let G be a finite group. Let further two irreducible representations ϑ and ϑ' of this group be given by their respective matrices $D(s)$ and $D'(s)$, i.e.,

$$\vartheta : s \rightarrow D(s) \quad \text{of dimension } n$$

$$\vartheta' : s \rightarrow D'(s) \quad \text{of dimension } n'$$

Finally, let U be some rectangular matrix not depending on s such that the matrix products $D'UD^{-1}$ are defined[1]. Hence U has length n and height n'.

Now we look at the following summation made over the group G:

$$W = \frac{1}{g} \sum_{s \in G} D'(s) U D(s)^{-1} \tag{5.1}$$

---

[1] To simplify notations, we shall frequently omit the argument s.

Here g is the order of G and summation of matrices is understood to be the summation of the corresponding matrix entries. Thus W also is a matrix of length n and height n'.

Let further a be a fixed element of the group G with

$$D(a) = A, \qquad D'(a) = A' \tag{5.2}$$

On inserting the identity matrix $E' = A'A'^{-1}$, we obtain

$$WA = \frac{1}{g}\sum_G [D'UD^{-1}A] = \frac{1}{g}\sum_G [A'(A'^{-1}D')U(D^{-1}A)] \tag{5.3}$$

whence

$$WA = A'\frac{1}{g}\sum_G [(A'^{-1}D')U(D^{-1}A)] \tag{5.4}$$

As s varies over the group G, so does $t = a^{-1}s$ (see Section 1.3). Therefore, (5.4) implies

$$WA = A'\frac{1}{g}\sum_G [D'UD^{-1}] = A'W \tag{5.5}$$

because $D(s^{-1}) = D(s)^{-1}$. Consequently,

$$WD(a) = D'(a)W \qquad \forall a \in G \tag{5.6}$$

Case I: ϑ and ϑ' are inequivalent. Schur's lemma (Section 2.1) implies

$$W = \text{null matrix} \tag{5.7}$$

By (5.1), an arbitrary matrix entry $w_{p\ell}$ of W satisfies

$$w_{p\ell} = \frac{1}{g}\sum_{s\in G}\sum_{\alpha,\beta} [d'_{p\alpha}(s)u_{\alpha\beta}d_{\beta\ell}(s^{-1})] = 0 \tag{5.8}$$

Let us now choose a special matrix U with entry 1 at position (q, k) and zeros elsewhere; that is,

$$u_{\alpha\beta} = \begin{cases} 1 & \text{if } \alpha = q \text{ and } \beta = k \\ 0 & \text{otherwise} \end{cases} \tag{5.9}$$

Thus (5.8) becomes

$$w_{p\ell} = \frac{1}{g}\sum_{s\in G} d'_{pq}(s) \cdot d_{k\ell}(s^{-1}) = 0 \qquad \forall p, q, k, \ell \tag{5.10}$$

Case II: ϑ and ϑ' are equivalent. By selecting a suitable coordinate system, say in the representation space of ϑ', we can achieve that

$$D(s) = D'(s) \qquad \forall s \in G \tag{5.11}$$

Equation (5.6) and Theorem 2.2 imply that

$$W = \lambda \cdot \text{identity matrix} \qquad (5.12)$$

Furthermore, by (5.1)

$$\text{tr}(W) = \lambda \cdot n = \text{tr}\left\{\frac{1}{g}\sum_G (DUD^{-1})\right\}$$

$$= \frac{1}{g}\sum_G \{\text{tr}(DUD^{-1})\} = \frac{1}{g}\sum_G \{\text{tr}(U)\} = \text{tr}(U)$$

because, by (5.128), $\text{tr}(XY) = \text{tr}(YX)$ holds for any two $n \times n$ matrices X, Y. Hence $\lambda = \frac{1}{n} \cdot \text{tr}(U)$, and by (5.12)

$$W = \frac{\text{tr}(U)}{n} \cdot \text{identity matrix} \qquad (5.13)$$

Now again, we choose a special (in this case a square) matrix (5.9) for U. The left equation in (5.10) and (5.13) written entrywise imply

$$w_{p\ell} = \frac{1}{g}\sum_{s \in G} d_{pq}(s) \cdot d_{k\ell}(s^{-1}) = \frac{\delta_{qk}}{n}\delta_{p\ell} \qquad (5.14)$$

where

$$\delta_{ij} = \begin{cases} 1 & \text{if } i = j \\ 0 & \text{otherwise} \end{cases}$$

Equations (5.10) and (5.14) are summarized as follows:

Theorem 5.1 *Let $D(s)$ and $D'(s)$ be the matrices of two inequivalent irreducible n- and n'-dimensional representations of the same finite group G. Then the matrix entries obey the following relations:*

$$\sum_{s \in G} d'_{pq}(s) \cdot d_{k\ell}(s^{-1}) = 0 \qquad \forall p, q, k, \ell \qquad (5.15)$$

$$\frac{1}{g}\sum_{s \in G} d_{pq}(s) \cdot d_{k\ell}(s^{-1}) = \frac{\delta_{qk}\delta_{p\ell}}{n} \qquad \forall p, q, k, \ell \qquad (5.16)$$

where g is the order of the group.

Theorem 5.1 is of central importance for group theory applications. We wish to reformulate it for the special case of a **unitary** representation ϑ. In this case[2]

$$D(s^{-1}) = \overline{D(s)}^T \qquad \forall s \in G \qquad (5.17)$$

Again ^{T} denotes the transpose and the bar complex conjugation. If we rewrite (5.17) entrywise, then Theorem 5.1 results in

---

[2]see Section 1.11.1.

Corollary 5.2 *Let $D(s)$ and $D'(s)$ be the matrices of two unitary inequivalent irreducible n- and n'-dimensional representations of the same finite group G. Then the following relations hold for the matrix entries:*

$$\sum_{s\in G} d'_{pq}(s) \cdot \overline{d_{\ell k}(s)} = 0 \qquad \forall k, \ell, p, q \tag{5.18}$$

$$\frac{1}{g} \sum_{s\in G} d_{pq}(s) \cdot \overline{d_{\ell k}(s)} = \frac{\delta_{qk}\delta_{p\ell}}{n} \qquad \forall k, \ell, p, q \tag{5.19}$$

Remark. Theorem 5.1 and Corollary 5.2 describe the so-called *orthogonality relations for irreducible representations* of G.

Geometric interpretation of Corollary 5.2. The numbers

$$d_{k\ell}(e), d_{k\ell}(a), d_{k\ell}(b), \ldots \tag{5.20}$$

assigned to the group elements $e, a, b, \ldots$ are regarded as g-dimensional vectors of $\mathbf{C}^g$ for fixed pairs k, ℓ. Equations (5.18), (5.19) say that all vectors (5.20) are orthonormal with respect to the inner product (1.80), regardless of whether they originate from the same or from different representations.

Since the maximal number of linearly independent g-dimensional vectors equals g, this implies the important fact that a **finite group** possesses only a **finite number** of irreducible and inequivalent **representations** ϑ_ρ. Let n_ρ be the dimension of ϑ_ρ. Since for a representation ϑ_ρ there are exactly n_ρ^2 vectors of the form (5.20), this further implies that

$$\sum_\rho n_\rho^2 \le |G| \tag{5.21}$$

In Section 1.8 we announced that even equality holds. A proof will be given in Section 5.6.

5.2 Algorithm for Symmetry Adapted Bases

Again let G first be a finite group and ϑ some n-dimensional representation of it. In this section, to avoid confusion, we shall employ sans serif type style T, D, P to denote the basis-independent notions of "transformation" and "operator". However, ordinary roman type style T, D, P is used to denote the corresponding basis-dependent matrices. That is, D(s) shall denote the linear transformation of the representation space V associated with the group element s.

Furthermore, let ϑ_ρ ($\rho = 1, 2, \ldots, N$) be the irreducible inequivalent n_ρ-dimensional representations of G that appear in ϑ with multiplicities $c_\rho \ge 0$.

Thus

$$\vartheta = c_1\vartheta_1 \oplus c_2\vartheta_2 \oplus \ldots \oplus c_N\vartheta_N \tag{5.22}$$

Accordingly, the representation space V of ϑ decomposes into[3]

$$V = V_1 \oplus V_2 \oplus \ldots \oplus V_N \tag{5.23}$$

The $c_\rho \cdot n_\rho$-dimensional subspaces V_ρ, called isotypic components (see Chapter 2), in turn decompose (for $c_\rho > 0$) into c_ρ irreducible subspaces V_ρ^1, V_ρ^2, ..., $V_\rho^{c_\rho}$ of dimension n_ρ:

$$V_\rho = V_\rho^1 \oplus V_\rho^2 \oplus \ldots \oplus V_\rho^{c_\rho} \tag{5.24}$$

In the following, we assume that all irreducible representations ϑ_ρ of G are given in the form of matrix tables. As we know, they are only determined up to equivalence (Section 1.5). However, in constructing an algorithm, it suffices to take an arbitrary fixed set of matrices for each ϑ_ρ. Let this set be denoted by

$$D_\rho(s) \qquad \forall s \in G \tag{5.25}$$

The corresponding matrix entries are denoted by

$$d_{\sigma\mu}^{(\rho)}(s) \qquad \forall s \in G, \qquad \sigma, \mu \in \{1, 2, \ldots, n_\rho\} \tag{5.26}$$

Our job is, based on this knowledge, to completely reduce the given representation ϑ; in other words, we shall try to determine the irreducible subspaces (5.24) and in addition, to determine a symmetry adapted basis. But prior to this, let us make some theoretical digressions.

There exists a special basis in V such that the corresponding representing matrices, denoted by $D^{sp}(s)$, of the linear transformations $D(s)$ of ϑ

---

[3]In the case $c_\rho = 0$, V_ρ is the null space.

have the following form:

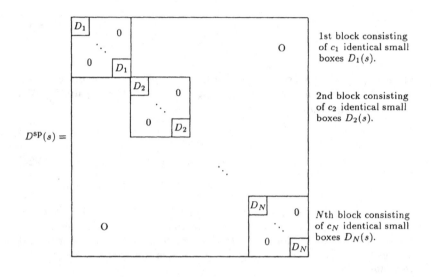

1st block consisting of c_1 identical small boxes $D_1(s)$.

2nd block consisting of c_2 identical small boxes $D_2(s)$.

Nth block consisting of c_N identical small boxes $D_N(s)$.

$$(5.27)$$

Here the ρth block is associated with the subspace V_ρ, and the c_ρ small boxes within, being identical with the fixed matrices given in (5.25), correspond to the decomposition (5.24). For the remainder of this section let j be an arbitrary, but fixed value of the index ρ with $c_j > 0$. As in Section 2.3 we shall label the special basis vectors of V_j mentioned above as follows:

$$
\begin{array}{llllll}
\text{for } V_j^1: & b_1^1 & b_2^1 & b_3^1 & \ldots & b_{n_j}^1 \\
\text{for } V_j^2: & b_1^2 & b_2^2 & b_3^2 & \ldots & b_{n_j}^2 \\
\vdots & \vdots & \vdots & \vdots & & \vdots \\
\text{for } V_j^{c_j}: & b_1^{c_j} & b_2^{c_j} & b_3^{c_j} & \ldots & b_{n_j}^{c_j}
\end{array}
\tag{5.28}
$$

For simplicity, we omit the subscript j of the basis vectors. Hence

$$
\mathsf{D}(s)b_\rho^i = \sum_{\sigma=1}^{n_j} d_{\sigma\rho}^{(j)}(s)b_\sigma^i
\tag{5.29}
$$

Now we consider the linear operators

$$
\mathsf{P}_{k\ell}^{(j)} = \frac{n_j}{g} \sum_{s\in G} d_{k\ell}^{(j)}(s^{-1})\mathsf{D}(s) \qquad k,\ell \in \{1,2,\ldots,n_j\}
\tag{5.30}
$$

It is worthwhile to point out that the numbers $d_{k\ell}^{(j)}(s^{-1})$ shall always mean fixed table values. Hence the matrix $\mathsf{P}_{k\ell}^{(j)\mathrm{sp}}$ describing the operator $\mathsf{P}_{k\ell}^{(j)}$ in our special coordinate system satisfies the equation

$$P_{k\ell}^{(j)\text{sp}} = \frac{n_j}{g} \sum_{s \in G} d_{k\ell}^{(j)}(s^{-1}) D^{\text{sp}}(s) \qquad k, \ell \in \{1, 2, \ldots, n_j\} \tag{5.31}$$

It will prove to be practical to regroup, according to (5.27), the matrix entries of $P_{k\ell}^{(j)\text{sp}}$ into blocks, which themselves contain small boxes along the main diagonal.

Carrying out the summation (5.31) entrywise yields that $P_{k\ell}^{(j)\text{sp}}$ has zero entries outside the small boxes, as has $D^{\text{sp}}(s)$.

In order to determine the small boxes in $P_{k\ell}^{(j)\text{sp}}$, we employ the orthogonality relations (5.15) and (5.16) from Theorem 5.1. Here ℓ, k, and j are fixed, while p and q are variable row and column numbers of the small boxes.

I. Equation (5.15) says that

$$\sum_{s \in G} d_{k\ell}^{(j)}(s^{-1}) \, d_{pq}^{(\rho)}(s) = 0 \qquad \forall p, q \quad \text{if } \rho \neq j \tag{5.32}$$

Hence all blocks of $P_{k\ell}^{(j)\text{sp}}$ with $\rho \neq j$ consist of zeros only.

II. Equation (5.16) applied to the jth block says

$$\frac{n_j}{g} \sum_{s \in G} d_{k\ell}^{(j)}(s^{-1}) \, d_{pq}^{(j)}(s) = \delta_{qk} \delta_{p\ell} \qquad \forall p, q, k, \ell \tag{5.33}$$

The term (5.33) is nonzero (namely $= 1$) if and only if

$$p = \ell \qquad \text{and} \qquad q = k$$

The only nonzero entry of an n_j-dimensional small box (inside the jth block) is found at the intersection of row no. ℓ and column no. k. Hence the jth block of $P_{k\ell}^{(j)\text{sp}}$ has the following form (5.34):

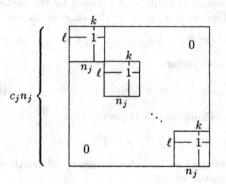

zeros throughout except
at the c_j marked spaces

$$\tag{5.34}$$

Thus the matrix $P_{k\ell}^{(j)\text{sp}}$ has exactly c_j entries 1, all others are 0. Hence its rank is c_j. It describes a linear operator with the following properties:

I. All basis vectors of the subspaces V_ρ with $\rho \neq j$ (see (5.22), (5.23)) are mapped into the zero vector. V_j is mapped onto itself. Hence

$$P_{k\ell}^{(j)} : V \to V_j \qquad (5.35)$$

II. For the basis vectors (5.28) in V_j, the block form (5.34) implies

$$\left.\begin{aligned} P_{k\ell}^{(j)} b_k^\alpha &= b_\ell^\alpha \\ P_{k\ell}^{(j)} b_{k'}^\alpha &= 0 \qquad \text{if } k \neq k' \end{aligned}\right\} \alpha = 1, 2, \ldots, c_j \qquad (5.36)$$

In words, this is summarized as follows:

$$\left.\begin{aligned} & P_{k\ell}^{(j)} \text{ maps the } c_j \text{ basis vectors of the } k\text{th column of (5.28)} \\ & \text{rowwise into those of the } \ell\text{th column. All other basis vectors} \\ & \text{of the } n\text{-dimensional representation space are annihilated,} \\ & \text{i.e., mapped into the zero vector.} \end{aligned}\right\} \quad (5.37)$$

Another important statement is an immediate consequence of (5.35) and (5.36) for the special case of $\ell = k$:

$$\left.\begin{aligned} & P_{kk}^{(j)} \text{ describes a projection of } V \text{ onto the } c_j\text{-dimensional} \\ & \text{subspace of } V_j \text{ spanned by the basis vectors in the } k\text{th col-} \\ & \text{umn of (5.28).} \end{aligned}\right\} \quad (5.38)$$

After these preparations we may proceed to the actual computation of **symmetry adapted basis vectors.** Let v be an arbitrary vector of the representation space V with the property

$$P_{11}^{(j)} v = v_1 \neq 0 \qquad (5.39)$$

This is possible only if the irreducible representation ϑ_j occurs with multiplicity $c_j > 0$.

By (5.38) the vector v_1 lies in the subspace spanned by the vectors of the first column of (5.28). Thus v_1 is a linear combination of the form

$$v_1 = \alpha_1 b_1^1 + \alpha_2 b_1^2 + \ldots + \alpha_{c_j} b_1^{c_j} \qquad \alpha_i \in \mathbf{C} \qquad (5.40)$$

Now we generate the additional vectors

$$v_\mu = P_{1\mu}^{(j)} v_1 \qquad \mu = 2, 3, \ldots, n_j \qquad (5.41)$$

Since $P_{k\ell}^{(j)} b_k^\alpha = b_\ell^\alpha$, Equations (5.40) and (5.41) imply that

$$v_\mu = \alpha_1 b_\mu^1 + \alpha_2 b_\mu^2 + \ldots + \alpha_{c_j} b_\mu^{c_j} \qquad \mu = 2, 3, \ldots, n_j \qquad (5.42)$$

These equations (5.42) reflect the fact that the vectors v_μ all are the same linear combination of the columns of (5.28).

Theorem 5.3 *The following holds:*

$$D(s)v_\mu = \sum_{\sigma=1}^{n_j} d_{\sigma\mu}^{(j)}(s)v_\sigma \qquad \mu = 1, 2, \ldots, n_j \tag{5.43}$$

In words (5.43) says: The vectors $v_1, v_2, \ldots, v_{n_j}$ span an irreducible invariant subspace of the representation ϑ. According to (5.29), the transformation behavior of the vectors v_i is exactly the same as that of an arbitrary row of vectors in (5.28).

Proof. Equations (5.29), (5.41), and (5.42) imply

$$D(s)v_\mu = \sum_{i=1}^{c_j} \alpha_i D(s) b_\mu^i = \sum_{i=1}^{c_j} \alpha_i \left(\sum_{\sigma=1}^{n_j} d_{\sigma\mu}^{(j)}(s) b_\sigma^i \right)$$

By interchanging summations, we obtain

$$D(s)v_\mu = \sum_{\sigma=1}^{n_j} d_{\sigma\mu}^{(j)}(s) \left(\sum_{i=1}^{c_j} \alpha_i b_\sigma^i \right) = \sum_{\sigma=1}^{n_j} d_{\sigma\mu}^{(j)}(s) v_\sigma$$

as was asserted in (5.43). $\qquad\qquad\qquad\qquad\qquad\qquad \Box$

Since the rank of P_{11}^j equals c_j, this procedure (Equations (5.39) through (5.42)) can be repeated with another vector v_1 which is linearly independent from the first, etc. This is done until c_j linearly independent vectors are determined.

Thus we proved the following algorithm; it is of central importance for applications:

Algorithm (Generating symmetry adapted basis vectors) *Given a finite group G. Assume that the irreducible pairwise inequivalent n_ρ-dimensional representations ϑ_ρ ($\rho = 1, 2, \ldots, N$) are available in the form of a table. More precisely, for each ϑ_ρ there exists a set of representing matrices whose entries are denoted as follows (σ = row number and μ = column number):*

$$d_{\sigma\mu}^\rho(s) \qquad \sigma, \mu \in \{1, 2, \ldots, n_\rho\}, \quad s \in G$$

Let ϑ be a given n-dimensional representation that assigns to each group element s the linear transformation $D(s)$. Let $D(s)$ be the matrices of the linear transformations $D(s)$ relative to some fixed basis $e_1, e_2, \ldots, e_n$ of the representation space V. Since ϑ is completely reducible, it reduces to

$$\vartheta = c_1 \vartheta_1 \oplus c_2 \vartheta_2 \oplus \ldots \oplus c_N \vartheta_N$$

Here the $c_\rho \geq 0$ denote the multiplicities with which the representations ϑ_ρ appear in ϑ. Correspondingly, V decomposes into

$$V = V_1 \oplus V_2 \oplus \ldots \oplus V_N$$

Now let j be some fixed value of the index ρ.

I. *Compute the matrix*

$$\pi^{(j)} = \sum_{s \in G} d_{11}^{(j)}(s^{-1}) D(s) \qquad (5.44)$$

associated with a multiple of the operator $\mathsf{P}_{11}^{(j)}$ *of rank* c_j.

II. *The matrix* $\pi^{(j)}$ *applied to the basis vectors of V yields vectors that are described by the columns of* $\pi^{(j)}$. *The latter span a c_j-dimensional subspace of V. If $c_j > 0$, we can find a basis of this subspace, denoted by*

$$v_1^1, v_1^2, \ldots, v_1^{c_j} \qquad (5.45)$$

by choosing c_j linearly independent columns among the n columns of the matrix $\pi^{(j)}$.

III. *Compute the matrices*

$$P_{1\mu}^{(j)} = \frac{n_j}{g} \sum_{s \in G} d_{1\mu}^{(j)}(s^{-1}) D(s) \qquad \mu = 2, 3, \ldots, n_j \qquad (5.46)$$

associated with the operators $\mathsf{P}_{1\mu}^{(j)}$ *of rank c_j (g is the group order).*

IV. *Generate symmetry adapted basis vectors in the irreducible subspaces* $V_j^1, V_j^2, \ldots, V_j^{c_j}$ *of V_j in the following way:*

$$v_\mu^i = P_{1\mu}^{(j)} v_1^i \qquad \mu = 2, 3, \ldots, n_j \qquad i = 1, 2, \ldots, c_j \qquad (5.47)$$

The n_j vectors $v_1^i, v_2^i, \ldots, v_{n_j}^i$ span V_j^i, as is conveniently remembered by the following array:

$$
\begin{array}{llccccc}
\textit{Basis for } V_j^1: & v_1^1 & v_2^1 & v_3^1 & \ldots & v_{n_j}^1 \\
\textit{Basis for } V_j^2: & v_1^2 & v_2^2 & v_3^2 & \ldots & v_{n_j}^2 \\
\vdots & \vdots & \vdots & \vdots & & \vdots \\
\textit{Basis for } V_j^{c_j}: & v_1^{c_j} & v_2^{c_j} & v_3^{c_j} & \ldots & v_{n_j}^{c_j}
\end{array}
\qquad (5.48)
$$

If this construction is carried out for each irreducible representation ϑ_j occurring in ϑ, then, with respect to the basis constructed above, the representing matrices have the desired form (5.27); in other words, the basis is symmetry adapted.

Remarks.

1. In Section 5.5, Equation (5.148), we shall deduce a formula to compute the multiplicities c_j, which will prove to be useful in applications. If $c_j = 0$, then evidently $\mathsf{P}_{11}^{(j)} =$ null matrix.

2. The linear maps

$$P_{11}^{(\rho)}, P_{12}^{(\rho)}, \ldots, P_{1n_\rho}^{(\rho)} \qquad \rho = 1, 2, \ldots, N$$

generate all of the desired basis vectors. For this reason they are called the **basis generators**.

3. Equations (5.44), (5.46) show that only the first row of the tabulated matrices of ϑ_ρ are used each time.

4. The decomposition of V_j into irreducible subspaces $V_j^1, V_j^2, \ldots, V_j^{c_j}$ is not unique, because the c_j linearly independent vectors in (5.45) were chosen arbitrarily.

5. From the derivation, it follows that in (5.44), (5.46), in place of

$$P_{11}^{(j)}, P_{12}^{(j)}, \ldots, P_{1n_j}^{(j)} \tag{5.49}$$

we could have used the basis generators

$$P_{kk}^{(j)}, P_{k1}^{(j)}, P_{k2}^{(j)}, \ldots, P_{k,k-1}^{(j)}, P_{k,k+1}^{(j)}, \ldots, P_{kn_j}^{(j)} \qquad k \in \{1, 2, \ldots, n_j\} \tag{5.50}$$

Here k can well take on varying values depending on the different values of j; that is, $k = k(j)$.

Finally, let us consider another

Modified algorithm using orthogonality: If an application involves unitary representations, then considerable simplification of the algorithm may be achieved: Let the tabulated representing matrices

$$D_\rho(s) = (d_{\sigma\mu}^{(\rho)}(s)) \qquad \sigma, \mu \in \{1, 2, \ldots, n_\rho\}, \qquad s \in G \tag{5.51}$$

used in the beginning of the algorithm be unitary. The representation property $D_j(s^{-1}) = D_j(s)^{-1}$ and (5.17) imply that the property

$$d_{1\mu}^{(j)}(s^{-1}) = \overline{d_{\mu 1}^{(j)}(s)} \qquad \mu = 1, 2, \ldots, n_j \tag{5.52}$$

holds for the equations (5.44) and (5.46). Here again the bar denotes complex conjugation.

Further simplification may result from

Theorem 5.4
Assumptions.

1. Let the irreducible representations $\vartheta_1, \vartheta_2, \ldots, \vartheta_N$ of the finite group G be tabulated by unitary matrices.

2. Let the representation $\vartheta : s \to \mathsf{D}(s)$ of G with the representation space $V = V_1 \oplus V_2 \oplus \ldots \oplus V_N$ (the V_j are isotypic components of type ϑ_j) be unitary with respect to a given inner product; i.e.,

$$\langle v, w \rangle = \langle \mathsf{D}(s)v, \mathsf{D}(s)w \rangle \qquad \forall v, w \in V, \quad \forall s \in G \tag{5.53}$$

Claims.

1. The isotypic components $V_1, V_2, \ldots, V_N$ are mutually orthogonal.

2. If in addition, the vectors in (5.45) are orthonormal[4], that is, if

$$\langle v_1^i, v_1^{i'} \rangle = \delta_{ii'} = \left\{ \begin{array}{ll} 1 & \text{if } i = i' \\ 0 & \text{otherwise} \end{array} \right. \tag{5.54}$$

holds for the various isotypic components, then the set of all vectors (5.48) constructed by the algorithm forms an orthonormal basis for the entire representation space V.

Proof. For v in (5.53) we select a vector v_k^i from the list (5.48) of ϑ_j and for w a vector $w_{k'}^{i'}$ from the list (5.48) of $\vartheta_{j'}$. Hence

$$\langle v_k^i, w_{k'}^{i'} \rangle = \langle \mathsf{D}(s)v_k^i, \mathsf{D}(s)w_{k'}^{i'} \rangle \tag{5.55}$$

Since (5.55) is valid for every element $s \in G$, this implies

$$\langle v_k^i, w_{k'}^{i'} \rangle = \frac{1}{g} \sum_{s \in G} \langle \mathsf{D}(s)v_k^i, \mathsf{D}(s)w_{k'}^{i'} \rangle$$

Substituting (5.43) into the right hand side and interchanging summations gives

$$\langle v_k^i, w_{k'}^{i'} \rangle = \frac{1}{g} \sum_{\sigma=1}^{n_j} \sum_{\sigma'=1}^{n_{j'}} \left(\langle v_\sigma^i, w_{\sigma'}^{i'} \rangle \sum_{s \in G} d_{\sigma k}^{(j)}(s) \overline{d_{\sigma'k'}^{(j')}(s)} \right) \tag{5.56}$$

By virtue of the orthogonality relations (5.18) and (5.19), summation over G in (5.56) yields

(a) the value 0, if $j \neq j'$, thus proving Claim I.

(b) the value $\delta_{\sigma\sigma'}\delta_{kk'}\frac{g}{n_j}$, if $j = j'$.

Substituted in (5.56), this in turn implies[5]

$$\langle v_k^i, v_{k'}^{i'} \rangle = \left\{ \begin{array}{ll} 0, & \text{if } k \neq k' \\ \frac{1}{n_j} \sum_{\sigma=1}^{n_j} \langle v_\sigma^i, v_{\sigma'}^{i'} \rangle, & \text{if } k = k' \end{array} \right. \tag{5.57}$$

---

[4]This can be effected, for instance, by the Schmidt orthogonalization method.
[5]Here $v = w$, since $j = j'$.

But the right side of (5.57) for $k = k'$ does not vary with k, thus by (5.54)

$$\langle v_k^i, v_k^{i'} \rangle = \langle v_1^i, v_1^{i'} \rangle = \delta_{ii'}$$

$\square$

Remark. *Theorem 5.4 is particularly interesting in the case where the representing matrices $D(s)$ are already related to an orthonormal basis, that is, when they are unitary (Section 1.11). Then the coordinate transformation between the original basis and the symmetry adapted basis is also described by a unitary matrix, which by*

$$S^{-1} = \overline{S}^T$$

is easily inverted. (T denotes the transpose and the bar complex conjugation.)

This plays a role, for instance, when solving simultaneous linear equations with symmetries, because then transforming the inhomogeneity does not cause any serious difficulties (see (3.30)).

5.3 Applications

By aid of the algorithm described in Section 5.2, eigenvalue problems and simultaneous linear equations

$$Mx = \lambda x, \qquad Mx = b$$

can be simplified by group theory methods if M possesses the symmetry of a completely reducible representation $\vartheta : s \to D(s)$, that is, if

$$MD(s) = D(s)M \qquad \forall s \in G$$

Because by the fundamental theorem of Section 2.3 the c_j-dimensional subspaces spanned by the columns of (5.48) are transformed into themselves under M; more still, each of them is transformed in exactly the same way. $\left.\begin{array}{c} \\ \\ \\ \\ \end{array}\right\}$ (5.58)

Algorithm (Simplification for Eigenvalue Computations) *By (5.58) it suffices to know just one column of (5.48). For instance, an eigenvalue λ appearing h times in the first column is also an eigenvalue appearing h times in the other columns; hence λ occurs in M at least[6] $(h \cdot n_j)$ times. The algorithm then reduces to the steps I and II.*

Finally another

---

[6] The same eigenvalue λ may occur in other isotypic components $V_{j'}$ $(j \neq j')$.

Remark on calculation of eigenvectors. By (5.58) again, *it suffices to calculate the eigenvectors in just* **one** *column.* For, if say,

$$x_1 = \sum_{i=1}^{c_j} \mu_i v_1^i \qquad \mu_i \in \mathbf{C}$$

is an eigenvector of M, then so are the corresponding vectors

$$x_\rho = \sum_{i=1}^{c_j} \mu_i v_\rho^i \qquad \rho = 2, 3, \ldots, n_j$$

of the remaining columns.

$$(5.59)$$

Example 5.1 (Waveguide junctions) Let us study a waveguide junction used in microwave circuits shown in Fig. 5.1.

The behavior of this junction against a certain type of high frequency electro-magnetic waves is described by a complex 3×3 matrix S (the so-called scattering matrix) with entries s_{ik} (Montgomery, Dicke, Purcell, 1965, Ch. 12). Then the complex amplitudes b_k of the emergent waves vary linearly with the complex amplitudes a_i of the incident waves; that is, $b_k = s_{k1}a_1 + s_{k2}a_2 + s_{k3}a_3$, $k = 1, 2, 3$, or simply

$$b = Sa \qquad (5.60)$$

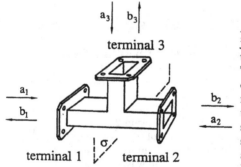

According to Fig. 5.1, the junction is symmetric about the plotted vertical plane. This reflection in the vertical plane denoted by σ corresponds to interchanging the terminals, or what is the same, to interchanging the subscripts 1 and 2 (3 is unaffected). The electrical behavior is unchanged, hence

Fig. 5.1

$$b_2 = s_{11}a_2 + s_{12}a_1 + s_{13}a_3$$
$$b_1 = s_{21}a_2 + s_{22}a_1 + s_{23}a_3 \qquad (5.61)$$
$$b_3 = s_{31}a_2 + s_{32}a_1 + s_{33}a_3$$

Comparison with (5.60) gives

$$s_{21} = s_{12}, \quad s_{22} = s_{11}, \quad s_{23} = s_{13}, \quad s_{32} = s_{31}$$

Thus S has the special form

$$S = \begin{bmatrix} \alpha & \beta & \gamma \\ \beta & \alpha & \gamma \\ \delta & \delta & \epsilon \end{bmatrix} \tag{5.62}$$

Therefore, instead of 9, we only need to measure 5 (complex) parameters in order to completely determine the scattering matrix S.

As the amplitudes of the standing waves are eigenvectors of S, we shall investigate the eigenvalue problem. In order to handle this by group theory methods, we note that with the aid of the permutation matrix

$$D(\sigma) = \begin{bmatrix} 0 & 1 & 0 \\ 1 & 0 & 0 \\ 0 & 0 & 1 \end{bmatrix}$$

Equation (5.61) can be written simply as $D(\sigma)b = SD(\sigma)a$. On substituting (5.60), we obtain

$$D(\sigma)S = SD(\sigma)$$

That is, S has the symmetry of the 3-dimensional representation

$$\vartheta : e \to E, \qquad \sigma \to D(\sigma)$$

of the cyclic group $C_2 = \{e = \sigma^2, \sigma\}$.

The complete list of irreducible representations of C_2 is given below:

	e	σ
ϑ_1	1	1
ϑ_2	1	-1

(5.63)

Step I in the algorithm "Generating symmetry adapted basis vectors" gives

$$\pi^{(1)} = E + D(\sigma) = \begin{bmatrix} 1 & 1 & 0 \\ 1 & 1 & 0 \\ 0 & 0 & 2 \end{bmatrix}, \quad \pi^{(2)} = E - D(\sigma) = \begin{bmatrix} 1 & -1 & 0 \\ -1 & 1 & 0 \\ 0 & 0 & 0 \end{bmatrix}$$

$$\text{rank of } \pi^{(1)} = c_1 = 2 \qquad\qquad \text{rank of } \pi^{(2)} = c_2 = 1$$

Consequently,

$$\vartheta = 2\vartheta_1 \oplus \vartheta_2 \tag{5.64}$$

Discussion of $\pi^{(2)}$. The corresponding subspace is 1-dimensional and is spanned by the vector $v_2^1 = (1, -1, 0)^T$. Action of S, given in (5.62), on this vector yields the eigenvalue $\alpha - \beta$.

Discussion of $\pi^{(1)}$. The corresponding subspace is 2-dimensional and is spanned by the vectors

$$v_1^1 = (1, 1, 0)^T \qquad \text{and} \qquad v_1^2 = (0, 0, 1)^T$$

which under the action of S are transformed as follows:

$$\begin{aligned} Sv_1^1 &= (\alpha + \beta)v_1^1 + 2\delta v_1^2 \\ Sv_1^2 &= \gamma v_1^1 + \epsilon v_1^2 \end{aligned}$$

Thus it remains to find the eigenvalues of the 2×2 matrix

$$\begin{bmatrix} \alpha + \beta & \gamma \\ 2\delta & \epsilon \end{bmatrix} \tag{5.65}$$

Remark. Equation (5.62) is a verification of Theorem 2.8. Here indeed $\sum_\rho c_\rho^2 = 2^2 + 1^2 = 5$, due to (5.64).

Example 5.2 (The membrane problem) We refer to Example 1.9. Its treatment involves the permutation group on the 9 lattice points shown in Fig. 5.2 originating from the dihedral group D_4 with the natural 9-dimensional representation ϑ_{nat} as described in (1.30).

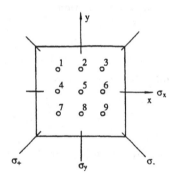

The vectors of the representation space correspond to the array of values taken on in the lattice; the ith basis vector has value 1 at lattice point no. i and 0 elsewhere.

We shall show that the algorithm (5.44) through (5.48) applied to the so constructed representation ϑ_{nat} of D_4 indeed generates symmetry adapted basis vectors, as already exhibited in (1.48). Here we use the following simplified method:

Fig. 5.2

Rather than writing out the permutation matrices, we *operate directly in the configuration of the permuted objects*, that means, here in the lattice of Fig. 5.2.

In the algorithm we choose as an example of the irreducible representation ϑ_ρ the 2-dimensional identity representation[7] of D_4. This representation is obtained by describing the rotations and reflections by matrices $(d_{\mu\nu}(s))$ relative to the cartesian coordinate system x, y. In the following table only the first columns of the matrices $(d_{\mu\nu}(s))$ are written down, since by $d_{\mu\nu}(s^{-1}) = \overline{d_{\nu\mu}(s)} = d_{\nu\mu}(s)$ only these are used.

	Id.	$+90°$	$-90°$	$180°$	σ_x	σ_y	σ_+	σ_-
1st row	1	0	0	-1	1	-1	0	0
2nd row	0	1	-1	0	0	0	1	-1

$$\tag{5.66}$$

---

[7]This representation is equivalent to ϑ_1^2 of (1.60).

Steps I and II of the algorithm can be summarized as follows: The action of

$$\pi = \sum_s d_{11}(s^{-1})D(s) = \sum_s d_{11}(s)D(s) \qquad (5.67)$$

on the ith basis vector simply means that one successively applies the 8 symmetry operators[8] of the dihedral group to the point no. i of Fig. 5.2 and then totals the thus obtained 8 lattice functions weighted by the numbers d_{11} from the first row of Table (5.66).

If, for example, we choose the third basis vector, then we obtain the array of values

$$\pi \cdot (3\text{rd basis vector}) = \begin{matrix} -1 & 0 & 1 \\ 0 & 0 & 0 \\ -1 & 0 & 1 \end{matrix} = u_2$$

Step III requires that the basis generator

$$P_{12} = \frac{1}{4}\sum_s d_{12}(s^{-1})D(s) = \frac{1}{4}\sum_s d_{21}(s)D(s) \qquad (5.68)$$

act on u_2; that is, one applies the 8 symmetry operators of D_4 to the array of values u_2 and then totals the so obtained 8 lattice functions, but now weighted by the numbers from the second row of Table (5.66).

We then obtain the array of values

$$P_{12}u_2 = \begin{matrix} 1 & 0 & 1 \\ 0 & 0 & 0 \\ -1 & 0 & -1 \end{matrix} = v_2$$

Thus the vectors u_2, v_2 are constructed, which in (1.48) were simply quoted without proof. We recommend that the reader construct the remaining arrays of values of (1.48) as an exercise.

Molecular Oscillations

Let us consider an arbitrary molecule consisting of N atoms oscillating about their equilibrium positions in the sense of classical mechanics. Here we assume that the displacements are small relative to the geometry of the molecule and that the forces involved are harmonic (that is, that they vary linearly with the displacement).

The linear approximations to the equations of motion for the kth atom are hence of the form

$$M_k \ddot{x}_{k\ell} = -\sum_{\kappa,\lambda} F_{k\ell,\kappa\lambda} x_{\kappa\lambda} \qquad \ell = 1,2,3 \qquad (5.69)$$

Summation extends over $\kappa = 1, 2, \ldots, N$ and $\lambda = 1, 2, 3$. M_k is the mass of the kth atom; the double dots stand for the second derivative in the time

---

[8]see Section 2.2.

variable t; $x_{k\ell}$ is the displacement of the kth atom relative to its equilibrium position in direction ℓ ($= 1, 2, 3$) of a cartesian coordinate system whose origin is the kth atom (see Example in Fig. 5.3); and $F_{k\ell,\kappa\lambda}$ are the force constants.

For better physical understanding of (5.69) we note the following: A displacement of atom no. k' in direction ℓ' results in a force component of magnitude $-F_{k\ell,k'\ell'}x_{k'\ell'}$ acting on atom no. k in direction ℓ.

Summation over all atoms k' and directions ℓ' yields the ℓth component of the resultant force on atom no. k caused by the displacements of the various atoms about their equilibrium positions.

Newton's principle **"action = reaction"** implies

$$F_{k\ell,k'\ell'} = F_{k'\ell',k\ell} \tag{5.70}$$

Our aim now is to find an approach to calculate the **eigenoscillations** and the **eigenfrequencies** of our molecule as a mechanical system. Separation of variables

$$x_{k\ell}(t) = \overset{o}{x}_{k\ell} \cdot e^{i\omega t} \qquad k = 1, 2, \ldots, N \qquad \ell = 1, 2, 3$$

reduces the system (5.69) of $3N$ linear differential equations to an algebraic problem

$$\omega^2 M_k \cdot \overset{o}{x}_{k\ell} = \sum_{\kappa,\lambda} F_{k\ell,\kappa\lambda} \overset{o}{x}_{\kappa\lambda} \tag{5.71}$$

In matrix form, this reads

$$\omega^2 \begin{bmatrix} M_1 \overset{o}{x}_{11} \\ M_1 \overset{o}{x}_{12} \\ M_1 \overset{o}{x}_{13} \\ M_2 \overset{o}{x}_{21} \\ M_2 \overset{o}{x}_{22} \\ \cdots \cdots \\ M_N \overset{o}{x}_{N3} \end{bmatrix} = \underbrace{\begin{bmatrix} F_{11,11} & F_{11,12} & F_{11,13} & F_{11,21} & F_{11,22} & \cdots & F_{11,N3} \\ F_{12,11} & F_{12,12} & F_{12,13} & F_{12,21} & F_{12,22} & \cdots & F_{12,N3} \\ F_{13,11} & F_{13,12} & F_{13,13} & F_{13,21} & F_{13,22} & \cdots & F_{13,N3} \\ F_{21,11} & F_{21,12} & F_{21,13} & F_{21,21} & F_{21,22} & \cdots & F_{21,N3} \\ F_{22,11} & \cdots & & & \cdots & \cdots & F_{22,N3} \\ \cdots & \cdots & & \cdots & \cdots & \cdots \\ F_{N3,11} & \cdots & \cdots & \cdots & \cdots & \cdots & F_{N3,N3} \end{bmatrix}}_{F} \begin{bmatrix} \overset{o}{x}_{11} \\ \overset{o}{x}_{12} \\ \overset{o}{x}_{13} \\ \overset{o}{x}_{21} \\ \overset{o}{x}_{22} \\ \cdots \\ \overset{o}{x}_{N3} \end{bmatrix}$$

$$\tag{5.72}$$

The **force matrix** F is symmetric by (5.70). Its entries may be calculated from the potential

$$\varphi = \frac{1}{2} \sum_{\kappa,\lambda} \sum_{\kappa',\lambda'} F_{\kappa\lambda,\kappa'\lambda'} \cdot x_{\kappa\lambda} \cdot x_{\kappa'\lambda'} = \frac{1}{2} x^T F x \tag{5.73}$$

In reducing (5.72) to an eigenvalue problem, we work with the normalized quantities

$$F'_{k\ell,\kappa\lambda} = \frac{1}{\sqrt{M_k M_\kappa}} F_{k\ell,\kappa\lambda} \qquad \overset{o}{x'}_{k\ell} = \sqrt{M_k} \, \overset{o}{x}_{k\ell} \tag{5.74}$$

Substituting these into (5.71) and multiplying by $1/\sqrt{M_k}$ gives a system of linear equations

$$\omega^2 \overset{o}{x}'_{k\ell} = \sum_{\kappa,\lambda} F'_{k\ell,\kappa\lambda} \cdot \overset{o}{x}'_{k\lambda} \tag{5.75}$$

This results in the eigenvalue problem

$$\det(F' - \omega^2 E) = 0 \tag{5.76}$$

where E is the $3N \times 3N$ identity matrix.

Thus the squares of the frequencies are eigenvalues of the matrix F', which again is symmetric.

The following representative example illustrates how in the case of **molecules** with certain **symmetries** the above problem can be simplified by utilizing basis generators.

Example 5.3 (Methane CH_4) The four hydrogen atoms are located at the vertices of a regular tetrahedron and the carbon atom at its centroid ($N = 5$), thus involving a 15×15 force matrix. We simplify notations by setting

$$x'_{m1} = x_m, \quad x'_{m2} = y_m, \quad x'_{m3} = z_m \qquad m = 1, 2, \dots, 5$$

The positions of the various coordinate systems are not prescribed. We choose them as in Fig. 5.3.

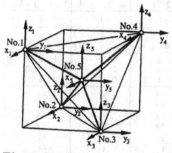

Obviously, the problem described above has symmetries: It is invariant under arbitrary permutations of the four hydrogen atoms. Each of the permutations of the atoms labeled 1 to 4 corresponds to a linear transformation of the 3-dimensional space which in turn corresponds to a linear transformation of the 15-dimensional space described by the coordinates x_m, y_m, z_m ($m = 1, 2, \dots, 5$).

Fig. 5.3

For example, the permutation

$$\left\downarrow \begin{array}{cccc} 1 & 2 & 3 & 4 \\ 1 & 3 & 2 & 4 \end{array} \right.$$

corresponds to a reflection in a plane that passes through the atoms labeled 1, 4, and 5. This reflection in turn produces the linear transformation

$$x_i \leftrightarrow -y_i \qquad z_i \leftrightarrow z_i \qquad i = 1, 4, 5$$
$$x_2 \leftrightarrow -y_3 \qquad x_3 \leftrightarrow y_2 \qquad z_2 \leftrightarrow z_3$$

$$(5.77)$$

	1 identity	8 ± 120° rotations about interior diagonals				
↓	1 2 3 4 1 2 3 4	1 2 3 4 3 1 2 4	1 2 3 4 2 3 1 4	1 2 3 4 4 1 3 2	1 2 3 4 2 4 3 1	1 2 3 4 4 2 1 3
ϑ_2^1	1	1	1	1	1	1
ϑ_1^2	1 0 0 1	$-1/2$ $\sqrt{3}/2$ $-\sqrt{3}/2$ $-1/2$	$-1/2$ $-\sqrt{3}/2$ $\sqrt{3}/2$ $-1/2$	$-1/2$ $-\sqrt{3}/2$ $\sqrt{3}/2$ $-1/2$	$-1/2$ $\sqrt{3}/2$ $-\sqrt{3}/2$ $-1/2$	$-1/2$ $\sqrt{3}/2$ $-\sqrt{3}/2$ $-1/2$
ϑ_1^3	1 0 0 0 1 0 0 0 1	0 -1 0 0 0 1 -1 0 0	0 0 -1 -1 0 0 0 1 0	0 0 -1 1 0 0 0 -1 0	0 1 0 0 0 -1 -1 0 0	0 1 0 0 0 1 1 0 0
ϑ	x_1	$-z_3$	$-y_2$	y_4	$-z_2$	z_4
	x_2	$-z_1$	$-y_3$	y_1	$-z_4$	z_2
	x_3	$-z_2$	$-y_1$	y_3	$-z_3$	z_1
	x_4	$-z_4$	$-y_4$	y_2	$-z_1$	z_3
	y_1	$-x_3$	z_2	$-z_4$	x_2	x_4
	y_2	$-x_1$	z_3	$-z_1$	x_4	x_2
	y_3	$-x_2$	z_1	$-z_3$	x_3	x_1
	y_4	$-x_4$	z_4	$-z_2$	x_1	x_3
	z_1	y_3	$-x_2$	$-x_4$	$-y_2$	y_4
	z_2	y_1	$-x_3$	$-x_1$	$-y_4$	y_2
	z_3	y_2	$-x_1$	$-x_3$	$-y_3$	y_1
	z_4	y_4	$-x_4$	$-x_2$	$-y_1$	y_3
	x_5	$-z_5$	$-y_5$	y_5	$-z_5$	z_5
	y_5	$-x_5$	z_5	$-z_5$	x_5	x_5
	z_5	y_5	$-x_5$	$-x_5$	$-y_5$	y_5

	6 reflections					
↓	1 2 3 4 2 1 3 4	1 2 3 4 1 3 2 4	1 2 3 4 1 2 4 3	1 2 3 4 3 2 1 4	1 2 3 4 4 2 3 1	1 2 3 4 1 4 3 2
ϑ_2^1	-1	-1	-1	-1	-1	-1
ϑ_1^2	$1/2$ $\sqrt{3}/2$ $\sqrt{3}/2$ $-1/2$	$1/2$ $-\sqrt{3}/2$ $-\sqrt{3}/2$ $-1/2$	$1/2$ $\sqrt{3}/2$ $\sqrt{3}/2$ $-1/2$	-1 0 0 1	$1/2$ $-\sqrt{3}/2$ $-\sqrt{3}/2$ $-1/2$	-1 0 0 1
ϑ_1^3	0 0 -1 0 1 0 -1 0 0	0 -1 0 -1 0 0 0 0 1	0 0 1 0 1 0 1 0 0	1 0 0 0 0 1 0 1 0	0 1 0 1 0 0 0 0 1	1 0 0 0 0 -1 0 -1 0
ϑ	$-z_2$	$-y_1$	z_1	x_3	y_4	x_1
	$-z_1$	$-y_3$	z_2	x_2	y_2	x_4
	$-z_3$	$-y_2$	z_4	x_1	y_3	x_3
	$-z_4$	$-y_4$	z_3	x_4	y_1	x_2
	y_2	$-x_1$	y_1	z_3	x_4	$-z_1$
	y_1	$-x_3$	y_2	z_2	x_2	$-z_4$
	y_3	$-x_2$	y_4	z_1	x_3	$-z_3$
	y_4	$-x_4$	y_3	z_4	x_1	$-z_2$
	$-x_2$	z_1	x_1	y_3	z_4	$-y_1$
	$-x_1$	z_3	x_2	y_2	z_2	$-y_4$
	$-x_3$	z_2	x_4	y_1	z_3	$-y_3$
	$-x_4$	z_4	x_3	y_4	z_1	$-y_2$
	$-z_5$	$-y_5$	z_5	x_5	y_5	x_5
	y_5	$-x_5$	y_5	z_5	x_5	$-z_5$
	x_5	z_5	x_5	y_5	z_5	$-y_5$

Representations of the symmetric group S_4

3   180° rotations about center of faces

1 2 3 4 3 2 4 1	1 2 3 4 1 4 2 3	1 2 3 4 1 3 4 2	1 2 3 4 2 1 4 3	1 2 3 4 3 4 1 2	1 2 3 4 4 3 2 1	row number
1	1	1	1	1	1	
$-1/2$ $-\sqrt{3}/2$ $\sqrt{3}/2$ $-1/2$	$-1/2$ $-\sqrt{3}/2$ $\sqrt{3}/2$ $-1/2$	$-1/2$ $\sqrt{3}/2$ $-\sqrt{3}/2$ $-1/2$	1 0 0 1	1 0 0 1	1 0 0 1	
0 0 1 1 0 0 0 1 0	0 0 1 -1 0 0 0 -1 0	0 -1 0 0 0 -1 1 0 0	-1 0 0 0 1 0 0 0 -1	1 0 0 0 -1 0 0 0 -1	-1 0 0 0 -1 0 0 0 1	
y_3	$-y_1$	z_1	$-x_2$	x_3	$-x_4$	1
y_2	$-y_4$	z_3	$-x_1$	x_4	$-x_3$	2
y_4	$-y_2$	z_4	$-x_4$	x_1	$-x_2$	3
y_1	$-y_3$	z_2	$-x_3$	x_2	$-x_1$	4
z_3	$-z_1$	$-x_1$	y_2	$-y_3$	$-y_4$	5
z_2	$-z_4$	$-x_3$	y_1	$-y_4$	$-y_3$	6
z_4	$-z_2$	$-x_4$	y_4	$-y_1$	$-y_2$	7
z_1	$-z_3$	$-x_2$	y_3	$-y_2$	$-y_1$	8
x_3	x_1	$-y_1$	$-z_2$	$-z_3$	z_4	9
x_2	x_4	$-y_3$	$-z_1$	$-z_4$	z_3	10
x_4	x_2	$-y_4$	$-z_4$	$-z_1$	z_2	11
x_1	x_3	$-y_2$	$-z_3$	$-z_2$	z_1	12
y_5	$-y_5$	z_5	$-x_5$	x_5	$-x_5$	13
z_5	$-z_5$	$-x_5$	y_5	$-y_5$	$-y_5$	14
x_5	x_5	$-y_5$	$-z_5$	$-z_5$	z_5	15

6 rotary reflections   (5.77)

1 2 3 4 2 3 4 1	1 2 3 4 2 4 1 3	1 2 3 4 3 1 4 2	1 2 3 4 4 1 2 3	1 2 3 4 3 4 2 1	1 2 3 4 4 3 1 2	row number
-1	-1	-1	-1	-1	-1	
-1 0 0 1	$1/2$ $-\sqrt{3}/2$ $-\sqrt{3}/2$ $-1/2$	$1/2$ $-\sqrt{3}/2$ $-\sqrt{3}/2$ $-1/2$	-1 0 0 1	$1/2$ $\sqrt{3}/2$ $\sqrt{3}/2$ $-1/2$	$1/2$ $\sqrt{3}/2$ $\sqrt{3}/2$ $-1/2$	
-1 0 0 0 0 -1 0 1 0	0 1 0 -1 0 0 0 0 -1	0 -1 0 1 0 0 0 0 -1	-1 0 0 0 0 1 0 -1 0	0 0 1 0 -1 0 -1 0 0	0 0 -1 0 -1 0 1 0 0	
$-x_2$	$-y_2$	y_3	$-x_4$	$-z_3$	z_4	1
$-x_3$	$-y_4$	y_1	$-x_1$	$-z_4$	z_3	2
$-x_4$	$-y_1$	y_4	$-x_2$	$-z_2$	z_1	3
$-x_1$	$-y_3$	y_2	$-x_3$	$-z_1$	z_2	4
z_2	x_2	$-x_3$	$-z_4$	$-y_3$	$-y_4$	5
z_3	x_4	$-x_1$	$-z_1$	$-y_4$	$-y_3$	6
z_4	x_1	$-x_4$	$-z_2$	$-y_2$	$-y_1$	7
z_1	x_3	$-x_2$	$-z_3$	$-y_1$	$-y_2$	8
$-y_2$	$-z_2$	$-z_3$	y_4	x_3	$-x_4$	9
$-y_3$	$-z_4$	$-z_1$	y_1	x_4	$-x_3$	10
$-y_4$	$-z_1$	$-z_4$	y_2	x_2	$-x_1$	11
$-y_1$	$-z_3$	$-z_2$	y_3	x_1	$-x_2$	12
$-x_5$	$-y_5$	y_5	$-x_5$	$-z_5$	z_5	13
z_5	x_5	$-x_5$	$-z_5$	$-y_5$	$-y_5$	14
$-y_5$	$-z_5$	$-z_5$	y_5	x_5	$-x_5$	15

We thus constructed a representation ϑ of the symmetric group S_4.

This representation is completely described in Table (5.77), as are all irreducible representations of S_4, except for ϑ_1^1, the unit representation. ϑ_2^1 denotes the alternating representation assigning the number 1 (-1) to every even (odd) permutation.

ϑ_1^3 is obtained by viewing the permutations of the 4 hydrogen atoms as rotations and rotary reflections of $\mathbf{R}^3$ (see head row of Table (5.77)). The representing matrices relate to the coordinate system x_5, y_5, z_5 in Fig. 5.3.

$\vartheta_2^3 = \vartheta_1^3 \otimes \vartheta_2^1$; thus all matrices of ϑ_2^3 associated with an odd permutation are multiplied by (-1), while the remaining 12 matrices remain unchanged (Section 1.10).

ϑ_1^2 is obtained by first assigning to each group element of S_4 the permutation of the nonoriented coordinate axes x_5, y_5, z_5 associated with ϑ_1^3 and then representing these permutations of S_3 (each appearing precisely 4 times) by those of the vertices of a triangle[9].

Table (5.77) is complete because $1^2 + 1^2 + 3^2 + 3^2 + 2^2 = 24$ (Theorem 1.4). To verify irreducibility of ϑ_1^3 and ϑ_2^3, see Problem 1.2.

The representing matrices of ϑ are given in a space saving form: The first row of the table shows the images of the first basis vector. In this row one can see, for example, that the coordinate x_1 is mapped to $-z_2$ under the representing matrix of

$$\left\downarrow \begin{array}{cccc} 1 & 2 & 3 & 4 \\ 2 & 4 & 3 & 1 \end{array}\right.$$

Labeling the 15 coordinates according to Table (5.77) from top to bottom shows that in the first column of the corresponding representing matrix all entries are zero except for the entry in the tenth row, which equals -1. All representing matrices of ϑ consist of exactly the elements $1, 0, -1$ and are orthogonal. Inspection of the table shows that the representation space decomposes in a natural way into a 12- and a 3-dimensional subspace. The latter transforms exactly after the manner of the irreducible representation ϑ_1^3, as is seen by comparing ϑ_1^3 with the bottom three rows of the table. In the following we treat the problem of decomposing the remaining 12-dimensional subspace by utilizing the algorithm "Generating symmetry adapted basis vectors".

Knowing the linearly independent column vectors k_1, k_2, ..., k_5 of the matrices (5.78), (5.79) is enough to find that for all occurring irreducible representations **steps I and II** ((5.44), (5.45)) are indeed completed, because to k_1 there is a 1-dimensional, to k_2 a 2-dimensional, and to each of k_3, k_4, k_5 there is a 3-dimensional irreducible subspace.

---

[9]See Example 1.8. Here, in contrast to Fig. 1.5, we place one of the vertices of the triangle on the x_2-axis.

$$\pi\!\begin{pmatrix}1\\1\end{pmatrix}=\begin{bmatrix}2&\cdots\cdots\\-2\\2\\-2\\[4pt]-2\\-2\\2\\2\\[4pt]2\\-2\\-2\\2&\cdots\cdots\end{bmatrix}\atop k_1 \qquad \pi\!\begin{pmatrix}2\\1\end{pmatrix}=\frac{3}{2}\begin{bmatrix}0&0&0&0&0&\cdots\\0&0&0&0&0\\ \cdot&\cdot&\cdot&\cdot&0\\ \cdot&\cdot&\cdot&\cdot&0\\[4pt]\cdot&\cdot&\cdot&\cdot&1\\ \cdot&\cdot&\cdot&\cdot&1\\ \cdot&\cdot&\cdot&\cdot&-1\\ \cdot&\cdot&\cdot&\cdot&-1\\[4pt]\cdot&\cdot&\cdot&\cdot&1\\ \cdot&\cdot&\cdot&\cdot&-1\\ \cdot&\cdot&\cdot&\cdot&-1\\0&0&0&0&1&\cdots\end{bmatrix}\atop k_2 \qquad(5.78)$$

$$\pi\!\begin{pmatrix}3\\1\end{pmatrix}=\begin{bmatrix}2&2&2&2&0&\cdots\\2&2&2&2&0\\2&2&2&2&0\\2&2&2&2&0\\[4pt]&&&&1\\&&&&-1\\&&&&-1\\&&&&1\\&&O&&\\&&&&-1\\&&&&-1\\&&&&1\\&&&&1&\cdots\end{bmatrix}\atop {k_3 \qquad k_4} \qquad \pi\!\begin{pmatrix}3\\2\end{pmatrix}=\begin{bmatrix}0&0&0&0&0&\cdots\\0&0&0&0&0\\ \cdot&\cdot&\cdot&\cdot&0\\ \cdot&\cdot&\cdot&\cdot&0\\[4pt]\cdot&\cdot&\cdot&\cdot&1\\ \cdot&\cdot&\cdot&\cdot&-1\\ \cdot&\cdot&\cdot&\cdot&-1\\ \cdot&\cdot&\cdot&\cdot&1\\[4pt]\cdot&\cdot&\cdot&\cdot&1\\ \cdot&\cdot&\cdot&\cdot&1\\ \cdot&\cdot&\cdot&\cdot&-1\\0&0&0&0&-1&\cdots\end{bmatrix}\atop k_5 \qquad(5.79)$$

These subspaces span all of the 12-dimensional subspace. Together with the 3-dimensional subspace already split off, this yields the representation

$$\vartheta = \vartheta_1^1 \oplus \vartheta_1^2 \oplus 3\vartheta_1^3 \oplus \vartheta_2^3 \qquad (5.80)$$

Since k_3 and k_4 are already orthogonal with respect to the inner product (1.80) and by Theorem 5.4, after having normalized the vectors k_1, k_2, ..., k_5, the algorithm produces an orthonormal basis of the 15-dimensional representation space. The normalized vectors are denoted as follows:

$$u = \frac{1}{\sqrt{48}}k_1, \quad v_1 = \frac{1}{\sqrt{8}}k_2, \quad w_1 = \frac{1}{\sqrt{8}}k_5, \quad x_1 = \frac{1}{4}k_3, \quad y_1 = \frac{1}{\sqrt{8}}k_4$$

$$(5.81)$$

Step III

$$P_{12}^{\left(\begin{smallmatrix}2\\1\end{smallmatrix}\right)} = \tfrac{1}{12}\sum_{s\in S_4}\overline{d}_{21}^{\left(\begin{smallmatrix}2\\1\end{smallmatrix}\right)}(s)\,D(s) \qquad P_{12}^{\left(\begin{smallmatrix}3\\1\end{smallmatrix}\right)} = \frac{3}{24}\sum_{s\in S_4}\overline{d}_{21}^{\left(\begin{smallmatrix}3\\1\end{smallmatrix}\right)}(s)\,D(s)$$

$$P_{13}^{\binom{3}{1}} = \tfrac{3}{24} \sum_{s\in S_4} \overline{d}_{31}^{\binom{3}{1}}(s)\, D(s)$$

$$P_{12}^{\binom{2}{1}} = \frac{\sqrt{3}}{24}
\left[\; O \quad
\begin{array}{rrrrrrrr}
2 & 2 & -2 & -2 & 2 & -2 & -2 & 2\\
-2 & -2 & 2 & 2 & -2 & 2 & 2 & -2\\
2 & 2 & -2 & -2 & 2 & -2 & -2 & 2\\
-2 & -2 & 2 & 2 & -2 & 2 & 2 & -2\\
1 & 1 & -1 & -1 & 1 & -1 & -1 & 1\\
1 & 1 & -1 & -1 & 1 & -1 & -1 & 1\\
-1 & -1 & 1 & 1 & -1 & 1 & 1 & -1\\
-1 & -1 & 1 & 1 & -1 & 1 & 1 & -1\\
-1 & -1 & 1 & 1 & -1 & 1 & 1 & -1\\
1 & 1 & -1 & -1 & 1 & -1 & -1 & 1\\
1 & 1 & -1 & -1 & 1 & -1 & -1 & 1\\
-1 & -1 & 1 & 1 & -1 & 1 & 1 & -1
\end{array}
\right] \tag{5.82}$$

$$P_{12}^{\binom{3}{1}} = \frac{1}{8}
\left[
\begin{array}{cccc|cccccccc}
 & O & & & 1 & -1 & -1 & 1 & -1 & -1 & 1 & 1\\
 & & & & -1 & 1 & 1 & -1 & 1 & 1 & -1 & -1\\
 & & & & -1 & 1 & 1 & -1 & 1 & 1 & -1 & -1\\
 & & & & 1 & -1 & -1 & 1 & -1 & -1 & 1 & 1\\
2 & 2 & 2 & 2 & & & & & & & &\\
2 & 2 & 2 & 2 & & & & O & & & &\\
2 & 2 & 2 & 2 & & & & & & & &\\
2 & 2 & 2 & 2 & & & & & & & &\\
 & O & & & 1 & -1 & -1 & 1 & -1 & -1 & 1 & 1\\
 & & & & -1 & 1 & 1 & -1 & 1 & 1 & -1 & -1\\
 & & & & 1 & -1 & -1 & 1 & -1 & -1 & 1 & 1\\
 & & & & -1 & 1 & 1 & -1 & 1 & 1 & -1 & -1
\end{array}
\right] \tag{5.83}$$

$$P_{13}^{\binom{3}{1}} = \frac{1}{8}
\left[
\begin{array}{cccc|cccccccc}
 & & & & -1 & 1 & 1 & -1 & 1 & 1 & -1 & -1\\
 & & & & -1 & 1 & 1 & -1 & 1 & 1 & -1 & -1\\
 & & & & 1 & -1 & -1 & 1 & -1 & -1 & 1 & 1\\
 & O & & & 1 & -1 & -1 & 1 & -1 & -1 & 1 & 1\\
 & & & & 1 & -1 & -1 & 1 & -1 & -1 & 1 & 1\\
 & & & & -1 & 1 & 1 & -1 & 1 & 1 & -1 & -1\\
 & & & & 1 & -1 & -1 & 1 & -1 & -1 & 1 & 1\\
 & & & & -1 & 1 & 1 & -1 & 1 & 1 & -1 & -1\\
2 & 2 & 2 & 2 & & & & & & & &\\
2 & 2 & 2 & 2 & & & & & & & &\\
2 & 2 & 2 & 2 & & & & O & & & &\\
2 & 2 & 2 & 2 & & & & & & & &
\end{array}
\right] \tag{5.84}$$

$$P_{12}^{\binom{3}{2}} = \frac{1}{8}
\left[
\begin{array}{cccccccccc}
O & -1 & 1 & 1 & -1 & -1 & -1 & 1 & 1\\
 & 1 & -1 & -1 & 1 & 1 & 1 & -1 & -1\\
 & 1 & -1 & -1 & 1 & 1 & 1 & -1 & -1\\
 & -1 & 1 & 1 & -1 & -1 & -1 & 1 & 1\\
O & & & & O & & & & O\\
 & 1 & -1 & -1 & 1 & 1 & 1 & -1 & -1\\
 & -1 & 1 & 1 & -1 & -1 & -1 & 1 & 1\\
O & 1 & -1 & -1 & 1 & 1 & 1 & -1 & -1\\
 & -1 & 1 & 1 & -1 & -1 & -1 & 1 & 1
\end{array}
\right] \tag{5.85}$$

$$P_{13}^{\binom{3}{2}} = \frac{1}{8}
\left[
\begin{array}{cccccccccc}
O & -1 & 1 & 1 & -1 & -1 & -1 & 1 & 1\\
 & -1 & 1 & 1 & -1 & -1 & -1 & 1 & 1\\
 & 1 & -1 & -1 & 1 & 1 & 1 & -1 & -1\\
 & 1 & -1 & -1 & 1 & 1 & 1 & -1 & -1\\
O & -1 & 1 & 1 & -1 & -1 & -1 & 1 & 1\\
 & 1 & -1 & -1 & 1 & 1 & 1 & -1 & -1\\
 & -1 & 1 & 1 & -1 & -1 & -1 & 1 & 1\\
 & 1 & -1 & -1 & 1 & 1 & 1 & -1 & -1\\
O & & & & O & & & & O
\end{array}
\right] \tag{5.86}$$

When calculating the basis generators (5.82) through (5.86), use the rank
(= multiplicity of the corresponding irreducible representation). This may
reduce the computational complexity. For instance, in the case of (5.82)
the rank is 1. Hence it suffices, from the sixth matrix column onward, to
just compute the entries of, say, the first row; the column vectors beyond
the fifth are then multiples of the fifth column vector.

Now we can calculate the remaining basis vectors, for example,

$$\left. \begin{array}{ll} x_2 = P_{12}^{\binom{3}{1}} x_1, & x_3 = P_{13}^{\binom{3}{1}} x_1 \\[2mm] w_2 = P_{12}^{\binom{3}{2}} w_1, & w_3 = P_{13}^{\binom{3}{2}} w_1 \end{array} \right\} \tag{5.87}$$

The complete symmetry adapted orthonormal basis (written out for the
15-dimensional representation space) hence is

$$\tag{5.88}$$

a) $\vartheta_1^1: u = \frac{1}{\sqrt{12}}[\ 1\ -1\ \ 1\ -1\ \ |\ -1\ -1\ \ 1\ \ 1\ \ |\ \ 1\ -1\ -1\ \ 1\ \ |\ 0\ \ 0\ \ 0]^{\mathrm{T}}$

b) $\vartheta_1^2: v_1 = \frac{1}{\sqrt{8}}[\ 0\ \ 0\ \ 0\ \ 0\ \ |\ \ 1\ \ 1\ -1\ -1\ \ |\ \ 1\ -1\ -1\ \ 1\ \ |\ 0\ \ 0\ \ 0]^{\mathrm{T}}$

$\quad\ v_2 = \frac{1}{\sqrt{24}}[\ \ 2\ -2\ \ \ 2\ -2\ \ |\ \ 1\ \ 1\ -1\ -1\ \ |\ -1\ \ 1\ \ 1\ -1\ \ |\ 0\ \ 0\ \ 0]^{\mathrm{T}}$

c) $\vartheta_2^3: w_1 = \frac{1}{\sqrt{8}}[\ 0\ \ 0\ \ 0\ \ 0\ \ |\ \ 1\ -1\ -1\ \ 1\ \ |\ \ 1\ \ 1\ -1\ -1\ \ |\ 0\ \ 0\ \ 0]^{\mathrm{T}}$

$\quad\ w_2 = \frac{1}{\sqrt{8}}[-1\ \ 1\ \ 1\ -1\ \ |\ 0\ \ 0\ \ 0\ \ 0\ \ |\ \ 1\ -1\ \ 1\ -1\ \ |\ 0\ \ 0\ \ 0]^{\mathrm{T}}$

$\quad\ w_3 = \frac{1}{\sqrt{8}}[-1\ -1\ \ 1\ \ 1\ \ |\ -1\ \ 1\ -1\ \ 1\ \ |\ 0\ \ 0\ \ 0\ \ 0\ \ |\ 0\ \ 0\ \ 0]^{\mathrm{T}}$

d) $\vartheta_1^3: x_1 = \frac{1}{2}[\ \ 1\ \ 1\ \ 1\ \ 1\ \ |\ 0\ \ 0\ \ 0\ \ 0\ \ |\ 0\ \ 0\ \ 0\ \ 0\ \ |\ 0\ \ 0\ \ 0]^{\mathrm{T}}$

$\quad\ x_2 = \frac{1}{2}[\ 0\ \ 0\ \ 0\ \ 0\ \ |\ \ 1\ \ 1\ \ 1\ \ 1\ \ |\ 0\ \ 0\ \ 0\ \ 0\ \ |\ 0\ \ 0\ \ 0]^{\mathrm{T}}$

$\quad\ x_3 = \frac{1}{2}[\ 0\ \ 0\ \ 0\ \ 0\ \ |\ 0\ \ 0\ \ 0\ \ 0\ \ |\ \ 1\ \ 1\ \ 1\ \ 1\ \ |\ 0\ \ 0\ \ 0]^{\mathrm{T}}$

$\quad\ y_1 = \frac{1}{\sqrt{8}}[\ 0\ \ 0\ \ 0\ \ 0\ \ |\ \ 1\ -1\ -1\ \ 1\ \ |\ -1\ -1\ \ 1\ \ 1\ \ |\ 0\ \ 0\ \ 0]^{\mathrm{T}}$

$\quad\ y_2 = \frac{1}{\sqrt{8}}[\ \ 1\ -1\ -1\ \ 1\ \ |\ 0\ \ 0\ \ 0\ \ 0\ \ |\ \ 1\ -1\ \ 1\ -1\ \ |\ 0\ \ 0\ \ 0]^{\mathrm{T}}$

$\quad\ y_3 = \frac{1}{\sqrt{8}}[-1\ -1\ \ 1\ \ 1\ \ |\ \ 1\ -1\ \ 1\ -1\ \ |\ 0\ \ 0\ \ 0\ \ 0\ \ |\ 0\ \ 0\ \ 0]^{\mathrm{T}}$

$\quad\ z_1 = \quad[\ 0\ \ 0\ \ \cdots\ \ |\ \cdots\cdots\ \ |\ \cdots\cdots\ 0\ \ |\ 1\ \ 0\ \ 0]^{\mathrm{T}}$

$\quad\ z_2 = \quad[\ 0\ \ 0\ \ \cdots\ \ |\ \cdots\cdots\ \ |\ \cdots\cdots\ 0\ \ |\ 0\ \ 1\ \ 0]^{\mathrm{T}}$

$\quad\ z_3 = \quad[\ 0\ \ 0\ \ \cdots\ \ |\ \cdots\cdots\ \ |\ \cdots\cdots\ 0\ \ |\ 0\ \ 0\ \ 1]^{\mathrm{T}}$

Symmetry properties of the force matrix F. In our example, the
force matrix, as appearing in (5.72), does not only satisfy (5.70), but also
other conditions, so that it commutes with the representing matrices of
ϑ. We now wish to establish such conditions. Here we shall make use of
the fact that the symmetric structure of the methane molecule causes the

potential φ in (5.73) to be invariant under the representation ϑ listed in (5.77).

If, for example, we choose the displacement

$$\left. \begin{array}{l} x_1 = 1 \\ \text{all other coordinates} = 0 \end{array} \right\} \qquad (5.89)$$

and the representing matrix described in the second column of (5.77), then the displacement (5.89) transforms to

$$\left. \begin{array}{l} z_3 = -1 \\ \text{all other coordinates} = 0 \end{array} \right\} \qquad (5.90)$$

Substituting each of (5.89) and (5.90) into (5.73) and equating the potentials yields

$$F_{x_1,x_1} = F_{z_3,z_3}$$

All other representing matrices of ϑ applied to the displacement (5.89) yield, according to the first row of Table (5.77),

$$F_{x_i,x_i} = F_{y_i,y_i} = F_{z_i,z_i} = A \qquad i = 1, 2, 3, 4 \qquad (5.91)$$

From the third-last row of Table (5.77) and by selecting the displacement

$$x_5 = 1$$

we see that

$$F_{x_5,x_5} = F_{y_5,y_5} = F_{z_5,z_5} = B \qquad (5.92)$$

Restrictions are imposed also on the elements off the main diagonal. For example, the displacement

$$\left. \begin{array}{l} x_1 = x_2 = 1 \\ \text{all other coordinates} = 0 \end{array} \right\} \qquad (5.93)$$

subjected to the various representing matrices of ϑ in (5.77) and (5.91) yield the following relations:

$$\left. \begin{array}{l} F_{x_1,x_2} = F_{z_3,z_1} = F_{y_2,y_3} = F_{y_4,y_1} = F_{z_2,z_4} = F_{x_4,x_3} = \\ F_{z_2,z_1} = F_{y_1,y_3} = F_{x_3,x_2} = F_{y_4,y_2} = F_{x_1,x_4} = F_{z_3,z_4} = C \end{array} \right\} \qquad (5.94)$$

Thus we have determined the (symmetric) force matrix F in parts. In order to obtain further relations, we select some matrix entry occurring in neither (5.91), (5.92), nor (5.94). For example, we select F_{x_1,x_3}, that is, the displacement

$$\left. \begin{array}{l} x_1 = x_3 = 1 \\ \text{all other coordinates} = 0 \end{array} \right\} \qquad (5.95)$$

Accordingly, the first and the third rows of ϑ in the list (5.77) imply the relations

$$F_{x_1,x_3} = F_{z_3,z_2} = F_{y_2,y_1} = F_{y_4,y_3} = F_{z_4,z_1} = F_{x_2,x_4} = D \qquad (5.96)$$

This procedure may be iterated for any of the unused matrix entries of F. By doing so successively, say, for the displacements

$$x_5 = x_1 = 1, \qquad y_5 = x_1 = 1, \qquad x_1 = y_1 = 1$$
$$x_3 = y_1 = 1, \qquad x_4 = y_1 = 1, \qquad x_5 = y_5 = 1$$

with the other coordinates being zero, we find that the matrix F must have the following structure:

x_1	x_2	x_3	x_4	y_1	y_2	y_3	y_4	z_1	z_2	z_3	z_4	x_5	y_5	z_5	
A	C	D	C	S	T	-T	U	-S	-U	T	-T	E	R	-R	x_1
	A	C	D	-T	-S	-U	T	-U	-S	-T	T	E	-R	-R	x_2
		A	C	T	-U	-S	-T	-T	T	S	U	E	-R	R	x_3
			A	U	-T	T	S	T	-T	U	S	E	R	R	x_4
				A	D	C	C	S	-T	U	T	R	E	R	y_1
					A	C	C	T	-S	-T	-U	-R	E	-R	y_2
						A	D	U	T	S	-T	-R	E	R	y_3
							A	-T	-U	T	-S	R	E	-R	y_4
								A	C	C	D	-R	R	E	z_1
									A	D	C	-R	-R	E	z_2
										A	C	R	R	E	z_3
											A	R	-R	E	z_4
												B	0	0	x_5
													B	0	y_5
														B	z_5

$$F_{ik} = F_{ki}$$

$$(5.97)$$

In particular, the 3×3 submatrix at the lower right hand corner is diagonal, because the second- and third-last rows of the list (5.77) imply

$$F_{x_5,y_5} = -F_{x_5,y_5} = F_{x_5,z_5} = -F_{x_5,z_5} = F_{y_5,z_5} = -F_{y_5,z_5} \qquad (5.98)$$

Now the question is whether or not the 9 numbers

$$A, B, C, D, E, R, S, T, U \qquad (5.99)$$

need to obey any further relations when considering some other displacements. This can be answered negatively by a modification of Theorem 2.6.

Proof. As mentioned at the end of Section 5.2, the coordinate transformation in the representation space transforming an orthonormal basis into a symmetry adapted orthonormal basis (with numbering along columns in each of the isotypic components (2.18)) is described by a unitary matrix S. If, as in our example, the bases are real, then S is even orthogonal. But then the following holds:

$$F \text{ real and symmetric} \quad \Leftrightarrow \quad S^T F S = \tilde{F} \text{ real and symmetric} \qquad (5.100)$$

An argument similar to that of Theorem 2.8 implies that, in an adjusted coordinate system, the set of operators commuting with ϑ as given in (2.20) is described by all matrices $\widetilde{F}$ containing arbitrary real symmetric $c_j \times c_j$ matrices $(j = 1, 2, \ldots, N)$, each of which occurs n_j times.

The set of all matrices $\widetilde{F}$ and hence *the set of all matrices F possesses a total of*

$$\sum_{j=1}^{N} \left(\frac{c_j^2}{2} + \frac{c_j}{2} \right) = \sum_{j=1}^{N} \frac{c_j(c_j + 1)}{2} \qquad (5.101)$$

free real parameters.

By (5.80), our example has the multiplicities 1, 1, 3, 1. Hence (5.101) says that for $N = 4$ the number of free real parameters is

$$\sum_{j=1}^{4} \frac{c_j(c_j + 1)}{2} = 1 + 1 + 6 + 1 = 9 \qquad (5.102)$$

As there are exactly the 9 values (5.99) that appear in matrix (5.97), we thus proved that these quantities are independent of each other.

The fundamental theorem in Section 2.3, applied to the representation (5.80), now tells us the following about the vectors (5.88):

1. u is an eigenvector of F to some eigenvalue α.
2. v_1, v_2 are eigenvectors of F, both to the same eigenvalue β.
3. w_1, w_2, w_3 are eigenvectors of F, all to the same eigenvalue γ.
4. The three subspaces spanned by the vectors x_i, y_i, z_i are transformed invariantly under F, and what is more, all three of them are transformed in exactly the same way.

In computing the eigenvalues it suffices to determine a nonzero component of the corresponding eigenvector. Equations (5.97) and (5.88) imply

$$\left. \begin{array}{llll} Fu & = \alpha u & = (A - 2C + D - 2S - 4T + 2U)u & \\ Fv_i & = \beta v_i & = (A - 2C + D + S + 2T - U)v_i & i = 1, 2 \\ Fw_i & = \gamma w_i & = (A - D + S - 2T - U)w_i & i = 1, 2, 3 \end{array} \right\} \qquad (5.103)$$

In order to determine, say, the first column of the symmetric 3×3 matrix occurring three times, it suffices by (5.88) to calculate the first, fifth, and the thirteenth component of Fx_1, and so forth.

$$\left. \begin{array}{llll} Fx_1 = & (A + 2C + D)x_1 + & \sqrt{2}(S + U)y_1 + & 2Ez_1 \\ Fy_1 = & (\ldots)x_1 + & (A + 2T + U - D - S)y_1 + & \sqrt{8}Rz_1 \\ Fz_1 = & (\ldots)x_1 + & (\ldots)y_1 + & Bz_1 \end{array} \right\} \qquad (5.104)$$

The 3×3 matrix occurring thrice in the partly diagonalized form $\widetilde{F}$ of F is

$$\begin{bmatrix} A + 2C + D & \sqrt{2}(S + U) & 2E \\ \sqrt{2}(S + U) & A + 2T + U - D - S & \sqrt{8}R \\ 2E & \sqrt{8}R & B \end{bmatrix} \qquad (5.105)$$

thus reducing the computation of the eigenvalues to solving a single cubic equation.

5.4 Similarity Classes of Groups

We consider an arbitrary finite or infinite group G, written multiplicatively.

Definition. Two elements $a, b \in G$ are said to be **similar to each other** if there exists an $s \in G$ such that $b = sas^{-1}$.

This relation is

1. reflexive, as a is similar to itself;
2. symmetric, because $b = sas^{-1}$ implies $a = s^{-1}bs$;
3. transitive, since $b = sas^{-1}$ and $c = tbt^{-1}$ imply $c = tsas^{-1}t^{-1} = (ts)a(ts)^{-1}$.

Hence the relation "similar to" is an **equivalence relation**. It induces in G a partitioning into equivalence classes, i.e., a decomposition into disjoint subsets T_ρ such that

$$\bigcup_\rho T_\rho = G \tag{5.106}$$

In our case of the similarity relation the T_ρ are called **similarity classes**.

Immediate consequences of the similarity definition for **arbitrary groups** are:

   I. *If $z \in G$ commutes with every element of the group G, that is, if*

$$zg = gz \qquad \forall g \in G \tag{5.107}$$

   then z forms a similarity class by itself, because $gzg^{-1} = gg^{-1}z = z$.

   II. *If a group G is abelian, then each element of G forms a class by itself.*

   III. *Every similarity class of a subgroup of G is a subset of a similarity class of G.*

Now we restrict our attention to **transformation groups**. We wish to analyze the relation between the two similar transformations a and $b = sas^{-1}$. The objects of the range W to be transformed are denoted by $X, Y, \ldots$ In the case of linear transformations, W is a vector space. In the case of permutations, W consists of the objects to be permuted, etc. From

$$X \xrightarrow{\ b\ } sas^{-1}(X) \qquad \forall X \in W \tag{5.108}$$

and

$$Y = s^{-1}(X) \tag{5.109}$$

it follows for the transformation $b = sas^{-1}$ that

$$s(Y) \overset{b}{\longrightarrow} s(a(Y)) \tag{5.110}$$

Formula (5.110) has the following interpretation:

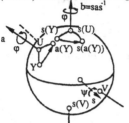

Fig. 5.4

Action of the similarity transformation s on an arbitrary object Y and on the image of Y under the transformation a yields elements that are associated with each other by the transformation $b = sas^{-1}$ as shown in Fig. 5.4.

Remark. We recall that as Y ranges through all of W, so does $s(Y)$, because s is a transformation. Thus (5.110) defines the correspondence for all elements of W.

We now apply these general similarity considerations to some special cases.

5.4.1   Subgroups of SO(3)

Subgroups of $O(n)$ are called *n*-**dimensional point groups**, as they leave the origin fixed. A point group can contain finitely or infinitely many elements. An example is the set of all rotations carrying a cube into itself. If two elements u, v of this set transform the cube into itself, then so does the composite $u \cdot v$ (v followed by u). Furthermore, all other group axioms are satisfied.

We now consider an arbitrary subgroup G of SO(3). Such a subgroup transforms the unit sphere into itself. Let further $a, s \in G$ be arbitrary rotations about oriented axes through rotation angles φ and ψ as shown in Fig. 5.5.

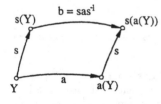

Given a, what is now the transformation $b = sas^{-1}$ similar to a ? Apply rotation s to the point U and to its image $a(U)$. As U lies on the rotation axis of a, we have $U = a(U)$. Hence by (5.110), $s(U)$ is a fixed point of the transformation b. The analog holds for V. Hence the *straight line passing through $s(U)$ and $s(V)$ is the rotation axis of b.*

Fig. 5.5

If we consider (5.110) for a point Y off the rotation axis of a as drawn in Fig. 5.5, then we find that b *has the same rotation angle φ as has* a. It is uniquely determined by the orientation of the rotation axis of b. Moreover, these considerations imply the following:

Two rotations through angles $\pm\varphi$ about the same oriented axis D are similar to each other if and only if in the group there exists a 180° rotation with rotation axis perpendicular to D.

Example 5.4 (Octahedral group O) This group consists of the proper rotations of $\mathbf{R}^3$ that carry a cube onto itself (Fig. 5.6).

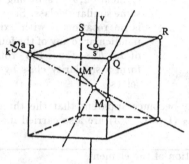

Let a be the 120° rotation about the interior diagonal k and s be the 90° rotation about the vertical axis v.

The two $\pm120°$ rotations about k are similar to each other. The reason is that the 180° rotation about an axis passing through the two midpoints M, M' of the edges is an element of O with axis perpendicular to k.

Fig. 5.6

The transformations sas^{-1} are the $\pm120°$ rotations about the interior diagonal passing through Q. The equivalence transformations s^2 and s^3 supply the $\pm120°$ rotations about the remaining two interior diagonals through R and S. These eight rotations form a similarity class since all other elements have a rotation angle different from $\pm120°$.

The reader may verify that the following collection of similarity classes of O is complete:

Class no.	Number of elements	Description of the elements
1	1	identity
2	8	$\pm120°$ rotations about the interior diagonals
3	6	180° rotations about the midpoints of edges
4	6	$\pm90°$ rotations about the midpoints of faces
5	3	180° rotations about the midpoints of faces

Similarity classes of the octahedral group O

$$(5.111)$$

Hence O is a group of 24 elements.

Example 5.5 (The group SO(3) of all proper rotations) Given two rotation axes, there always exists a rotation s that maps one axis into the other. Consequently, *in the group SO(3) all rotations with angles equal in magnitude form a similarity class.*

Example 5.6 (The tetrahedral group T) This group consists of all proper rotations that carry a regular tetrahedron into itself. Let s be the 120° rotation about the vertical axis (in Fig. 5.7 drawn in direction of the projection) through the vertex D, and let a be the 120° rotation about the axis BF_1.

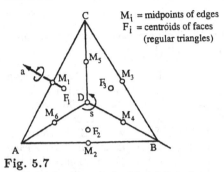

M_i = midpoints of edges
F_i = centroids of faces (regular triangles)

Fig. 5.7

$b = sas^{-1}$ is the 120° rotation about CF_2, and $b' = sbs^{-1}$ is the 120° rotation about AF_3. For symmetry reasons the four elements a, b, b', s belong to the same similarity class. Since there is no rotation T with axis perpendicular to any of the four axes of a, b, b', s, the $-120°$ rotations form a similarity class by themselves.

Furthermore, from Fig. 5.7 it becomes apparent that the three 180° rotations about the axes $M_i M_{i+3}$ ($i = 1, 2, 3$) are also carried into each other by s or s^2. We thus obtain

Class no.	Number of elements	Description of the elements
1	1	identity
2	4	120° rotations about the midpoints of faces
3	4	$-120°$ rotations about the midpoints of faces
4	3	180° rotations about the midpoints of edges

Similarity classes of the tetrahedral group T

$$(5.112)$$

Hence T is a group of 12 elements. If to each element of T we associate the corresponding even permutation of the four vertices of the tetrahedron, then we find that the correspondence is one-to-one: *The alternating group A_4 is isomorphic with* T.

Furthermore, T is a subgroup of O; for, a regular tetrahedron can be inscribed into a cube (Fig. 1.11). Hence a congruence transformation of the tetrahedron also carries a cube into itself.

5.4.2 Permutation Groups

By Section 1.3 the permutation groups are subgroups of the symmetric group S_N, which itself consists of all $N!$ permutations of N objects. Without loss of generality, we may choose as the N objects the natural numbers $1, 2, \ldots, N$.

For $N = 3$ the symmetric group consists of the 6 elements

$$
\left.
\begin{array}{lll}
e = \left| \begin{array}{ccc} 1 & 2 & 3 \\ 1 & 2 & 3 \end{array} \right| &
a = \left| \begin{array}{ccc} 1 & 2 & 3 \\ 2 & 3 & 1 \end{array} \right| &
b = \left| \begin{array}{ccc} 1 & 2 & 3 \\ 3 & 1 & 2 \end{array} \right| \\[3mm]
c = \left| \begin{array}{ccc} 1 & 2 & 3 \\ 1 & 3 & 2 \end{array} \right| &
d = \left| \begin{array}{ccc} 1 & 2 & 3 \\ 3 & 2 & 1 \end{array} \right| &
f = \left| \begin{array}{ccc} 1 & 2 & 3 \\ 2 & 1 & 3 \end{array} \right|
\end{array}
\right\}
\quad (5.113)
$$

In studying similarity classes it is practical to use the **cycle notation**: e.g., (312) denotes the permutation mapping 3 to 1, 1 to 2, and 2 to 3. This is the permutation a from (5.113). Evidently,

$$a = (312) = (231) = (123) \tag{5.114}$$

That is, the number sequence is determined only up to cyclic order.

The following theorem is readily verified:

Theorem 5.5 *Each permutation is a product of disjoint cycles.*

For example, the following element of S_8 may be written as

$$\left\downarrow \begin{array}{cccccccc} 1 & 2 & 3 & 4 & 5 & 6 & 7 & 8 \\ 4 & 2 & 5 & 6 & 3 & 1 & 8 & 7 \end{array} \right. = (146)(2)(35)(78) \tag{5.115}$$

The objects in a cycle of length 1 remain fixed under the permutation in question. *Clearly, disjoint cycles commute* without changing the permutation. Thus the permutation in (5.115) can be rewritten as follows:

$$(78)(146)(2)(35) = (2)(146)(35)(78) \tag{5.116}$$

Now we introduce the **type** of a permutation. It is specified by the lengths of the cycles appearing in the disjoint decomposition. Thus, if we rearrange the cycles by decreasing lengths, the permutation (5.116) is of type

$$(...)(..)(..)(.)$$

More precisely, for an arbitrary permutation, let

$$\alpha_j = \text{ number of cycles of length } j \qquad j = 1, 2, \ldots, N \tag{5.117}$$

with the convention that $\alpha_\ell = 0$ if cycles of length ℓ fail to occur in the permutation. The **type of the permutation** hence is characterized by N nonnegative integers $\alpha_1, \alpha_2, \ldots, \alpha_N$ with the property

$$\sum_{k=1}^{N} k \cdot \alpha_k = N \tag{5.118}$$

(Clearly, it suffices to give the positive α_k only.)

Example. The two permutations (132)(45) and (412)(35) of S_5 are of the same type.

Elementary combinatorial considerations lead to the following statement: *There exist exactly*

$$\frac{N!}{\displaystyle\prod_{k=1}^{N}(\alpha_k! \cdot k^{\alpha_k})} \tag{5.119}$$

permutations of the symmetric group S_N of type $[\alpha_1, \alpha_2, \ldots, \alpha_N]$ (see Problem 5.1(b)).

Finally, as an example we consider the **type table** of S_5 (5.120). The table lists the cycles by decreasing lengths:

Type	$\alpha_j=$ number of cycles of length j					Number of permutations
	α_1	α_2	α_3	α_4	α_5	
(.....)	0	0	0	0	1	24
(....)(.)	1	0	0	1	0	30
(...)(..)	0	1	1	0	0	20
(...)(.)(.)	2	0	1	0	0	20
(..)(..)(.)	1	2	0	0	0	15
(..)(.)(.)(.)	3	1	0	0	0	10
(.)(.)(.)(.)(.)	5	0	0	0	0	1
						total 120=5!

$$(5.120)$$

The last column is obtained from (5.119).

Similarity Classes of Permutation Groups.   As a representative example we consider

$$a = (146)(35)(2) \quad \text{and} \quad s = \begin{vmatrix} 1 & 2 & 3 & 4 & 5 & 6 \\ 3 & 6 & 5 & 4 & 2 & 1 \end{vmatrix} \quad (5.121)$$

as elements of the permutation group on 6 objects. How do we obtain the permutation $b = sas^{-1}$ similar to a? As in (5.110) we apply the permutation s to $Y \in \{1, 2, \ldots, 6\}$ and to its image $a(Y)$. The permutation b is then obtained by assigning

$$\underset{s(Y)}{\circ} \xrightarrow{\;b\;} \underset{s(a(Y))}{\circ} \qquad (5.122)$$

Denoting by s_i the image of the number i under the permutation s, we obtain

$$b = (s_1 s_4 s_6)(s_3 s_5)(s_2) = (341)(52)(6) \qquad (5.123)$$

Evidently, a and b are of the same type.

It is obvious that in the case of the full permutation group, that is, of the symmetric group S_N, the converse also holds:

If two permutations a, b are of the same type, then there exists an $s \in S_N$ such that $b = sas^{-1}$.

Example.

$$\left. \begin{array}{l} a = (251)(436) \\ b = (234)(156) \end{array} \right\} \qquad (5.124)$$

Then

$$s = \begin{vmatrix} 1 & 2 & 3 & 4 & 5 & 6 \\ 4 & 2 & 5 & 1 & 3 & 6 \end{vmatrix} \qquad (5.125)$$

obeys the condition mentioned above.

The properties found in these examples can easily be generalized:

Theorem 5.6 *For an arbitrary permutation group, a necessary condition for two elements to be similar is that they be of the same type. For the symmetric group, this condition is also sufficient.*

Remark. This again confirms property III mentioned after (5.107).

A consequence of Theorem 5.6 and definition (5.117), (5.118) of the permutation type is

Theorem 5.7 *The number of similarity classes of S_N equals the number of all N-tuples $\alpha_1, \alpha_2, \ldots, \alpha_N$ with the property*

$$\sum_{k=1}^{N} k \cdot \alpha_k = N \qquad \text{where} \qquad \alpha_k \in \{0, 1, 2, 3, \ldots, N\} \qquad (5.126)$$

5.5 Characters

5.5.1 General Properties

The trace tr of an $n \times n$ matrix $A = (a_{ij})$ is known to be the sum of all diagonal elements:

$$\text{tr}(A) = \sum_{i=1}^{n} a_{ii} \qquad (5.127)$$

That

$$\text{tr}(AB) = \text{tr}(BA) \qquad (5.128)$$

holds for any two $n \times n$ matrices A and B, follows from

$$\text{tr}(AB) = \sum_i \sum_j a_{ij} b_{ji} = \sum_j \sum_i b_{ji} a_{ij} = \text{tr}(BA)$$

By (5.128), with B nonsingular, we obtain

$$\text{tr}(BAB^{-1}) = \text{tr}(B(AB^{-1})) = \text{tr}((AB^{-1})B) = \text{tr}(A) \qquad (5.129)$$

Hence the trace is preserved under coordinate and similarity transformations. This fact enables us to characterize representations in a coordinate-independent way.

Definition. Let G be an arbitrary group and $\vartheta : s \rightarrow D(s), s \in G$, be a representation of it by matrices. The complex-valued function

$$\chi(s) = \text{tr}(D(s)) \qquad s \in G \qquad (5.130)$$

is called the **character** of the representation ϑ of G.

Properties of characters χ

1. *Equivalent representations have the same character.*

   Proof. Given two equivalent representations $\vartheta : s \rightarrow D(s)$ and $\vartheta' : s \rightarrow D'(s)$, there exists a coordinate transformation T such that $D'(s) = T^{-1}D(s)T$. But then $\chi'(s) = \mathrm{tr}(T^{-1}D(s)T) = \mathrm{tr}(D(s)) = \chi(s)$. □

   Remark. It should be noted that for finite groups the converse also holds (Section 5.6). However, it does not hold in general, as is exemplified by the following **counterexample:** For the multiplicative group of all real numbers $x \neq 0$, consider the representation assigning to every x the 2×2 identity matrix as opposed to the representation (1.46) of the same group.

2. *The dimension of a representation is equal to $\chi(e)$, where e is the identity element of the group.*
   The proof is clear and therefore omitted.

3. *Characters are class functions; that is,*

   $$\chi(a) = \chi(sas^{-1}) \qquad \forall s \in G \tag{5.131}$$

   In other words, χ takes on a constant value on each similarity class.

   Proof. $\chi(sas^{-1}) = \mathrm{tr}(SAS^{-1}) = \mathrm{tr}(A) = \chi(a)$, where $S = D(s)$ and $A = D(a)$. □

4. *If a representation ϑ decomposes into $\vartheta = \vartheta_1 \oplus \vartheta_2 \oplus \ldots \oplus \vartheta_m$, then*

   $$\chi(s) = \chi_1(s) + \chi_2(s) + \ldots + \chi_m(s) \qquad \forall s \in G \tag{5.132}$$

   Proof. Use a coordinate system which is suitably adjusted to the decomposition of ϑ, so that the representing matrices take the form of boxes. □

   Remark. The $\vartheta_1, \vartheta_2, \ldots, \vartheta_m$ need not be irreducible.

5. *Let χ_1 and χ_2 be the characters of the two representations ϑ_1 and ϑ_2 of some arbitrary group G. Then the Kronecker product representation $\vartheta_1 \otimes \vartheta_2$ has the character*

   $$\chi(s) = \chi_1(s) \cdot \chi_2(s) \qquad \forall s \in G \tag{5.133}$$

   Proof. Let $\vartheta_1 : s \rightarrow D_1(s)$ be of dimension m and $\vartheta_2 : s \rightarrow D_2(s)$ be of dimension n. In a suitable coordinate system, the representing matrices of $\vartheta_1 \otimes \vartheta_2$ have the form (see (1.65))

   $$D_1 \otimes D_2 = \begin{bmatrix} d_{11}^{(1)}D_2 & d_{12}^{(1)}D_2 & \ldots & d_{1m}^{(1)}D_2 \\ d_{21}^{(1)}D_2 & d_{22}^{(1)}D_2 & \ldots & d_{2m}^{(1)}D_2 \\ \vdots & \vdots & & \vdots \\ d_{m1}^{(1)}D_2 & d_{m2}^{(1)}D_2 & \ldots & d_{mm}^{(1)}D_2 \end{bmatrix} \tag{5.134}$$

Hence

$$\chi(s) = \operatorname{tr}(D_1 \otimes D_2) = (d_{11}^{(1)} + d_{22}^{(1)} + \ldots + d_{mm}^{(1)}) \cdot \operatorname{tr}(D_2)$$
$$= \operatorname{tr}(D_1) \cdot \operatorname{tr}(D_2) = \chi_1(s) \cdot \chi_2(s)$$

□

6. *If $\chi(s)$ is the character of an n-dimensional representation $\vartheta : s \to D(s)$ of some group G, then*

$$\chi(s) = \text{ sum of the } n \text{ eigenvalues of } D(s) \qquad (5.135)$$

This is a consequence of basic theorems in linear algebra (triangular form of a matrix, invariance of the characteristic polynomial under coordinate transformation).

7. *For every representation ϑ of a finite group G and for every unitary representation of an arbitrary group G*

$$\chi(s^{-1}) = \overline{\chi(s)} \qquad \forall s \in G \qquad (5.136)$$

The bar denotes complex conjugation.

Proof. By Theorem 1.6 and Property 1 of characters we may restrict our attention to unitary representing matrices $D(s)$. Their eigenvalues are of magnitude 1, thus of the form

$$\lambda_k = e^{i\varphi_k(s)}$$

On the other hand, the eigenvalues of inverses $D(s)^{-1} = D(s^{-1})$ are

$$\lambda_k^{-1} = e^{-i\varphi_k(s)} = \overline{\lambda_k}$$

This together with Property 6 proves the assertion.   □

Properties 3 and 7 imply

8. *If along with the assumptions made in Property 7, the group G satisfies the additional condition that every element be similar to its inverse (i.e., $\forall a \in G \; \exists b \in G$ such that $a^{-1} = bab^{-1}$), then the character χ is a real-valued function.*

Examples. The symmetric group S_N, the octahedral group O (see (5.111)).

9. *If χ is the character of a representation by permutation matrices[10], then $\forall s \in G$*

$$\chi(s) = \text{ number of basis vectors fixed under } D(s) \qquad (5.137)$$

Proof. Let $D(s)$ map the ith basis vector into the jth basis vector. Then the ith column of the matrix $D(s)$ contains exactly one nonzero element. This element is located in the jth row and equals 1. It has a contribution to the trace only if $i = j$.   □

----
[10] See text after (1.28).

Character table. In this table we collect the characters of all irreducible representations belonging to a certain group. Unlike the tables of representations, however, it gives the characters by classes only.

From (5.77) or from Problem 5.8, we obtain the following character table for the group S_4:

Similarity class	$(.)(.)(.)(.)$	$(...)(.)$	$(....)$	$(..)(..)$	$(..)(.)(.)$
Number of elements per class	1	8	6	3	6
χ_1^1 of the identity rep. ϑ_1^1	1	1	1	1	1
χ_2^1 of the alternating rep. ϑ_2^1	1	1	-1	1	-1
χ_1^3 of the representation ϑ_1^3	3	0	-1	-1	1
$\chi_2^3 = \chi_1^3 \cdot \chi_2^1$ of the rep. $\vartheta_1^3 \otimes \vartheta_2^1$	3	0	1	-1	-1
χ_1^2 of the representation ϑ_1^2	2	-1	0	2	0

$$(5.138)$$

The first column also contains the dimension of the representation in question. The number of elements per class was obtained from Equation (5.119).

5.5.2   Orthogonality Relations of Characters

They are derived from Theorem 5.1, i.e., from the orthogonality relations of irreducible representations. With the special choice of indices $\ell = k$, $p = q$, Equations (5.15) and (5.16) imply

$$\sum_{s \in G} d'_{pp}(s) d_{\ell\ell}(s^{-1}) = 0 \qquad (5.139)$$

$$\frac{1}{g} \sum_{s \in G} d_{pp}(s) d_{\ell\ell}(s^{-1}) = \frac{\delta_{p\ell}\delta_{p\ell}}{n} = \frac{\delta_{p\ell}}{n} \qquad (5.140)$$

Summation over ℓ and p and Property 7 of characters yield

Theorem 5.8 (Orthogonality of characters) *Let $\chi(s)$ and $\chi'(s)$ be the characters of two inequivalent and irreducible representations of a finite group G, then*

$$\sum_{s \in G} \chi'(s)\overline{\chi(s)} = 0 \qquad (5.141)$$

i.e., χ and χ' are orthogonal to each other with respect to the summation $\frac{1}{g}\sum_{s \in G}$ and

$$\sum_{s \in G} \chi(s)\overline{\chi(s)} = 1 \qquad (5.142)$$

i.e., χ has norm[11] 1 with respect to the summation $\frac{1}{g} \sum\limits_{s \in G}$.

5.5.3 A Composition Formula for Characters

Another consequence of Section 5.1 is the so-called **multiplication theorem**. For any $n \times n$ matrix U, Equations (5.1) and (5.13) imply

$$W = \frac{1}{g} \sum_{s \in G} D(s) U D(s^{-1}) = \frac{\mathrm{tr}(U)}{n} E \qquad (5.143)$$

Here $D(s)$ are the matrices of an irreducible n-dimensional representation ϑ, and E is the $n \times n$ identity matrix.

Now we choose $U = D(a)$, some fixed representing matrix of ϑ. Then

$$\frac{1}{g} \sum_{s \in G} D(s) D(a) D(s^{-1}) = \frac{\chi(a)}{n} E \qquad (5.144)$$

By linearity of the sum and by multiplying (5.144) by some arbitrary representing matrix $D(b)$, we obtain

$$\frac{1}{g} \sum_{s \in G} D(b) D(s) D(a) D(s^{-1}) = \frac{\chi(a)}{n} D(b)$$

Linearity of the trace implies

Theorem 5.9 (Multiplication Theorem) *The character χ of an n-dimensional irreducible representation of a finite group G satisfies*

$$\frac{1}{g} \sum_{s \in G} \chi(bsas^{-1}) = \frac{\chi(a)\chi(b)}{n} \qquad \forall a, b \in G \qquad (5.145)$$

5.5.4 Fundamental Theorems about Characters

Every repesentation ϑ of a finite group G is completely reducible (see Theorem 1.6). Hence

$$\vartheta = c_1 \vartheta_1 \oplus c_2 \vartheta_2 \oplus \ldots \oplus c_N \vartheta_N \qquad (5.146)$$

Here the ϑ_j denote the irreducible inequivalent representations of G. The values $c_j \geq 0$ are the multiplicities with which the ϑ_j occur in ϑ. If a

---

[11]The notions "orthogonal" and "norm" are justified, because $(\varphi, \psi) = \frac{1}{g} \sum\limits_{s \in G} \varphi(s) \cdot \overline{\psi(s)}$ defines an inner product in the vector space of complex-valued functions on G.

representation ϑ_k does not occur in ϑ, then $c_k = 0$. We shall calculate c_j by use of the character χ of ϑ.

Property 4 (Equation (5.132)) implies

$$\chi(s) = c_1\chi_1(s) + c_2\chi_2(s) + \ldots + c_N\chi_N(s) \qquad s \in G \qquad (5.147)$$

where χ_j is the character associated with the irreducible representation ϑ_j. Thus by Theorem 5.8

$$\frac{1}{g}\sum_{s\in G}\chi(s)\overline{\chi_j(s)} = \frac{1}{g}\sum_{s\in G}(c_1\chi_1 + c_2\chi_2 + \ldots + c_N\chi_N)\overline{\chi_j} = c_j$$

This proves the following

Algorithm (for computing multiplicities) *Let ϑ (with character χ) be a representation of a finite group G of order g. Then the irreducible representation ϑ_j (with character χ_j) occurs in ϑ exactly c_j times, where*

$$c_j = \frac{1}{g}\sum_{s\in G}\overline{\chi_j(s)}\chi(s) \qquad c_j \in \{0,1,2,\ldots\} \qquad (5.148)$$

Example 5.7 Let us calculate the multiplicities of the representation ϑ of the membrane problem (Example 5.2). From (1.59) and (1.60) we obtain the following *character table for the group D_4* ($g = 8$, χ_i^d denotes the character of ϑ_i^d, the elements are denoted according to Fig. 5.2 by (5.65)):

Similarity class	identity	$\pm 90°$ rot.	$180°$ rot.	σ_x, σ_y	σ_+, σ_-
Number of elements	1	2	1	2	2
χ_1^1	1	1	1	1	1
χ_2^1	1	1	1	-1	-1
χ_3^1	1	-1	1	1	-1
χ_4^1	1	-1	1	-1	1
χ_1^2	2	0	-2	0	0

$$(5.149)$$

The value of the character of ϑ equals the number of points in Fig. 5.2 that are fixed under the transformation in question of D_4, thus

$$\chi: \qquad 9, 1, 1, 3, 3$$

Hence, by (5.148) the multiplicity of, say, ϑ_1^2 equals

$$\frac{1}{8}(2 \cdot 9 - 2 \cdot 1) = 2$$

Calculating all multiplicities, we obtain

$$\vartheta = 3\vartheta_1^1 \oplus \vartheta_3^1 \oplus \vartheta_4^1 \oplus 2\vartheta_1^2 \qquad (5.150)$$

This can also be obtained from (3.11) with $i = 0$, $h = k = \delta = 1$.

Example 5.8 (Abelian groups) In an abelian group, each element is itself a similarity class. Furthermore, by Theorem 2.3 all irreducible representations are of dimension 1. Thus the character table is identical with the table of all irreducible representations. For the cyclic group it is described in (1.52).

Remark. Equation (5.148) for computing multiplicities will be of interest for the algorithm "Generating symmetry adapted basis vectors" (end of Section 5.2). Namely, from the c_j we obtain the **truncation criteria** for computing the matrices I and II (which we know to be of rank c_j).

Theorem 5.10 *A representation of a finite group is irreducible if and only if the corresponding character χ has the norm 1, that is, if*

$$\frac{1}{g} \sum_{s \in G} \chi(s)\overline{\chi(s)} = 1 \qquad (5.151)$$

Proof. If we substitute $\chi(s) = \sum c_j \chi_j(s)$ into (5.151), then we obtain the orthogonality relations (5.141) and (5.142): $\sum c_j^2 = 1$. Therefore, there is precisely one nonzero value of c_j, which must equal 1. $\qquad \square$

The sums appearing in the character values can be simplified since they are class functions. Thus

$$\frac{1}{g} \sum_{s \in G} \chi(s)\overline{\chi(s)} = \frac{1}{g} \sum_i g_i \chi(i)\overline{\chi(i)} \qquad (5.152)$$

Here $\chi(i)$ denotes the value of the character in class no. i, and g_i denotes the number of group elements belonging to class no. i.

With Theorem 5.10 at hand, the question of irreducibility is easily answered. We shall do this for the representations $\vartheta_1^3, \vartheta_2^3$ of the symmetric group S_4. By virtue of the character table (5.138) and with (5.151)

$$\left. \begin{array}{l} \frac{1}{g} \sum \chi_1^3(s)\overline{\chi_1^3(s)} = \frac{1}{24}(1 \cdot 9 + 8 \cdot 0 + 6 \cdot 1 + 3 \cdot 1 + 6 \cdot 1) = 1 \\[2mm] \frac{1}{g} \sum \chi_2^3(s)\overline{\chi_2^3(s)} = \frac{1}{24}(1 \cdot 9 + 8 \cdot 0 + 6 \cdot 1 + 3 \cdot 1 + 6 \cdot 1) = 1 \end{array} \right\} \qquad (5.153)$$

Hence the two representations are irreducible, as the reader may already have shown in Problem 1.2(a).

Theorem 5.11 *Two representations of a finite group G are equivalent to each other if and only if their characters are identical.*

Proof.

(a) Equivalence implies identical characters, due to Property 1.

(b) Conversely, let ϑ, ϑ' be representations with identical characters, i.e.,

$$\chi(s) = \chi'(s) \qquad \forall s \in G$$

Equation (5.148) applied to ϑ and ϑ' gives

$$c_j = c'_j \qquad \forall j \tag{5.154}$$

That is, for each j the irreducible representation ϑ_j occurs in ϑ and ϑ' the same number of times. This implies equivalence of ϑ and ϑ'. $\square$

5.5.5 Projectors

This subsection supplies a tool for the decomposition of a representation into its isotypic components (so-called *canonical decomposition*). We again start from the situation described in Section 5.2 and consider the operator (5.30) in the special case $k = \ell$:

$$\mathsf{P}^{(j)}_{kk} = \frac{n_j}{g} \sum_{s \in G} d^{(j)}_{kk}(s^{-1})\mathsf{D}(s) \qquad k \in \{1, 2, \ldots, n_j\} \tag{5.155}$$

Now we form a new operator by summing over all k and by using (5.136):

$$\mathsf{P}^{(j)} = \sum_{k=1}^{n_j} \mathsf{P}^{(j)}_{kk} = \frac{n_j}{g} \sum_{k=1}^{n_j} \sum_{s \in G} d^{(j)}_{kk}(s^{-1})\mathsf{D}(s) \tag{5.156}$$

$$= \frac{n_j}{g} \sum_{s \in G} \sum_{k=1}^{n_j} d^{(j)}_{kk}(s^{-1})\mathsf{D}(s) = \frac{n_j}{g} \sum_{s \in G} \overline{\chi_j(s)}\mathsf{D}(s)$$

In order to find its properties, we reconsider $\mathsf{P}^{(j)}$ in a special coordinate system relative to which the representing matrices have the form (5.27). There, by (5.33), the corresponding matrix

$$P^{(j)\mathrm{sp}} = \sum_{k=1}^{n_j} P^{(j)\mathrm{sp}}_{kk} \tag{5.157}$$

has the following form (see (5.34) with $k = \ell$): The jth block is a $c_j n_j \times c_j n_j$ identity matrix. All other entries outside this block are zeros. Hence this is the **projection of V onto V_j**. In fact, the projection property

$$[P^{(j)\mathrm{sp}}]^2 = P^{(j)\mathrm{sp}} \tag{5.158}$$

holds.

This together with (5.156) implies

Theorem 5.12 *Let $\vartheta : s \to D(s)$ be a representation of a finite group G. Assume that the irreducible n_j-dimensional representation ϑ_j with character $\chi_j(s)$ occurs exactly c_j times. Then, with $g = |G|$,*

$$\mathsf{P}^{(j)} = \frac{n_j}{g} \sum_{s \in G} \overline{\chi_j(s)} \mathsf{D}(s) \tag{5.159}$$

is the **projection operator** *or the* **projector** *of the representation space V onto the $c_j n_j$-dimensional subspace V_j, whose c_j irreducible subspaces are transformed after the manner of ϑ_j.*

Remarks.

1. In order to determine the $c_j n_j$-dimensional subspaces V_j, one uses the table of characters rather than the table of irreducible representations.

2. For the special case of $c_j = 1$, step I (Equation (5.44)) of the algorithm "Generating symmetry adapted basis vectors" can be modified as follows: In place of $\pi^{(j)}$ use the matrix corresponding to (5.159).

(Chan, 1975) exhibits an application of the *projector method* in connection with membrane problems that do not have the classical symmetry, and which in (Hersch, 1966-67) and (Hersch, 1982-83) were approached *analytically.*

5.5.6 Summary

With the findings of this chapter, we can treat the problems that have the symmetry of a representation $\vartheta : s \to D(s)$ of a finite group G by matrices in the following way:

A. *Determine the multiplicities c_j of the various irreducible representations ϑ_j of ϑ, for instance by the algorithm "Computing multiplicities" (5.148).*

B. (a) *If $c_j = 1$: Calculate the matrix associated with the projector (5.159)*

$$P^{(j)} = \frac{n_j}{g} \sum_{s \in G} \overline{\chi_j(s)} D(s) \tag{5.160}$$

(The coefficient is inessential.) By selecting n_j linearly independent columns of $P^{(j)}$ one obtains a basis for the irreducible subspace of ϑ_j.

(b) *If $c_j > 1$: Use the algorithm "Generating symmetry adapted basis vectors" (Equations (5.44) through (5.48)). To find an orthonormal basis, use Theorems 5.4 and 1.6. If only the eigenvalues of an operator with symmetries are to be calculated, then steps I and II of the algorithm will do.*

C. *Use the fundamental theorem in Section 2.3. In order to solve a system of linear equations use (3.30), and to calculate the eigenvectors observe the considerations made in connection with (5.59).*

5.6   Representation Theory of Finite Groups

5.6.1   The Regular Representation

The regular representation ϑ_{reg} of a finite group G of order g is defined as follows:

I. To each group element of G we assign a unique basis vector of a g-dimensional complex vector space V.

II. Each element of G is then associated with the $g \times g$ matrix that permutes these basis vectors in accordance with the group table of G (Section 1.3).

The representation property is guaranteed by Theorem 1.2. Moreover, ϑ_{reg} is faithful by Theorem 1.3, and it is a representation by permutation matrices. The corresponding permutations form a subgroup of S_g of order g.

Example 5.9 (Regular representation of S_3) We consider the group table (1.19).

I. To each group element we assign a basis vector in the order of the head row. Thus e corresponds to the first, a to the second, ..., f to the sixth basis vector.

II. The representing matrix of b, for example, is

$$b \xrightarrow{\vartheta_{reg}} \begin{bmatrix} 0 & 1 & 0 & 0 & 0 & 0 \\ 0 & 0 & 1 & 0 & 0 & 0 \\ 1 & 0 & 0 & 0 & 0 & 0 \\ 0 & 0 & 0 & 0 & 0 & 1 \\ 0 & 0 & 0 & 1 & 0 & 0 \\ 0 & 0 & 0 & 0 & 1 & 0 \end{bmatrix} \tag{5.161}$$

Let us return to the general theory: If we multiply a group element $s \in G$ by a fixed element $f \neq e$ from the left, then

$$f \cdot s \neq s \qquad \forall s \in G \tag{5.162}$$

For, if on the contrary $f \cdot s = s$, then multiplying from the right by s^{-1} would yield $f = e$, a contradiction.

On the other hand, $e \cdot s = s \ \forall s \in G$.

Thus by (5.137), the character of ϑ_{reg} satisfies

$$\chi_{reg}(s) = \begin{cases} \text{group order } g & \text{if } s = e \\ 0 & \text{otherwise} \end{cases} \tag{5.163}$$

Let again χ_j be the character of some irreducible n_j-dimensional representation ϑ_j of the group G. By (5.163) and Property 2 of characters (Section 5.5), Equation (5.148) yields the multiplicity of ϑ_j in ϑ_{reg}

$$c_j = \frac{1}{g} \sum_{s \in G} \chi_{\text{reg}}(s)\overline{\chi_j(s)} = \frac{1}{g}\chi_{\text{reg}}(e)\overline{\chi_j(e)} = \frac{1}{g} \cdot g \cdot n_j = n_j$$

Thus we proved

Theorem 5.13 *The multiplicity with which the irreducible representation ϑ_j occurs in the regular representation ϑ_{reg} equals the dimension of ϑ_j; thus*

$$\vartheta_{\text{reg}} = \bigoplus_\rho n_\rho \vartheta_\rho \tag{5.164}$$

And by comparing dimensions in (5.164), we find

$$group \ order \ g = \sum_\rho n_\rho^2 \tag{5.165}$$

Summation is made over all irreducible representations ϑ_ρ of the group G.

An easy consequence of this is that $n_\rho \leq \sqrt{|G|}$ for every ρ. So, for example, in the case of $G = S_3$ this implies $n_\rho \leq \sqrt{6}$, hence $n_\rho \leq 2$. From Example 1.13 we know that equality holds.

Remark. This result was already stated in Theorem 1.4. It is the strong version of (5.21).

5.6.2 The Character Matrix

Let the similarity classes of an arbitrary finite group G of order g be designated by

$$(1), (2), \ldots, (K) \tag{5.166}$$

Hence K is the number of distinct classes in G. Let (1) contain the identity element e. The identity element e forms a class by itself. Let the number of group elements in (i) be g_i. Let further

$$\vartheta_1, \vartheta_2, \ldots, \vartheta_N \tag{5.167}$$

be the list of **all** irreducible inequivalent representations of G and $\chi_1, \chi_2, \ldots, \chi_N$ be their characters. Let $\chi_\rho(i)$ denote the value of χ_ρ taken on in class (i).

The **character matrix of a finite group** is a character table modified by factors of the following form:

Similarity classes	(1)	(2)	$\cdots$	(K)
Number of elements	$g_1 = 1$	g_2		g_K
associated with χ_1	$\sqrt{\frac{g_1}{g}} \cdot 1$	$\sqrt{\frac{g_2}{g}} \cdot 1$	$\cdots$	$\sqrt{\frac{g_K}{g}} \cdot 1$
associated with χ_2	$\sqrt{\frac{g_1}{g}} \cdot \chi_2(1)$	$\sqrt{\frac{g_2}{g}} \cdot \chi_2(2)$		$\sqrt{\frac{g_K}{g}} \cdot \chi_2(K)$
$\vdots$	$\vdots$	$\vdots$		$\vdots$
associated with χ_1	$\sqrt{\frac{g_1}{g}} \cdot \chi_N(1)$	$\sqrt{\frac{g_2}{g}} \cdot \chi_N(2)$	$\cdots$	$\sqrt{\frac{g_K}{g}} \cdot \chi_N(K)$

$$(5.168)$$

Hence the matrix entry in the jth row and ith column is $\sqrt{\frac{g_i}{g}}\chi_j(i)$. The relations (5.141), (5.142) are summarized in

$$\frac{1}{g} \sum_{s \in G} \overline{\chi_\rho(s)}\chi_\sigma(s) = \delta_{\rho\sigma} = \begin{cases} 1, & \text{if } \rho = \sigma \\ 0, & \text{if } \rho \neq \sigma \end{cases} \qquad (5.169)$$

Therefore summation over classes (according to (5.152)) yields

$$\frac{1}{g} \sum_{(i)} g_i \overline{\chi_\rho(i)}\chi_\sigma(i) = \sum_{(i)} \left(\sqrt{\frac{g_i}{g}}\overline{\chi_\rho(i)}\right) \cdot \left(\sqrt{\frac{g_i}{g}}\chi_\sigma(i)\right) = \delta_{\rho\sigma} \qquad (5.170)$$

The last equation in (5.170) says that the **rows** of the character matrix form an orthonormal set.

Thus

$$N \leq K \qquad (5.171)$$

Now let us show that the columns are also pairwise orthonormal: The multiplication theorem (5.145) and property (5.136) imply, upon replacing b by b^{-1}, that

$$\chi_\rho(a)\overline{\chi_\rho(b)} = n_\rho \cdot \frac{1}{g} \sum_{s \in G} \chi_\rho(b^{-1}sas^{-1}) \qquad \forall a, b \in G \qquad (5.172)$$

Summation over ρ and Equations (5.172) and (5.164) yield

$$\sum_{\rho=1}^{N} \chi_\rho(a)\overline{\chi_\rho(b)} = \frac{1}{g} \sum_{s \in G} \sum_{\rho=1}^{N} n_\rho \chi_\rho(b^{-1}sas^{-1}) = \frac{1}{g} \sum_{s \in G} \chi_{\text{reg}}(b^{-1}sas^{-1}) \qquad (5.173)$$

Now let a, b be two elements belonging to distinct similarity classes $(i), (j)$, respectively. Hence $sas^{-1} \neq b$ $\forall s \in G$. Thus

$$b^{-1}sas^{-1} \neq e \qquad \forall s \in G \qquad (5.174)$$

Hence by (5.174) and (5.163), the right-most sum in (5.173) equals zero. Therefore,

$$\sum_{\rho=1}^{N} \chi_\rho(i)\overline{\chi_\rho(j)} = 0 \qquad i \neq j \tag{5.175}$$

As $\chi(e) \neq 0$, the norm of any character χ can never vanish. Thus

$$K \leq N \tag{5.176}$$

whence by (5.171)

$$K = N \tag{5.177}$$

This is summarized in

Theorem 5.14 *For a finite group G there are exactly as many irreducible inequivalent representations as there are similarity classes of G. Thus the character matrix is a square matrix; moreover, it is unitary.*

The reader may verify by use of Tables (5.138) and (5.149) that the corresponding character matrices are unitary.

5.6.3 Completeness

Now we consider the vector space of all complex-valued class functions $f(s)$ on G. They satisfy

$$f(a) = f(sas^{-1}) \qquad \forall s \in G \tag{5.178}$$

Orthogonality of characters (5.141), (5.142) and Theorem 5.14 immediately imply

Theorem 5.15 *For a finite group G the characters of all irreducible representations form a basis in the vector space of complex-valued class functions.*

5.7 Extension to Compact Lie Groups

Important results of this chapter can be extended to the so-called **compact Lie groups**. To develop the general definition of an abstract compact Lie group is beyond the scope of this book. The following well-known statement serves to understand the subsequent considerations:

Every group isomorphic to a subgroup of $U(n)$ *is a compact Lie group.*

The class of compact Lie groups includes $U(n)$, $SU(n)$, $O(n)$, $SO(n)$, and all finite groups; the latter because of the regular representation, which describes an isomorphism between the finite group in question and a group of permutation matrices of $O(n)$.

Examples of non-compact groups are $GL(n, \mathbf{C})$, $GL(n, \mathbf{R})$, $SL(n, \mathbf{C})$, $SL(n, \mathbf{R})$, and the proper positive Lorentz group L.

The following theorem holds (Bredon,1972), (Serre, 1977):

Let G be a compact Lie group. Then there exists a unique real-valued functional $\underset{s \in G}{\mathcal{M}} f(s)$ defined for continuous [12] real-valued functions f on G such that the following four properties are satisfied:

I. *Linearity:*

$$\underset{s \in G}{\mathcal{M}}\{\alpha f(s) + \beta g(s)\} = \alpha \underset{s \in G}{\mathcal{M}} f(s) + \beta \underset{s \in G}{\mathcal{M}} g(s) \qquad \forall \alpha, \beta \in \mathbf{C} \quad (5.179)$$

II. *Positivity:*

$$f(s) \geq 0 \Rightarrow \underset{s \in G}{\mathcal{M}} f(s) \geq 0 \qquad\qquad (5.180)$$

Equality holds if and only if $f = 0$.

III. *Normalization:*

$$f(s) = 1 \ \forall s \in G \Rightarrow \underset{s \in G}{\mathcal{M}} f(s) = 1 \qquad\qquad (5.181)$$

IV. *Invariance:*

$$\underset{s \in G}{\mathcal{M}} f(s) = \underset{s \in G}{\mathcal{M}} f(as) = \underset{s \in G}{\mathcal{M}} f(sa) = \underset{s \in G}{\mathcal{M}} f(s^{-1}) \qquad a \in G \quad (5.182)$$

Property IV involves the so-called **left and right translations** of a group:

$$s \to as, \qquad s \to sa, \qquad \forall s \in G, \qquad a \in G \text{ fixed} \qquad (5.183)$$

If s runs through the group exactly once, then so do as and sa (see Problem 1.7).

$\mathcal{M}$ is called the **Haar measure** or the **invariant measure**. We adopt Hermann Weyl's notation and call it more suitably the **averaging over a group** (Weyl, 1946). Namely, for the special case of a finite group G, we have

$$\underset{s \in G}{\mathcal{M}} f(s) = \frac{1}{g} \sum_{s \in G} f(s) \qquad\qquad (5.184)$$

where g is the group order.

An existence and uniqueness proof for $\mathcal{M}$ due to Neumann is found in (Pontrjagin, 1946).

A classification of the Lie groups can be found in (Sagle, 1973) or (Serre, 1965).

---

[12] because, as in Chapter 4, we restrict our attention to continuous representations of such groups. Also observe the statements made at the beginning of Section 7.1.

Example 5.10 (Averaging over the group SU(2)) By (4.8) the elements of this group have the form

$$A = \begin{bmatrix} a & b \\ -\bar{b} & \bar{a} \end{bmatrix} \qquad a, b \in \mathbf{C} \text{ with } \det A = a\bar{a} + b\bar{b} = 1$$

In the notation

$$a = a_0 + ia_1, \qquad b = a_2 + ia_3 \qquad a_i \in \mathbf{R} \tag{5.185}$$

the determinant condition reads

$$a_0^2 + a_1^2 + a_2^2 + a_3^2 = 1 \tag{5.186}$$

The a_i may be thought of as coordinates in $\mathbf{R}^4$. Thus (5.186) means that the group manifold of SU(2) is the 3-dimensional unit sphere $S^3 \subset \mathbf{R}^4$.

Now let us analyze the transformations involved in Property IV of $\mathcal{M}$. The left translation

$$S \to A \cdot S = T \qquad \forall S \in SU(2), \quad A \text{ fixed} \tag{5.187}$$

is described by

$$\begin{bmatrix} a_0 + ia_1 & a_2 + ia_3 \\ -a_2 + ia_3 & a_0 - ia_1 \end{bmatrix} \cdot \begin{bmatrix} s_0 + is_1 & s_2 + is_3 \\ -s_2 + is_3 & s_0 - is_1 \end{bmatrix} = \begin{bmatrix} t_0 + it_1 & t_2 + it_3 \\ -t_2 + it_3 & t_0 - it_1 \end{bmatrix} \tag{5.188}$$

where

$$\begin{aligned} t_0 &= a_0 s_0 - a_1 s_1 - a_2 s_2 - a_3 s_3 & t_2 &= a_2 s_0 + a_3 s_1 + a_0 s_2 - a_1 s_3 \\ t_1 &= a_1 s_0 + a_0 s_1 - a_3 s_2 + a_2 s_3 & t_3 &= a_3 s_0 - a_2 s_1 + a_1 s_2 + a_0 s_3 \end{aligned} \tag{5.189}$$

Hence it may also be described by

$$\begin{bmatrix} t_0 \\ t_1 \\ t_2 \\ t_3 \end{bmatrix} = \underbrace{\begin{bmatrix} +a_0 & -a_1 & -a_2 & -a_3 \\ +a_1 & +a_0 & -a_3 & +a_2 \\ +a_2 & +a_3 & +a_0 & -a_1 \\ +a_3 & -a_2 & +a_1 & +a_0 \end{bmatrix}}_{A'} \cdot \begin{bmatrix} s_0 \\ s_1 \\ s_2 \\ s_3 \end{bmatrix} \tag{5.190}$$

The real 4×4 matrix A' is orthogonal.

Correspondingly, by interchanging s and a in (5.189), the right translation $S \to T = SA$ of the group yields a matrix A'', which differs from A' only by its sign and is also orthogonal.

Likewise, in passing to the inverse $S \to S^{-1}$, we have

$$s_0 \to s_0, \qquad s_i \to -s_i \qquad i = 1, 2, 3 \tag{5.191}$$

Thus this transformation is also described by an orthogonal matrix (namely, by the diagonal matrix with diagonal entries $1, -1, -1, -1$).

Therefore, all three cases in IV involve a rotation of the 3-dimensional unit sphere $S^3 \subset \mathbf{R}^4$ in the Euclidean sense.

Consequently, in the class of continuous functions on S^3 there exists an **average by the Euclidean volume measure** of S^3, because this measure is invariant under the rotations of S^3.

For better understanding and as a preparation for investigating the irreducibility of representations of the group SU(2), we shall explicitly carry out the **averaging for continuous class functions**.

This involves functions that are constant for all elements of the same similarity class, that is, functions with the property

$$f(A) = f(SAS^{-1}) \qquad \forall S \in \mathrm{SU}(2) \tag{5.192}$$

As we know, every matrix of SU(2) is similar to one of the form

$$\begin{bmatrix} e^{i\varphi} & 0 \\ 0 & e^{-i\varphi} \end{bmatrix} \tag{5.193}$$

Here φ can be restricted to the interval

$$0 \leq \varphi \leq \pi \tag{5.194}$$

because

$$\begin{bmatrix} 0 & 1 \\ -1 & 0 \end{bmatrix} \begin{bmatrix} e^{i\varphi} & 0 \\ 0 & e^{-i\varphi} \end{bmatrix} \begin{bmatrix} 0 & -1 \\ 1 & 0 \end{bmatrix} = \begin{bmatrix} e^{-i\varphi} & 0 \\ 0 & e^{i\varphi} \end{bmatrix} \tag{5.195}$$

Every similarity class of SU(2) is represented by a unique matrix (5.193), where $0 \leq \varphi \leq \pi$.

Moreover, *matrices of* SU(2) *are similar to each other if and only if they have the same traces*

$$a + \bar{a} = 2a_0 = 2\cos\varphi \tag{5.196}$$

that is, if they have the same value a_0 *or* φ, *because the function* $2\cos\varphi$ *is strictly monotone on the interval* $0 \leq \varphi \leq \pi$.

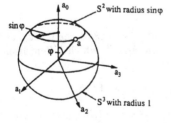

Now we consider the space $\mathbf{R}^4$ (Fig. 5.8). A point a of the 3-dimensional unit sphere S^3 corresponds to a unique element A of SU(2), and vice versa. Hence the similarity classes consist of 2-dimensional unit spheres S^2 centered at $(\cos\varphi, 0, 0, 0)$ with radius $\sin\varphi$.

Fig. 5.8

Therefore, the contribution of the similarity class to the average is

$$f(A) \cdot (\text{surface of } S^2) \cdot d\varphi = f(A) \cdot 4\pi \sin^2 \varphi \cdot d\varphi$$

Averaging over all similarity classes and taking into account the normalization (5.181) yields

$$\underset{A \in SU(2)}{\mathcal{M}} f(A) = \frac{2}{\pi} \int_0^\pi f(\varphi) \sin^2 \varphi \, d\varphi \qquad (5.197)$$

for continuous class functions on SU(2).

Let us return to the general theory: *It is of essential importance that the algorithm "Generating symmetry adapted basis vectors" and the following statements hold for compact Lie groups likewise:*

(α) A more general version of Theorem 1.6: *Every continuous representation ϑ of a compact Lie group G with respect to an appropriate inner product is unitary and thus completely reducible.*

(β) All of the current Chapter 5 holds, except for the geometric interpretation of Equation (5.20) and Section 5.6.

The reason is that in the proofs the summation is just replaced by the average, or more precisely,

$$\frac{1}{g} \sum_{s \in G} \rightarrow \underset{s \in G}{\mathcal{M}} \qquad (5.198)$$

Moreover, the words "finite group" are replaced by "compact Lie group".

Example 5.11 (Irreducibility and completeness of the representations ϑ_ℓ of SU(2) in Theorem 4.3)

(α) **Irreducibility.** In order to prove irreducibility, we use a theorem corresponding to (5.151): *A representation of a compact Lie group is irreducible if and only if its character χ satisfies*

$$\underset{s \in G}{\mathcal{M}} \chi(s)\overline{\chi(s)} = 1 \qquad (5.199)$$

Now we consider the representations ϑ_ℓ of SU(2), as described in Section 4.3: In order to determine the character of ϑ_ℓ, it suffices to consider a representative element of each of the various similarity classes. By construction of ϑ_ℓ in Section 4.3.1 we have

$$\begin{bmatrix} e^{i\varphi} & 0 \\ 0 & e^{-i\varphi} \end{bmatrix} \xrightarrow{\vartheta_\ell} \begin{bmatrix} e^{-i2\ell\varphi} & & & 0 \\ & e^{-i2(\ell-1)\varphi} & & \\ 0 & & \ddots & \\ & & & e^{+i2\ell\varphi} \end{bmatrix} \qquad (5.200)$$

Thus the associated character

$$\chi_\ell(\varphi) = 2 \left[\cos 2\ell\varphi + \cos 2(\ell - 1)\varphi + \ldots + \left\{ \begin{array}{ll} \frac{1}{2} & \text{if } \ell \text{ integer} \\ \cos\varphi & \text{otherwise} \end{array} \right. \right] \tag{5.201}$$

We calculate this sum:

$$\chi_\ell = \sum_{k=-\ell}^{\ell} e^{ik2\varphi}; \qquad e^{i2\varphi}\chi_\ell = \sum_{k=-\ell+1}^{\ell+1} e^{ik2\varphi} \tag{5.202}$$

Subtraction yields

$$(e^{i2\varphi} - 1)\chi_\ell = e^{i2(\ell+1)\varphi} - e^{-i2\ell\varphi} \tag{5.203}$$

Multiplying by $e^{-i\varphi}$ gives

$$(e^{i\varphi} - e^{-i\varphi})\chi_\ell = e^{i(2\ell+1)\varphi} - e^{-i(2\ell+1)\varphi}$$

This results in an equation equivalent to (5.201):

$$\chi_\ell(\varphi) = \frac{\sin(2\ell+1)\varphi}{\sin\varphi} \tag{5.204}$$

A verification of the dimension of ϑ_ℓ is immediately obtained from (5.201) or analytically by passing to the limit in (5.204): $\chi_\ell(0) = 2\ell + 1$.

Now we use the criterion (5.199). Equations (5.197) and (5.204) imply

$$\left. \begin{array}{ll} \underset{SU(2)}{\mathcal{M}} \chi_\ell(A)\overline{\chi_\ell(A)} & = \frac{2}{\pi} \int_0^\pi \chi_\ell(\varphi)\overline{\chi_\ell(\varphi)} \sin^2 \varphi \, d\varphi \\[2mm] & = \frac{2}{\pi} \int_0^\pi \sin^2(2\ell+1)\varphi \, d\varphi = 1 \end{array} \right\} \tag{5.205}$$

Thus ϑ_ℓ is irreducible.

(β) **Completeness.** Suppose that there is a continuous irreducible representation ϑ other than ϑ_ℓ. Then by (5.141), (5.197), and (5.204), its character χ should satisfy

$$\underset{SU(2)}{\mathcal{M}} (\chi\overline{\chi_\ell}) = \frac{2}{\pi} \int_0^\pi (\chi \sin\varphi) \sin(2\ell+1)\varphi \, d\varphi = 0 \quad \forall \ell \in \{0, \frac{1}{2}, 1, \ldots\} \tag{5.206}$$

Since the set of functions $\sin\varphi, \sin 2\varphi, \sin 3\varphi, \ldots$ of the space of all continuous functions f on the interval $[0, \pi]$ with $f(0) = f(\pi)$ is complete, continuity implies that $\chi \sin\varphi = 0$, whence $\chi = 0$. But a character cannot vanish identically $(\chi(e) \neq 0)$. □

Finally, let us mention that this also proves Theorem 4.3.

Example 5.12 (Circle group SO(2)) Likewise, for SO(2) there exists an average for continuous functions f on the circle, namely,

$$\underset{s\in SO(2)}{\mathcal{M}} f(s) = \frac{1}{2\pi} \int_0^{2\pi} f(s)\, ds \qquad (5.207)$$

We leave it to the reader to verify the properties (5.179) through (5.182).

This also proves that every continuous representation of SO(2) is completely reducible.

PROBLEMS

Problem 5.1 (Permutations) (a) Using the cycle notation (read from right to left)

$$(j_1 j_2 \ldots j_\ell) = (j_1 j_2)(j_2 j_3)\ldots(j_{\ell-1} j_\ell) \qquad (5.208)$$

(a1) formulate a criterion to determine whether a given permutation is even or odd;

(a2) show that the so-called adjacent transpositions $(12), (23), \ldots, (N\text{-}1\ N)$ form a generating set of the symmetric group S_N;

(a3) show that the two elements (12) and $(123\ldots N)$ already form a generating set of S_N.

(b) Prove Equation (5.119).

Problem 5.2 (The dihedral group D_h)

(a) Determine the similarity classes of the group D_h and also its character table (in particular, do this also for $D_3 \cong S_3$).
    Hint: Distinguish even and odd values of h. Use the relations (1.56) and (1.57).

(b) Using (5.148) verify the multiplicities given in (3.11) of the permutation representation ϑ_{per}.

Problem 5.3 According to (5.80), the representation ϑ_2^3 in Example 5.3 (methane molecule) appears with multiplicity 1. Using Tables (5.77) and (5.138) calculate the associated 3-dimensional irreducible subspace together with the projector (5.159).

Problem 5.4 Verify the irreducibility of the representations of D_4 listed in (5.149) by use of the corresponding character equation.

Problem 5.5

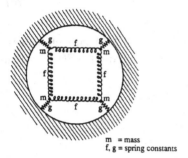

What are the eigenfrequencies of the mechanical system consisting of 4 mass points arranged in the vertices of a square, as indicated in Fig. 5.9? The spring constants are f and g, respectively. Calculate the frequencies for the linear approximation (small displacements).

Fig. 5.9

Problem 5.6 (Representation theory of the circle group SO(2)) Prove that the solution to Problem 2.4 contains **all** continuous irreducible representations (proof of completeness).

Problem 5.7 Prove the **Clebsch-Gordan series** (4.81) by aid of the theory of characters. **Hint:** Use (5.148). For the character of the Kronecker product, use both (5.201) and (5.204).

Problem 5.8 (Character table of the symmetric group S_4) The first four rows of Table (5.138) are easily written down if one considers the fact that the trace of the rotation matrix of $\mathbf{R}^3$ through angle φ is $1 + 2\cos\varphi$. Find the last row by employing Theorem 5.14.

Chapter 6

Various Topics of Application

In conformity with the purpose of this book, we shall here discuss the topics in which group theory methods play a vital part. However, we shall have to confine ourselves to just giving a brief outline of the concepts used in the various fields of application. We shall attempt to present these ideas in a manner such that even nonexpert readers will find them palatable and refer the interested reader to the literature for more detailed information.

6.1 Bifurcation and A New Technique

The committed reader will find all necessary basic concepts described in (Golubitsky, Stewart, Schaeffer, 1988, vol. II), unless otherwise specified.

For bifurcation problems with symmetries, group theory methods furnish powerful techniques, *techniques that have not yet been fully explored and with the help of which most intriguing results are thought to be found.*

For convenience, we divide this section into the following parts:

- Introductory examples related to bifurcations;

- Further notions and statements of group theory that are necessary for understanding bifurcation problems with symmetry;

- Presentation of a new technique for bifurcation problems with symmetry and explicit application to the so-called Brusselator.

6.1.1 Introduction

By **bifurcation** we essentially mean the number of solutions of a nonlinear problem as a function of some parameter.

For certain problems, including partial and ordinary differential equations, it can be shown that the study of bifurcations, by utilizing some of the techniques of bifurcation theory, can be formulated by aid of an equation

$$\varphi(x, \lambda) = 0 \qquad (6.1)$$

of a single scalar x, the **state variable**, with λ being a real parameter, the **bifurcation parameter**.

The set of points (x, λ) satisfying (6.1) constitutes the **bifurcation diagram** and is also called the **solution set** of φ. A point (x_0, λ_0) is called a **static bifurcation point** if the number of solutions changes as λ varies in the neighborhood of λ_0.

Example 6.1 (Pitchfork bifurcation) We consider the equation

$$x^3 - \lambda x = 0$$

Here $(0,0)$ is the only bifurcation point. The bifurcation diagram is shown in Fig. 6.1.

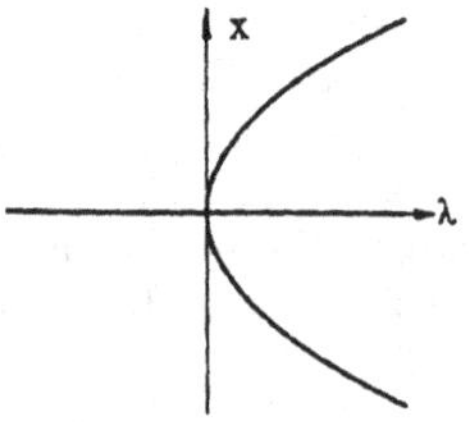

The solutions are

$$\begin{aligned} x_1 &= 0 \quad \text{for} \quad -\infty < \lambda < \infty \\ x_{2,3} &= \pm\sqrt{\lambda} \quad \text{for} \quad \lambda \geq 0 \end{aligned}$$

The number of solutions in the neighborhood of $(0,0)$ changes from 1 to 3.

Fig. 6.1 Pitchfork bifurcation

6.1.2   The Hopf Bifurcation

The term **Hopf bifurcation** refers to a phenomenon where a steady state of an evolution equation evolves into a periodic orbit as the bifurcation parameter λ varies.

Let us consider an autonomous system of ordinary differential equations

$$\frac{d\vec{y}}{dt} = \varphi(\vec{y}, \lambda) \qquad (6.2)$$

where $\varphi : \mathbf{R}^n \times \mathbf{R} \to \mathbf{R}^n$ is C^∞ and λ is the bifurcation parameter. Suppose that

$$\varphi(0, \lambda) = \vec{0} \qquad \forall \lambda$$

Then for every λ the vector $\vec{y} = \vec{0}$ is a **steady-state solution**. Hopf showed that a one-parameter set of periodic solutions to (6.2) emanating from $(\vec{y}, \lambda) = (\vec{0}, 0)$ can be found if along the steady-state solutions the $n \times n$ Jacobian matrices of $\varphi(\vec{y}, \lambda)$ denoted by

$$J(\lambda) = \left[\frac{\partial \varphi}{\partial \vec{y}} \right]_{\vec{y}=\vec{0}, \lambda}$$

satisfy the following assumptions:

(i) $J(0)$ has exactly two simple eigenvalues. They are complex conjugates located on the imaginary axis ($\pm \alpha i, \alpha \neq 0$).

(ii) The matrices $J(\lambda)$ in a neighborhood of $\lambda = 0$ have the simple eigenvalues $f_1(\lambda) + i f_2(\lambda)$, where $f_1(\lambda)$ and $f_2(\lambda)$ are real and imaginary parts and

$$\frac{df_1}{d\lambda}(0) \neq 0$$

holds. In dynamical terms this means that the paths of the two eigenvalues in the Gaussian plane properly **intersect** the imaginary axis at "time" $\lambda = 0$ with nonvanishing "speed".

A simple but instructive example is given by

$$J(\lambda) = \begin{bmatrix} \lambda & -1 \\ 1 & \lambda \end{bmatrix} \tag{6.3}$$

with eigenvalues $\lambda \pm i$ and $f_1'(0) = 1$.

We first discuss the *linear problem*

$$\frac{d\vec{y}}{dt} = \begin{bmatrix} \lambda & -1 \\ 1 & \lambda \end{bmatrix} \cdot \vec{y}$$

with inital condition $\vec{y}(0) = (r, 0)^T$. The solution is then

$$\vec{y}(t) = r \cdot e^{\lambda t} \begin{bmatrix} \cos t \\ \sin t \end{bmatrix}$$

describing logarithmic spirals if $\lambda \neq 0$ and concentric circles if $\lambda = 0$. The steady state $\vec{y} = \vec{0}$ is

(i) **stable if $\lambda < 0$:** spiral directed towards the origin.

(ii) **unstable if $\lambda > 0$:** spiral directed away from the origin.

(iii) **neutrally stable if $\lambda = 0$** and *each* of the orbits is 2π-periodic.

Hopf's finding is that this set of periodic solutions persists in a certain sense even when higher order terms of $\vec{y}$ and λ are added. However, this one-parameter family (parametrized by the amplitude r) need not remain in the plane $\lambda = 0$. In fact, *when higher order terms are added, then for each fixed λ in a neighborhood of a Hopf bifurcation point there exists at most one periodic solution.*

We now discuss the *nonlinear problem*

$$\varphi(\vec{y}, \lambda) = J(\lambda) \cdot \vec{y} - (y_1^2 + y_2^2) \cdot \vec{y}$$

with $J(\lambda)$ from (6.3).

As the nonlinear part points directly towards the origin, we obtain the following:

(i) For $\lambda < 0$ the qualitative behavior in the linear case is strengthened (spiraling towards the origin);

(ii) For $\lambda > 0$ a circle with amplitude $r = \sqrt{y_1^2 + y_2^2} = \sqrt{\lambda}$ is a possible path.

As $|\vec{y}|$ increases, the direction points more and more towards the origin.

However, for small $|\vec{y}|$ the linear part prevails; that is, in this case the spiral behavior away from the origin will be weakened only slightly.

For $|\vec{y}| = \sqrt{\lambda}$ = amplitude, the solution will be a **circle**. There, evidently, the repulsing and the attracting forces are in equilibrium. Apparently, the solution curves inside and outside approach the circle. This circle is an example of a so-called **stable limit cycle**.

The phase portraits for the nonlinear example are shown in Fig. 6.2.

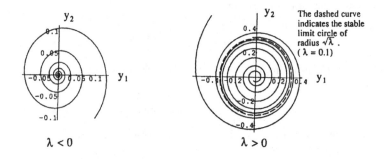

Fig. 6.2 Phase portraits

Fig. 6.3 shows the bifurcation diagrams for both the linear and the nonlinear example.

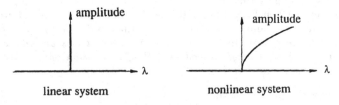

Fig. 6.3 Bifurcation diagrams

A new phenomenon appears in the nonlinear case; namely, for each $\lambda > 0$ there is a unique *stable* periodic solution. All solutions (except 0) approach this periodic solution.

6.1.3 Symmetry-Breaking

Bifurcation in general is considered in connection with **nonlinear** maps of the form

$$\varphi(\vec{y}, \lambda) : \mathbf{R}^n \times \mathbf{R} \to \mathbf{R}^n$$

Let G be a compact Lie group and $\vartheta : s \in G \to D(s)$ its representation by $n \times n$ matrices $D(s)$.
If

$$\varphi(D(s) \cdot \vec{y}, \lambda) = D(s) \cdot \varphi(\vec{y}, \lambda) \qquad \forall s \in G, \vec{y} \in \mathbf{R}^n, \forall \lambda \qquad (6.4)$$

we say that φ **commutes with the actions of the group** G', where $G' = \{D(s) | s \in G\}$ or φ is G'-**equivariant** or φ **has the symmetry of** G'.

Remarks.

  (i) G' itself is a compact Lie group.

 (ii) Because φ takes $\mathbf{R}^{n+1}$ into $\mathbf{R}^n$, the representing matrices $D(s)$ are all **real**.

(iii) G' can be regarded as a subgroup of the orthogonal group $O(n)$ if one chooses a suitable basis in the representation space (observe (α) after (5.198)).

(iv) Often $G \cong G'$ (i.e., ϑ is faithful), so that we may reword (6.4) as follows: φ **commutes with** G or φ is G-**equivariant** or φ **has the symmetry of** G.

Example 6.2 Let $\varphi(\vec{y}, \lambda)$ be C_2-equivariant. Then, as $C_2 \cong \{E_n, -E_n\}$, the following relation holds:

$$\varphi(-\vec{y}, \lambda) = -\varphi(\vec{y}, \lambda) \qquad \forall \vec{y}$$

Thus φ is an **odd** function of $\vec{y}$.

Next we consider the autonomous problem (6.2) with the property that φ *has the symmetry of some given group* of orthogonal $n \times n$ matrices.

Steady-state solutions or equilibria satisfy $\dfrac{d\vec{y}}{dt} = \vec{0}$; that is,

$$\varphi(\vec{y}, \lambda) = \vec{0} \qquad (6.5)$$

In order to study the structure of steady-state solutions, we need to introduce further elementary definitions. The **orbit of the group** G' **through** $\vec{y}$ **with respect to** $\mathbf{R}^n$ is the set

$$G' \cdot \vec{y} = \{D(s) \cdot \vec{y} \mid s \in G\} \subset \mathbf{R}^n$$

If $\vec{y}$ is a solution of (6.5), then so is the vector $D(s)\vec{y}$. This follows immediately from (6.4):

$$\varphi(D(s) \cdot \vec{y}, \lambda) = D(s) \cdot \varphi(\vec{y}, \lambda) = D(s) \cdot \vec{0} = \vec{0}$$

Thus, if φ vanishes, then it does so on the union of the orbits of G'.

To measure the amount of symmetry of a steady state $\vec{y}$, we introduce the **isotropy subgroup of G' associated with $\vec{y}$:**

$$S_{\vec{y}} = \{D(s) \in G' \mid D(s)\vec{y} = \vec{y}\}$$

We say that $\vec{y}$ **has the symmetry of $S_{\vec{y}}$.**

Example 6.3 Let φ be a linear map that commutes with the representation $\vartheta : s \rightarrow D(s), s \in G$. If ϑ contains an isotopic component V_1 of the unit representation type $(s \rightarrow 1 \; \forall s \in G)$, then each vector $\vec{v} \in V_1$ has the maximal symmetry, namely, the symmetry of the linear map φ:

$$S_{\vec{v}} = G'$$

Example 6.4 Let us again study the 9-dimensional linear map $\varphi = M$ in (1.44) with the symmetry of $\vartheta(D_4)$ described in (1.30). Then for vectors in (1.48), we have

$$S_{x_i} \cong D_4 \qquad \text{and} \qquad S_y \cong S_z \cong S_{u_i} \cong S_{v_i} \cong C_2$$

Therefore, each of the vectors in (1.48) has a certain amount of symmetry, the vectors x_i having the maximal possible symmetry.

There are vectors with no symmetry at all (the other extreme), for example, $y + u_1$ or $u_1 + 2u_2$. However, for every nontrivial linear combination $u = \alpha u_1 + \beta u_2$, $v = \alpha v_1 + \beta v_2$, we have

$$S_u \cong S_v \cong C_2$$

Two points $D(s_1)\vec{y}$, $D(s_2)\vec{y}$ belonging to the same orbit are identical if and only if $D(s_2)^{-1}D(s_1) \cdot \vec{y} = \vec{y}$, or equivalently, if $D(s_2)^{-1}D(s_1) \in S_{\vec{y}}$.

In particular, if $|G'|$ is finite, then the number of elements in the orbit of G' through $\vec{y}$ is

$$|G' \cdot \vec{y}| = \frac{|G'|}{|S_{\vec{y}}|}$$

Let us verify this equation through Example 6.4: Indeed, the orbits of D_4 with respect to the vector space $\mathbf{R}^9$

  containing the vector x_i is the 1-element set $\{x_i\}$;
  containing the vector y is the 2-element set $\{y, -y\}$;
  containing the vector u_i is the 4-element set $\{u_i, -u_i, v_i, -v_i\}$.

Let Σ be a subgroup of G': $\Sigma \subset G'$. The **fixed point space** of Σ is a linear subspace of $\mathbf{R}^n$ defined by

$$\text{Fix}(\Sigma) = \{\vec{y} \in \mathbf{R}^n \mid D(s) \cdot \vec{y} = \vec{y} \; \forall D(s) \in \Sigma\}$$

For the trivial subgroup $\Sigma = \{e\}$, we have $\text{Fix}(\Sigma) = \mathbf{R}^n$.

Example 6.5 We are given the fixed point subspaces of some subgroups of D_4 with respect to (1.48):

$$\begin{aligned}
\text{Fix}(D_4) &= \text{Fix}(C_4) = [x_1, x_2, x_3] \\
\text{Fix}(\sigma, e) &= [x_1, x_2, x_3, y, v_1, v_2] \\
\text{Fix}(\mu, e) &= [x_1, x_2, x_3, z]
\end{aligned}$$

Here $[x_1, x_2, x_3]$, for example, denotes the 3-dimensional subspace spanned by the vectors x_1, x_2, x_3; the letter σ denotes the reflection in the vertical line and μ the reflection in the diagonal.

A remarkable fact is stated in

Theorem 6.1 *Fixed point subspaces are invariant under the G'-equivariant map φ, even if φ is nonlinear.*

Proof. To show:

$$\varphi(\text{Fix}(\Sigma), \lambda) \subset \text{Fix}(\Sigma)$$

Let $D(s) \in \Sigma$ and $\vec{y} \in \text{Fix}(\Sigma)$. Then

$$\varphi(\vec{y}, \lambda) = \varphi(D(s)\vec{y}, \lambda) = D(s) \cdot \varphi(\vec{y}, \lambda) \tag{6.6}$$

The first equation is true by definition of $\text{Fix}(\Sigma)$. The second equation holds, because φ has the symmetry of G'.
From (6.6), it follows that

$$D(s)\varphi(\vec{y}, \lambda) = \varphi(\vec{y}, \lambda) \qquad \forall \vec{y} \in \text{Fix}(\Sigma)$$

$$\square$$

Theorem 6.1 provides a strategy for finding solutions of (6.5):
Restrict the problem (6.5) to $\vec{y} \in \text{Fix}(\Sigma)$ and solve it in this (low dimensional) subspace.

In the following, we limit ourselves to just giving a brief sketch of the further procedure in order to get more information on the structure of bifurcations of steady-state solutions: The following **equivariant branching lemma** due to (Vanderbauwhede, 1980) and (Cicogna, 1981) plays an important role. It states the existence of a unique branch of solutions provided that Σ meets certain conditions, mainly that of

$$\dim \text{Fix}(\Sigma) = 1 \tag{6.7}$$

(see bifurcation subgroup introduced in the next section 6.1.4). In practice, this condition is often satisfied. Problem (6.5) relates to "spontaneous symmetry-breaking" as follows: Assume that for each λ, the autonomous system (6.2) has the trivial solution $\vec{y} = \vec{0}$ (with isotropy subgroup G').

Furthermore, assume that it is asymptotically stable (see (Golubitsky et al., 1988, vol. I)) for $\lambda < 0$ but ceases to be so at $\lambda = 0$. Typically, a loss of asymptotical stability is associated with a bifurcation point, at which the number of solutions depending on the bifurcation parameter λ changes.

New branches of solutions $\vec{y}(\lambda) \neq \vec{0}$ to (6.5) emerge from the trivial branch at $\lambda = 0$. Such solutions often have isotropy subgroups Σ smaller than G'.

We also say that **the solution spontaneously breaks symmetry** from G' to Σ, where "spontaneously" means that the *equation* (6.5) still has the symmetry of G'; but instead of a single G'-symmetric solution $\vec{y} = \vec{0}$, there is an additional family of solutions $\vec{y} \neq \vec{0}$ with symmetry Σ (or conjugate to Σ).

It is worthwhile to point out that the Hopf bifurcation can also be treated as a steady-state bifurcation with the circle group symmetry S^1.

6.1.4 A New Approach

The content of this subsection up to the end of Section 6.1 is mainly based on the work (Werner, 1989), which also includes other results.

We study the parameter-dependent nonlinear dynamical system (6.2). It has the symmetry of a group of orthogonal $n \times n$ matrices; that is, (6.4) holds with $G' \subset O(n)$.

Assume that we know a branch of G'-symmetric equilibria denoted by $\vec{y_0}(\lambda)$ with λ varying over an interval of $\mathbf{R}$.

Along a branch of G'-symmetric steady states, the Jacobian matrices

$$J(\lambda) = \left[\frac{\partial \varphi}{\partial \vec{y}} \right]_{\vec{y} = \vec{y_0}(\lambda), \lambda}$$

share the G'-invariance with the nonlinear map φ:

$$D(s) \cdot J(\lambda) = J(\lambda) \cdot D(s) \qquad \forall D(s) \in G'$$

The eigenvalues of the Jacobians $J(\lambda)$ account for steady-state and Hopf bifurcations; namely, steady-state bifurcations correspond to zero eigenvalues, while Hopf bifurcations correspond to purely imaginary eigenvalues. Therefore, we may utilize a **symmetry adapted basis in $\mathbf{R}^n$**. It has the property that all matrices $J(\lambda)$ simultaneously have block diagonal form as described in Section 2.3.

This approach has not yet been used for bifurcation problems in the past. But it will prove to be very useful from both the theoretical and the numerical point of view.

Underlying symmetries often force the Jacobians $J(\lambda)$ to have eigenvalues of high multiplicities. Let us recall a consequence of the fundamental theorem in Section 2.3:

If an eigenvector $\vec{v}$ of $J(\lambda)$ with eigenvalue μ is an element of the isotypic component V_j of type ϑ_j, then

$$\text{the algebraic multiplicity of } \mu \geq \text{ the dimension of } \vartheta_j \qquad (6.8)$$

For example, the multiplicity of the eigenvalue 4 in (2.41) is 3, but the dimension of the unit representation is 1. For the remaining eigenvalues $4 \pm 2\sqrt{2}$, $4 \pm \sqrt{2}$, however, equality holds in (6.8).

Because of the presence of eigenvalues with algebraic multiplicity > 1, standard bifurcation results cannot be applied directly to simple steady-state and Hopf bifurcations.

On the other hand, underlying symmetries enable us to classify the bifurcation phenomena. To do so, we first introduce some definitions. For simplicity, we shall restrict our attention to so-called **real type** representations (Golubitsky, et al., 1988, vol. II, pp. 41). We state without proof that our examples are such representations.

Definition. An **eigenvalue** μ of a G'-equivariant matrix J is called G'-**simple** if in (6.8) equality holds for μ in place of λ.

Remark. μ is G'-simple means that μ is associated with one and only one isotypic component V_j of G'.

In the example just mentioned, all eigenvalues of M, except 4, are D_4-simple.

Definitions. A point $(\vec{y_0}, \lambda_0)$ is a G'-**simple bifurcation point of symmetry type** ϑ_j if there exists a G'-simple eigenvalue $\mu(\lambda)$ of $J(\lambda)$ in the neighborhood of $\lambda = \lambda_0$ that intersects the imaginary axis in the Gaussian plane with nonzero "speed" (with respect to λ).

If $\mu(\lambda_0) = 0$, the point $(\vec{y_0}, \lambda_0)$ is called a **steady-state bifurcation point**; otherwise it is called a **Hopf bifurcation point**.

A word of caution: The notation G'-*simple bifurcation point* has to be taken with care, as it is not proven yet that a bifurcation does take place at that point.

Definition. A subgroup Σ of G' is called a **bifurcation subgroup** for ϑ_j if for each irreducible subspace of ϑ_j the fixed point space Fix(Σ) *is 1-dimensional.*

For solutions with at least the symmetry of Σ, (Dellnitz, Werner, 1989) showed that G'-*simple bifurcation points* of real type representations ϑ_j (with multiplicity = dimension of ϑ_j) *can be reduced to simple bifurcation points* provided Σ is a bifurcation subgroup for ϑ_j.

For Hopf bifurcation points the concept of bifurcation subgroups is insufficient to describe all bifurcating branches; however, the symmetry type ϑ_j of a bifurcation point seems to give the essential information needed to describe the various bifurcations of steady-state or periodic solutions; see (Golubitsky, et al., 1988) and (Dellnitz, 1989).

6.1.5   The Brusselator

The **Brusselator** (created in Brussels) is the name given to the simplest model that complies with chemical kinetics and exhibits oscillatory behavior like the Belusov-Zhabotinsky reaction (Prigogine and Lefever, 1974).

We consider a 6-box Brusselator problem. Let us assume that we have six identical chemical reactors (boxes) connected in a hexagonal configuration.

For the state vector $[x(t), y(t)]^T$ of one box, the reaction law is given by

$$\left.\begin{array}{l}\varphi_1(x,y) = a - (b+1)x + x^2 y \\ \varphi_2(x,y) = \quad\quad bx - x^2 y\end{array}\right\}$$

with the constants a and b.

Let $[x_i(t), y_i(t)]^T$ be the state vector for box no. i. If the diffusion rates clockwise and counterclockwise are the same, then the dynamics of the whole system is modeled by the following 12 nonlinear differential equations ($i = 1, 2, \ldots, 6$), where λ controls the size of diffusion:

$$\left.\begin{array}{l}\dot{x}_i = \varphi_1(x_i, y_i) + \frac{d_1}{\lambda^2}(x_{i+1} - 2x_i + x_{i+1}) \\[2mm] \dot{y}_i = \varphi_2(x_i, y_i) + \frac{d_2}{\lambda^2}(y_{i+1} - 2y_i + y_{i+1})\end{array}\right\} \tag{6.9}$$

Here the subscript 7 is to be replaced by 1 and the subscript 0 by 6. The quantities d_1 and d_2 are the diffusion coefficients. The model (6.9) is obviously D_6-equivariant, thus reflecting the symmetry of the physical situation.

There exists a D_6-symmetric branch of equilibria of (6.9):

$$x_i = a, \qquad y_i = b/a \qquad i = 1, 2, \ldots, 6 \tag{6.10}$$

The 12×12 Jacobians $J(\lambda)$ of the right hand side of (6.9) commute with the faithful representation

$$\vartheta : s \to P(s) \qquad s \in D_6$$

where each group element s is represented by the corresponding 6×6 block permutation matrix and each block itself is the 2×2 identity matrix.

Therefore, we have

$$P(s) \cdot J(\lambda) = J(\lambda) \cdot P(s) \qquad \forall s \in D_6 \tag{6.11}$$

The reader may verify (6.11) by explicit calculations. The *eigenvalues of the Jacobians $J(\lambda)$ are responsible for steady-state and Hopf bifurcations* on the branch (6.10).

With (5.148) or directly with the double values of (3.20) (remember we have 2×2 blocks), we obtain

$$\vartheta = 2\vartheta_1^1 \oplus 2\vartheta_3^1 \oplus 2\vartheta_1^2 \oplus 2\vartheta_2^2 \tag{6.12}$$

The irreducible representations of D_6 are described in (1.59) and (1.60). The superscripts indicate the dimensions of the irreducible constituents.

With the fundamental theorem of Section 2.3 at hand, the set of the Jacobians $J(\lambda)$ can be transformed *simultaneously into block diagonal form* by use of the symmetry adapted basis which is explicitly tabulated in Table (3.7).

The first two summands in (6.12) each yield one 2×2 block. The remaining two each yield two identical 2×2 blocks. Thus we have a total of four different 2×2 blocks.

For the following **numerical results**, we used the values

$$a = 2 \qquad b = 5.9 \qquad d_1 = 1 \qquad d_2 = 10$$

Our objective now is to calculate the eigenvalues of $J(\lambda)$ for *a large number of* λ-values.

In pursuing this, we notice that knowing the symmetry adapted basis does not only serve us in our theoretical reasonings, but also considerably facilitates numerical computations, since we expect that for almost all values of λ there are two D_6-simple eigenvalues of each of the four different symmetry types in (6.12).

These eigenvalues were computed for numerous λ-values by solving the eigenvalue problem of four 2×2 matrices. This is where the symmetry adapted basis comes into play.

The paths (as functions of λ) of the eigenvalues of the three symmetry types $\vartheta_3^1, \vartheta_1^2, \vartheta_2^2$ in the complex plane are qualitatively the same.

The paths of such a pair of eigenvalues are described in Fig. 6.4:

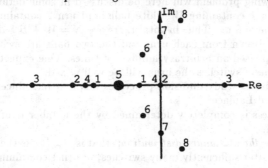

Fig. 6.4 Paths of two eigenvalues of $J(\lambda)$

The labels refer to the paths as the bifurcation parameter λ varies.

(i) For $\lambda = 0$ (infinite diffusion), the two eigenvalues are negative real. They are located at the two points labeled 1 in Fig. 6.4.

(ii) For increasing $\lambda > 0$, the two eigenvalues move along the real axis, one to the left, the other to the right. The one moving to the right crosses the imaginary axis at the origin. This instant is pictured by the two points labeled 2 (**steady-state bifurcation**).

(iii) For some larger values of λ, the two eigenvalues still move along the real axis. Then at position no. 3 they change directions. The eigenvalue moving to the right again crosses the imaginary axis at the origin (another **steady-state bifurcation**) before it joins the other eigenvalue at collision point no. 5.

(iv) After colliding at point no. 5 both eigenvalues bifurcate into the complex plane at infinite "speed" (as a function of λ) passing through the points numbered 6, before they intersect the imaginary axis at the points labeled 7 (**Hopf bifurcation**) and proceed to the points labeled 8.

Thus, for all symmetry types in (6.12), except the unit representation ϑ_1^1, *we have found two steady-state bifurcation points and one Hopf bifurcation point.*

6.2 A Diffusion Model in Probability Theory

6.2.1 Introduction

The subject treated in this section is mainly based on (Diaconis, Shashahani, 1987), which also includes a collection of examples where this theory applies. For more applications, the reader may also consult the recent work of (Diaconis, Graham, Morrison, 1990).

The following problem will here be discussed in some detail. Consider two urns, urn 1 containing W white balls and urn 2 containing B black balls at the beginning. Thus in total there are $N = W + B$ balls.

A ball is drawn from each urn, and the two balls are switched. This process is repeated an arbitrary number of times. One expects that after sufficiently many switches the urns will be well mixed.

This model to study diffusion processes was first introduced by Daniel Bernoulli[1] and Laplace[2].

The process is completely determined by the number w of white balls in, say, urn 1.

Let P_∞ be the *stationary distribution*; that is, $P_\infty(w)$ is the asymptotic probability (after sufficiently many switches) of urn 1 containing exactly w white balls.

---

[1] Bernoulli, Daniel (1700-1782). Son of Johann Bernoulli, moved to Petersburg in 1725, worked in Basel, Switzerland, from 1733 on. Treated the Ricatti differential equation. Separated the notion of mathematical expectation from that of moral expectation and treated probability questions analytically for large numbers. One of the founders of hydrodynamics (Bernoulli's equation). Made objections to unjustified limit considerations by Euler.

[2] Laplace, Pierre Simon (1749-1827). Studied in Caen, taught in Paris at the military academy from 1772 (Napoleon was his student), Minister of the Interior and Chancellor. Investigated partial differential equations (Laplace operator). Studied probability theory; in celestial mechanics, periods of Jupiter and Saturn and those of the three inner Jupiter satellites, also the stability of the solar system, moon theory. Created the famous cosmological theory and thus was also a great philosopher.

The existence and uniqueness conditions of P_∞ are satisfied; see (Hoel, Port, Stone, 1972, Theorems 5 and 7 in Ch. 2) or (Kemeny, Snell, 1965, Theorems 4.1.4. and 4.16).

$P_\infty(w)$ is given by the ratio of the number of ways of selecting W balls out of N balls such that among them there are exactly w white balls (and thus $W - w$ black balls) relative to the total number of combinations of N balls taken W at a time.

$$P_\infty(w) = \frac{\binom{W}{w} \cdot \binom{N-W}{W-w}}{\binom{N}{W}} \qquad 0 \le w \le W \qquad (6.13)$$

This is also called the hyper-geometric distribution.

Our main question now is:

How long or how many switches does it take to reach stationary distribution?

Let P_k be the (unknown) distribution after k switches or after time k, assuming that in each time unit exactly one exchange takes place. Thus $P_k(w)$ is the probability of having exactly w white balls in urn 1 after k switches.

A measure for the deviation of P_k relative to the stationary distribution P_∞ is the **variation distance**

$$d_k = \|P_k - P_\infty\| = \frac{1}{2} \cdot \sum_{w=0}^{W} |P_k(w) - P_\infty(w)| \qquad k = 0, 1, 2, \ldots \qquad (6.14)$$

The right hand side tends to 0 as $k \to \infty$. Furthermore, $d_k \le 1$, since by the triangle inequality

$$\sum_w |P_k(w) - P_\infty(w)| \le \sum_w [P_k(w) + P_\infty(w)] = 2$$

By virtue of

$$P_0(w) = \begin{cases} 1 & \text{if } w = W \\ 0 & \text{if } w < W \end{cases}$$

and because $P_\infty(w)$ is small, it follows that the variation distance $d_0 = \|P_0 - P_\infty\|$ is close to 1.

Let us analyze the Bernoulli-Laplace process by a corresponding *nearest neighbor random walk* (which is a *Markov chain*) in the following sense:

We identify the elements of $Q = S_N/(S_W \times S_B)$ with the $\binom{N}{W}$ combinations consisting of W elements taken from the total set of N elements. S_N is the symmetric group on the set $\{1, 2, \ldots, N\}$, thus $|S_N| = N!$. The subgroup $S_W \times S_B \subset S_N$ consists of the permutations of S_N that separately permute the elements of the subset $\{1, 2, \ldots, W\}$ and those of the complement $\{W + 1, W + 2, \ldots, N\}$ among themselves.

Indeed, the order $|S_N/(S_W \times S_B)| = \dfrac{N!}{W!(N-W)!} = \begin{pmatrix} N \\ W \end{pmatrix}$.

Let us define the metric

$$\mu(q,r) = W - |q \cap r| \qquad q, r \in Q$$

The nearest neighbor random walk on Q consists in moving from $q \in Q$ to $r \in Q$ by randomly selecting a ball in q and a ball in the complement $\bar{q}$ and then switching the two balls to obtain the new subset $r \in Q$.

6.2.2   General Considerations

In order to analyze our main question it is favorable to work in a more general setting: Let G be a finite group, H a subgroup of G, and $Q = G/H$ be the associated cosets of G modulo H.

We choose as fixed coset representatives $q_0 = e, q_1, \ldots, q_m$; thus

$$G = H \cup q_1 H \cup \ldots \cup q_m H$$

We shall often identify Q with the set $\{q_0, q_1, \ldots, q_m\}$ of our representatives.

Let P be an H **bi-invariant probability** on G; that is,

$$P(h_1 g h_2) = P(g) \qquad \forall h_1, h_2 \in H, \quad g \in G \tag{6.15}$$

The probability distribution P induces a **random walk** or a **Markov chain** on the homogeneous space Q by defining

$$P(q,r) = P(q^{-1} r H) \tag{6.16}$$

where $P(q,r)$ denotes the probability of moving from $q \in Q$ to $r \in Q$ in one single step.

We obtain (6.16) as follows: Each choice of $g \in G$ effects a transition from $q \in G$ to $r = q \cdot g$. Hence the problem amounts to finding $g \in G$ so that $rH = q \cdot gH$; thus $P(q,r) = P\{g \in G \mid qgH = rH\} = P(q^{-1} r H)$.

Observe that this is not the random walk induced by the left action $x \to gx$ of G on Q, whose hermitian matrix would have been given by $\check{P}(q,r) = P(rHq^{-1})$.

An immediate consequence of (6.16) is

$$P(q,r) = P(gq, gr) \qquad \forall g \in G \tag{6.17}$$

The 2-step transition matrix $P(q,r)^2$ of the Markov chain $P(q,r)$ is obtained as follows ($e = $ identity):

$$\begin{aligned}
P(q,r)^2 &= P(r^{-1}q, e)^2 = \sum_{q_i} P(r^{-1}q, q_i) \cdot P(q_i, e) \\
&= \sum_{q_i} P(q^{-1} r q_i H) \cdot P(q_i^{-1} H) \tag{6.18}
\end{aligned}$$

Here we used first (6.17), then the definition of the matrix product, and finally (6.16).

In comparing (6.18) with (3.43), we see that

$$P(q,r)^2 = P \star P(q^{-1}rH)$$

Thus $P(q,r)^2$ is **the convolution of the associated bi-invariant probability P with itself.**

Complete induction yields

$$P(q,r)^k = \underbrace{(P \star P \star P \ldots \star P)}_{k \text{ times}}(q^{-1}rH) \tag{6.19}$$

Another simple fact is that the variation distance between P^n and the stationary distribution on Q equals that between P^n and the uniform distribution $U(g) = 1/|G|$ on G if, in the sense above, P is interpreted as a transition probability on Q or G, respectively.

Now we consider the vector space $\mathcal{F}(G/H)$ of all complex-valued functions φ on G that are constant on each of the left cosets of $H \subset G$. This is expressed in the notation $\varphi(qH)$, where q is an arbitrary representative of the coset qH.

The following facts are used in dealing with probability distributions φ. Let $\vartheta : g \to D(g)$ be the representation obtained by letting g act on the functions of $\mathcal{F}$; that is,

$$D(g) : \varphi(qH) \to \varphi(gqH) = \varphi(q'H) \tag{6.20}$$

Undoubtedly, ϑ is a homomorphism satisfying the representation properties.

An upper bound for the variation distance between P and the uniform distribution U can be given: Utilizing the Cauchy-Schwarz inequality in the first step and the Plancherel theorem in the theory of Fourier transforms (Serre, 1979, Sec. 6.2) in the second step, we obtain

$$\| P - U \|^2 = \frac{1}{4}\left[\sum_{g \in G} |P(g) - U(g)|\right]^2 \leq \frac{|G|}{4}\sum_{g \in G} |P(g) - U(g)|^2$$

$$= \frac{1}{4}\sum_j n_j \cdot \mathrm{tr}\left[\hat{P}(\vartheta_j) \cdot \hat{P}(\vartheta_j)^H\right] \tag{6.21}$$

where in the last expression the sum extends over all irreducible representations of G occurring in ϑ and ^{H} means hermitian conjugate or conjugate transpose ($M^H = \overline{M}^T$). In Plancherel's theorem we made use of the fact that for $U(g)$ the Fourier transform

$$\hat{U}(\vartheta_j) = \begin{cases} 1 & \text{if } \vartheta_j \text{ is the unit representation} \\ \text{null matrix} & \text{otherwise} \end{cases}$$

because $\hat{U}(\vartheta_j) = \frac{1}{|G|} \sum_{s \in G} D_j(s)$ can be thought of as the projection of the representation ϑ_j onto the subspace that is transformed identically. But since for each irreducible representation ϑ_j different from the unit representation there is no such nontrivial subspace, the 0 follows readily.

The reason for the summation in (6.21) to extend only over the irreducible representations of G that indeed appear in our representation (6.20) is the following:

The so-called **reciprocity theorem** of Frobenius (see (Hall, 1976, Sec. 16.7) or (Serre, 1977, Theorem 13)) implies that every irreducible constituent ϑ_k of a representation ϑ of G occurs with multiplicity c_k, where c_k equals the dimension of the restriction to H of the constituent ϑ_k of the subspace transformed identically (whence transformed like the unit representation).

In other words: The multiplicity c_k with which the irreducible constituent ϑ_k occurs in the representation ϑ of G equals the multiplicity with which the unit representation occurs in the representation ϑ_k restricted to the subgroup H.

Therefore, *if ϑ_k fails to appear*, then the restriction of ϑ_k to H does not contain the unit representation. Because in

$$\hat{P}(\vartheta_k) = \sum_{q_i} P(q_i H) \cdot D_k(q_i) \cdot \sum_{h \in H} D_k(h) \qquad (6.22)$$

summation over H yields the null matrix. This is due to the orthogonality of the irreducible representations (5.18), where one chooses one of the two representations to be the unit (trivial) representation. Or in the parlance of Chapter 5, the sum over H describes, up to a scalar factor, the projection onto the subspace S that transforms after the manner of the unit representation, i.e., that transforms identically. But, as $c_k = 0$, the reciprocity theorem implies that S is the null space. Thus the Fourier transform $\hat{P}(\vartheta_k)$ of P at ϑ_k is the null matrix.

From the findings of the preceding chapters we may infer that the case where each multiplicity of the irreducible constituents of ϑ is less than or equal to one is particularly simple.

In the following, we restrict ourselves to the special cases of this kind.

Definition. The pair (G, H) is called a **Gelfand pair** if each irreducible constituent of the representation (6.20) has multiplicity ≤ 1.

For a Gelfand pair (G, H), let

$$\mathcal{F}(G/H) = V_0 \oplus V_1 \oplus \ldots \oplus V_J \qquad (6.23)$$

be the decomposition of $\mathcal{F}$ into its isotypic components, each of which is an irreducible subspace of ϑ.

Let V_0 be the subspace that is transformed after the manner of the unit reppresentation, i.e., transformed identically (dim $V_0 = 1$, V_0 is spanned by the constant function.)

Remark: If the more general diffusion problem with three urns is to be solved, then the framework described is needed, where in the decomposition (6.23) multiplicities > 1 will occur.

From the reciprocity theorem again we may infer that each V_j has exactly one 1-dimensional subspace of functions $s_j(qH)$ which is transformed identically under the restriction to H of the representation (6.20).

Let the function $s_j(qH)$ be so normalized that $s_j(H) = 1$. Such a function is called the jth **spherical function**. Spherical functions have been explicitly computed for many Gelfand pairs.

Theorem 6.2 *If (G, H) is a Gelfand pair and P is an H bi-invariant probability distribution, then the following estimate holds for the variation distance between P and the uniform distribution $U(g) = 1/|G|$:*

$$\parallel P - U \parallel^2 \le \frac{1}{4} \sum_{j=1}^{J} n_j |\hat{P}_j|^2$$

Here the sum extends only over those irreducible representations that occur in ϑ and differ from the unit representation, where

$$\hat{P}_j = \sum_{g \in G} P(g) \cdot s_j(g)$$

Proof. We consider a fixed subspace V_k in (6.23) with the irreducible representation ϑ_k of G. In V_k we choose a basis such that the normalized spherical function $s_k(qH)$ is the first basis vector. Relative to this basis the Fourier transform of an H bi-invariant function P on G at ϑ_k has the form (6.22).

Now let us consider the representation ϑ_k *restricted to the subgroup* H (which is reducible). The inner sum over H in (6.22) is the $|H|$-fold projection of V_k onto the 1-dimensional subspace spanned by s_k.

Thus, for $M = \sum_{h \in H} D_k(h)$, we have

$$m_{ij} = \left\{ \begin{array}{ll} |H| & \text{if } i = j = 1 \\ 0 & \text{otherwise} \end{array} \right. \tag{6.24}$$

This is a direct consequence of (5.159), where the character of the unit representation is inserted.

Now the product of the matrix of the outer sum of Equation (6.22) with M furnishes for the matrix $\hat{P}(\vartheta_k) = (p_{ij}^{(k)})$ the following result: Except for the first column, all elements of $\hat{P}$ are zero.

Let us recall that, in order to obtain (6.22), only the right invariance $P(q_i H) = P(q_i)$ is used from the H bi-invariance of P. If (6.22) is reworded analogously by the left invariance $P(H q_i) = P(q_i)$, then the factor M of the matrix product for $\hat{P}$ goes to the left; thus yielding the result that, except for the first row of $\hat{P}$, all elements of $\hat{P}$ are zero.

Comparing the two statements regarding column and row finally yields

$$p_{ij}^{(k)} = \begin{cases} |H| \cdot \sum_{q_i} P(q_i H) \cdot s_k(q_i H) & \text{if } i = j = 1 \\ 0 & \text{otherwise} \end{cases}$$

Since the trace norm[3] $\| \hat{P}(\vartheta_j) \| = \sqrt{\text{tr}[\hat{P}(\vartheta_j) \cdot \hat{P}(\vartheta_j)^H]}$ is invariant under unitary coordinate transformations, the assertion of Theorem 6.2 follows from (6.21). □

6.2.3 The Bernoulli-Laplace Model

Let us return to the Bernoulli-Laplace diffusion model, where initially W white balls are in one and B black balls in the other urn with a total number of $N = W + B$ balls. Then the following holds for the variation distance defined in (6.14):

Theorem 6.3 *If the number of switches is*

(a) $k = \frac{W}{2}(1 - \frac{W}{N})(\log N + c)$ *for* $c > 0$, *then there exists a constant* a, *independent of* N, *such that*

$$\| P_k - P_\infty \| \leq a e^{-c} \qquad \text{(upper bound)}$$

(b) $k = \frac{W}{2}(1 - \frac{W}{N})(\log N - \lambda)$ *for* $0 \leq \lambda \leq \frac{1}{4} \log N$, *then there exists a universal constant* b *such that*

$$\| P_k - P_\infty \| \geq 1 - b e^{-4\lambda} \qquad \text{(lower bound)}$$

We confine ourselves to **proving statement (a)**:

Without loss of generality we may assume that $W \leq N/2$. As was proven by (James, 1978, p. 52), the pair $(S_N, S_W \times S_B)$ is a **Gelfand pair** where in (6.23) the subscript J is replaced by W and

$$n_j = \dim(V_j) = \binom{N}{j} - \binom{N}{j-1} \qquad 0 \leq j \leq W$$

The spherical functions needed here are described in (Stanton, 1984). They are in fact classically studied functions known as the **dual Hahn polynomials**.

---

[3]Trace norm (or Frobenius norm): the norm on matrices that arises by treating a matrix as a vector and using the Euclidean norm of that vector: $\|A\|^2 = \sum_{i=1}^{n} \sum_{j=1}^{m} |a_{ij}|^2$.

This quantity is also the trace of $A \cdot A^H = A \cdot \bar{A}^T$.

The function $s_k(q_i)$ on $Q = S_N/(S_W \times S_B)$ varies only with the distance $d(q_i, q_0)$ and is the polynomial

$$s_k(d) = \sum_{j=0}^{k} \frac{(-k)_j \cdot (k - N - 1)_j \cdot (-d)_j}{(W - N)_j \cdot (-W)_j \cdot j!} \qquad 0 \le k \le W$$

Here $(\alpha)_0 = 1$ and $(\alpha)_j = \alpha(\alpha + 1)\ldots(\alpha + j - 1)$ for $j > 0$. Thus in particular

$$
\begin{aligned}
s_0(d) &= 1, \qquad s_1(d) = 1 - \frac{dN}{W(N - W)} \\
s_2(d) &= 1 - \frac{2d(N - 1)}{W(N - W)} - \frac{(N - 1)(N - 2)d(d - 1)}{(N - W)(N - W - 1)W(W - 1)}
\end{aligned}
$$

The underlying probability distribution P in this problem may be regarded as the uniform distribution on the $W(N - W)$ sets having distance $d = 1$ from the set $\{1, 2, \ldots, W\}$. Hence, by Theorem 6.2, it follows that

$$\hat{P}_j = s_j(1) = 1 - \frac{j(N - j + 1)}{W(N - W)} \qquad 0 \le j \le W$$

whence

$$\| P_k - P_\infty \|^2 \le \frac{1}{4} \sum_{j=1}^{W} \left\{ \binom{N}{j} - \binom{N}{j-1} \right\} \cdot \left(1 - \frac{j(N - j + 1)}{W(N - W)} \right)^{2k} \tag{6.25}$$

In the latter sum, let us first consider the term with $j = 1$:

$$(N - 1) \left(1 - \frac{N}{W(N - W)} \right)^{2k} < \exp \left(\log N - \frac{2kN}{W(N - W)} \right) \tag{6.26}$$

The inequality (6.26) is true because

$$\exp(-2ky) > (1 - y)^{2k} \quad \text{for } 0 < y < 1.2 \tag{6.27}$$

the latter being a consequence of $\exp(-y) > 1 - y$.

The argument of exp is strictly negative for

$$k = \frac{W}{2} \left(1 - \frac{W}{N} \right) (\log N + c) \qquad \text{where } c > 0 \tag{6.28}$$

If (6.28) is substituted into (6.25) with

$$\binom{N}{j} - \binom{N}{j-1} < \binom{N}{j} < \frac{N^j}{j!} = \exp(j \log N - \log(j!))$$

and (6.27) with $y = \frac{j(N-j+1)}{W(N-W)}$ is used, then the problem reduces to that of bounding

$$\sum_{j=1}^{W} e^{a(j)+b(j)} \qquad (6.29)$$

where

$$a(j) = c \cdot j \cdot \frac{j-1-N}{N} \qquad b(j) = \frac{j(j-1)\log N}{N} - \log(j!)$$

For $c > 0$, the points with coordinates $(j, a(j))$ lie on a parabola that opens upward and has zeros at $j = 0$ and $j = N + 1$.

Thus for $c \geq 0$,

$$a(j) \leq a(1) = -c \qquad \forall j \in [2, N/2]$$

Hence in the sum (6.29) the term e^{-c} can be factored out. It remains to show that there exists a constant C, independent of N, such that

$$\sum_{j=1}^{N/2} e^{b(j)} \leq C \qquad (6.30)$$

Because

$$\log(j!) = \sum_{\alpha=1}^{j} \log(\alpha) > \int_{1}^{j} \log \alpha \cdot d\alpha = j \log j - j$$

we have $b(j) < j^2 \log N/N - j \log j + j$.

But now the right hand side is $< -j$ for $j \geq 21$; or equivalently,

$$\frac{\log N}{N} < \frac{\log j - 2}{j}$$

for the following reason:

Let us consider the function $f(x) = \dfrac{\log x - 2}{x}$. Then its derivative $f'(x) = \dfrac{3 - \log x}{x^2} < 0$ for $x > e^3 = 20.08...$ Thus for $j > 21 > e^3$,

$$\frac{\log j - 2}{j} > \frac{\log(N/2) - 2}{N/2}$$

As the right hand side is $> \dfrac{\log N}{N}$ whenever $N \geq e^{4+2\log 2}$, it follows that

$$\sum_{j=22}^{N/2} e^{b(j)} \leq \sum_{j=22}^{N/2} e^{-j} < 1$$

This guarantees the existence of a universal constant C in (6.30).

For the proof of part (b) regarding the lower bound, we refer to (Diaconis, Shashahani, 1987). $\qquad \square$

6.2.4 Discussion and Applications

Theorem 6.3 provides the number of switches, $k_0 = \dfrac{W}{2}\left(1 - \dfrac{W}{N}\right)\log N$, needed to reach stationary distribution *in a sharp sense*:

For slightly fewer switches, the variation distance is close to its maximum value of 1; for slightly more switches, however, the variation distance tends to 0 exponentially. In other words, there is a rather abrupt change at k_0 switches.

In the special case where $W = B = N/2$ (N even), we have $k_0 = \frac{1}{8}N \cdot \log N$. Therefore, the time needed to reach stationary distribution is unexpectedly short. The problem has been computer-simulated for $N = 100$ and the pairs $(W, B) = (50,50)$; $(80,20)$; $(20,80)$ with $k_0 \approx 58$ and $k_0 \approx 37$ (see Fig. 6.5).

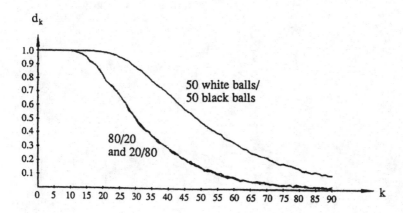

Fig. 6.5

Remark. Error correcting codes is another attractive application of symmetry considerations. See, for example, (Peterson, 1961, Section 11.2).

PROBLEMS

Problem 6.1 In Example 6.4 verify a statement analogous to $S_u \cong S_v$ for the general case of a symmetry adapted basis in an isotypic component.

Problem 6.2 Verify (6.11) by explicit calculations.

Chapter 7

Lie Algebras

The theory of Lie algebras has its main application in theoretical physics. We shall attempt to describe the most important constructive methods.

As in Chapter 4, we confine our treatment to the groups listed in Table (4.3) in order to facilitate proofs. But we shall still use the term "continuous matrix groups" for the entirety of these types. In the theory of Lie[1] groups this notion is used in a broader sense. All statements made in this chapter about continuous matrix groups are also valid in this larger class.

7.1 Infinitesimal Operator and Exponential Mapping

7.1.1 Infinitesimal Operators

Let G be a continuous $n \times n$ matrix group (see Section 4.1). By a **one-parameter family** $A(t) = (a_{ik}(t))$ we mean a subset of G, defined for real values of t in a neighborhood of $t = 0$, differentiable in t and such that

$$A(0) = E = \text{identity matrix} \tag{7.1}$$

The proper spherical rotation group SO(3) (see Example 1.5) serves as an introductory example. In it we choose as a one-parameter family the rotations

$$A(t) = \begin{bmatrix} \cos t & -\sin t & 0 \\ \sin t & \cos t & 0 \\ 0 & 0 & 1 \end{bmatrix} \tag{7.2}$$

---

[1] Lie, Sophus (1842-1899). He succeeded Felix Klein in his professorship in Leipzig and was mainly involved in transformation groups. Shortly before his death he returned to Norway.

about the x_3-axis, the parameter t being the rotation angle. By the **infinitesimal operator** of the family we mean the matrix

$$J = \left.\frac{dA(t)}{dt}\right|_{t=0} = \begin{bmatrix} 0 & -1 & 0 \\ 1 & 0 & 0 \\ 0 & 0 & 0 \end{bmatrix} \tag{7.3}$$

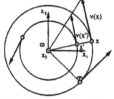

Orthogonal projection of the x_3-axis

Its geometric meaning becomes apparent if we apply J to the position vector x of some point of $\mathbf{R}^3$ and attach the resulting vector $v(x) = Jx = (-x_2, x_1, 0)^T$ to the point x (see Fig. 7.1).

Fig. 7.1

In the terminology of mechanics, $v(x)$ is the field of velocity vectors when rotating about the x_3-axis, and t is the time variable. Here ω is the vector of angular velocity (of length 1) in direction of the x_3-axis. As a matter of fact,

$$v(x) = Jx = \omega \times x$$

That is, it equals the cross product of the rotation vector and the position vector. If the rotation axis has an arbitrary position and $\omega = (\omega_1, \omega_2, \omega_3)$ is the vector of angular velocity of arbitrary length $\sqrt{\omega_1^2 + \omega_2^2 + \omega_3^2}$ (= angular speed), then the velocity field satisfies

$$v(x) = \omega \times x = \begin{bmatrix} \omega_2 x_3 - \omega_3 x_2 \\ \omega_3 x_1 - \omega_1 x_3 \\ \omega_1 x_2 - \omega_2 x_1 \end{bmatrix}; \text{ thus } J = \begin{bmatrix} 0 & -\omega_3 & \omega_2 \\ \omega_3 & 0 & -\omega_1 \\ -\omega_2 & \omega_1 & 0 \end{bmatrix} \tag{7.4}$$

This small argumentation is ingenious in that it is not necessary to explicitly establish the one-parameter family of rotations about the inclined axis, thus suggesting that the infinitesimal operators of a group G are often more easily manipulated than the transformations of G.

We now generally define the following:

Let $A(t)$ be a one-parameter family of a continuous $n \times n$ matrix group G. Then

$$J = \left.\frac{dA(t)}{dt}\right|_{t=0} \tag{7.5}$$

is said to be **the infinitesimal operator of the family** $A(t)$ of G.

The Taylor expansion of the family starts with

$$A(t) = E + Jt + \frac{1}{2}\left.\frac{d^2A}{dt^2}\right|_{t=0} \cdot t^2 + \ldots \tag{7.6}$$

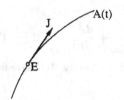

J is the "tangent vector" to the family at the point $A(0) = E$ (see Fig. 7.2).

Fig. 7.2

Let us now determine the infinitesimal operators in the matrix groups of Table (4.3):

1. $GL(n, \mathbf{C})$ and $GL(n, \mathbf{R})$: For an arbitrary regular or singular complex (or real) $n \times n$ matrix J and for sufficiently small t-values, $A(t) = E + t \cdot J$ is a **one-parameter family** in $GL(n, \mathbf{C})$ or $GL(n, \mathbf{R})$. Hence the following holds:

Theorem 7.1 *The infinitesimal operators in $GL(n, \mathbf{C})$ or $GL(n, \mathbf{R})$ comprise all of the complex or real $n \times n$ matrices.*

2. $SL(n, \mathbf{C})$ and $SL(n, \mathbf{R})$: The matrices A of these groups have determinants $\det A = 1$. To bring to light the implications of this condition, we expand the characteristic polynomial of an arbitrary $n \times n$ matrix M:

$$\det(M + \lambda E) = \lambda^n \pm \lambda^{n-1} \cdot \operatorname{tr}(M) + \ldots + \det M \qquad (7.7)$$

where again $\operatorname{tr}(M) = \sum_{i=1}^{n} m_{ii}$ is the trace of M. Next we set $M = t \cdot N$ and $\lambda = 1$. Thus (7.7) becomes

$$\det(tN + E) = 1 \pm t \cdot \operatorname{tr}(N) + t^2(\ldots) + \ldots + t^n \cdot \det N \qquad (7.8)$$

If $A(t)$ is a family in $SL(n, \mathbf{C})$ or $SL(n, \mathbf{R})$, then the Taylor formula for $A(t)$ with remainder $R(t)$ is

$$A(t) = E + tJ + t^2 R(t) = t \cdot (J + t \cdot R(t)) + E \qquad (7.9)$$

In (7.8) we choose $N = J + t \cdot R(t)$. Then (7.9) implies

$$\det A(t) = 1 \pm t \cdot \operatorname{tr}(J + tR) + t^2(\ldots) + \ldots + t^n \det(J + tR) = 1$$

Differentiation with respect to the parameter t at $t = 0$ finally gives $\operatorname{tr}(J) = 0$. Thus the following holds:

Theorem 7.2 *The infinitesimal operators in $SL(n, \mathbf{C})$ or $SL(n, \mathbf{R})$, are the complex or real $n \times n$ matrices with trace zero*[2].

---

[2]The still missing part of the proof that **all** of the listed matrices are infinitesimal operators will be supplied in the next subsection 7.1.2. The same is true for the Theorems 7.3, 7.4, and 7.5.

3. U(n): A family $A(t)$ of U(n) satisfies $A^T \bar{A} = E$ by (4.3). Differentiating this matrix equation and setting $t = 0$ gives, by $\bar{A}(0) = A^T(0) = E$,

$$\frac{dA^T}{dt}\bigg|_{t=0} \cdot \bar{A}\bigg|_{t=0} + A^T \frac{d\bar{A}}{dt}\bigg|_{t=0} = \frac{dA^T}{dt}\bigg|_{t=0} + \frac{d\bar{A}}{dt}\bigg|_{t=0} = 0 \qquad (7.10)$$

Therefore, the infinitesimal operator J of the family $A(t)$ satisfies

$$J^T + \bar{J} = 0 \qquad (7.11)$$

In appealing to the definition (4.18) of hermitian matrices, a matrix J is said to be **skew-hermitian** if it satisfies property (7.11). Its diagonal elements are purely imaginary. The following holds:

Theorem 7.3 *The infinitesimal operators in U(n) are the skew-hermitian $n \times n$ matrices.*

4. SU(n): Theorems 7.2 and 7.3 imply

Theorem 7.4 *The infinitesimal operators in SU(n) are the skew-hermitian $n \times n$ matrices with trace zero.*

We verify Theorem 7.4 for the example of SU(2). By (4.8) a one-parameter family of this group has the form

$$A(t) = \begin{bmatrix} a(t) & b(t) \\ -\bar{b}(t) & \bar{a}(t) \end{bmatrix} \quad \text{where} \quad a\bar{a} + b\bar{b} = 1 \quad \text{and} \quad \begin{matrix} a(0) = 1 \\ b(0) = 0 \end{matrix} \qquad (7.12)$$

Writing

$$\alpha = \frac{da(t)}{dt}\bigg|_{t=0}, \qquad \beta = \frac{db(t)}{dt}\bigg|_{t=0}$$

we obtain the derivative of (7.12) at $t = 0$:

$$J = \begin{bmatrix} \alpha & \beta \\ -\bar{\beta} & -\alpha \end{bmatrix} \quad \text{where} \quad \alpha \text{ is purely imaginary} \quad (\bar{\alpha} = -\alpha) \quad (7.13)$$

due to the constraint $\alpha + \bar{\alpha} = 0$.

5. O(n) and SO(n): By (4.3), a family $A(t)$ of O(n) satisfies $A^T A = E$ with A real. Differentiation at $t = 0$ shows that the infinitesimal operator J of $A(t)$ satisfies

$$J^T + J = 0 \qquad (7.14)$$

in analogy with (7.10). A matrix with the property (7.14) is said to be **skew-symmetric**. Its diagonal elements are all zero. In addition, a neighborhood of the identity matrix E of the group O(n) also lies in SO(n) ($\det(A) = 1$), so that the following holds:

Theorem 7.5 *The infinitesimal operators in O(n) and SO(n) are the real skew-symmetric $n \times n$ matrices.*

7.1.2 The Exponential Mapping

As a converse of the foregoing Section 7.1.1, we construct a one-parameter family $A(t)$ of a continuous matrix group G from a given infinitesimal operator J of G via the prescription

$$A(t) = e^{tJ}$$

where the exponential map is defined by the matrix series

$$e^{tJ} = E + tJ + \frac{t^2}{2!}J^2 + \ldots + \frac{t^k}{k!}J^k + \ldots \qquad (7.15)$$

In justifying this construction (that is, in proving that the e^{tJ} are elements of the matrix group G in question), we confine our discussion to the examples of Table (4.3).

1. $GL(n, \mathbf{C})$ and $GL(n, \mathbf{R})$: The series (7.15) converges for any J and t. We recommend that the reader prove this statement as an exercise.

   Let now $\lambda_1, \lambda_2, \ldots, \lambda_n$ be the eigenvalues of the matrix J. Then the matrix formed by the first $(k + 1)$ summands of (7.15) has the eigenvalues

   $$1 + t\lambda_i + \frac{t^2}{2!}\lambda_i^2 + \ldots + \frac{t^k}{k!}\lambda_i^k \qquad (i = 1, 2, \ldots, n)$$

   On passing to the limit $k \to \infty$, we obtain

   Theorem 7.6 *If $\lambda_1, \lambda_2, \ldots, \lambda_n$ are the eigenvalues of an $n \times n$ matrix J, then $e^{\lambda_1 t}, e^{\lambda_2 t}, \ldots, e^{\lambda_n t}$ are the eigenvalues of the matrix e^{tJ}.*

   Hence the determinant satisfies

   $$\det(e^{tJ}) = \prod_{i=1}^{n} e^{\lambda_i t} \qquad (7.16)$$

   It differs from zero, whence e^{tJ} indeed belongs to $GL(n,\mathbf{C})$ or $GL(n,\mathbf{R})$. Let us consider the product of two series (7.15)

   $$e^{tJ}e^{tK} = E + t(J + K) + \frac{t^2}{2!}(J^2 + K^2 + 2JK) + \ldots \qquad (7.17)$$

   Clearly, $e^{tJ}e^{tK} = e^{t(J+K)}$ if and only if J and K commute with respect to multiplication. We may then use the same combination rules as for ordinary numbers.

2. $SL(n, \mathbf{C})$ and $SL(n, \mathbf{R})$: Let J be an arbitrary (complex or real) $n \times n$ matrix with trace zero. Then its eigenvalues satisfy

   $$\lambda_1 + \lambda_2 + \ldots + \lambda_n = 0 \qquad (7.18)$$

Combining this with (7.16), we obtain

$$\det(e^{tJ}) = 1$$

Hence e^{tJ} belongs to SL(n, **C**) or SL(n, **R**), respectively.

3. U(n): Let now J be an arbitrary skew-hermitian matrix. Then (7.11) and commutativity of $\bar{J}$ with $-\bar{J}$ imply

$$(e^{tJ})^T(\overline{e^{tJ}}) = e^{t(J^T)}e^{t\bar{J}} = e^{-t\bar{J}}e^{t\bar{J}} = E \qquad (7.19)$$

Thus e^{tJ} belongs to U(n).

4. For the remaining groups SU(n), O(n), SO(n), equations similar to (7.19) are true. We thus showed the converse of Theorem 7.2, as announced in the footnote. Moreover, this also proves that the exponential map makes sense for the groups in (4.3).

Let us point out that the one-parameter family e^{tJ} of G is in fact even a one-parameter subgroup of G; for,

$$e^{t_1 J} \cdot e^{t_2 J} = e^{(t_1 + t_2)J} \qquad (7.20)$$

Addition of t-values corresponds to group multiplication. Thus the one-parameter subgroup is abelian. The inverses are easily found due to (7.20):

$$(e^{tJ})^{-1} = e^{-tJ} \qquad (7.21)$$

7.2 Lie Algebra of a Continuous Group

If J and K are infinitesimal operators of a continuous matrix group G associated with the one-parameter families $A(t) = E + tJ + \ldots$ and $B(t) = E + tK + \ldots$, then $J + K$ is also an infinitesimal operator, namely, that associated with the one-parameter family $A(t)B(t) = E + t(J + K) + \ldots$ Moreover, if J is an infinitesimal operator of G, then so is rJ (r real). To prove this we introduce a new parameter τ by $t = r \cdot \tau$ in the family $A(t) = E + tJ + \ldots$ associated with J:

$$A(r \cdot \tau) = E + (rJ)\tau + \ldots \qquad r \in \mathbf{R} \qquad (7.22)$$

Hence the infinitesimal operator of this family with parameter τ is

$$\frac{d}{d\tau}A(r \cdot \tau)\Big|_{\tau=0} = rJ \qquad (7.23)$$

This may be summarized in

Theorem 7.7 *The infinitesimal operators of a matrix group G form a real vector space $\mathcal{L}$.*

In the group $SU(n)$ this is manifested in that each linear combination of skew-hermitian matrices with trace zero is again a skew-hermitian matrix with trace zero.

Remark. By virtue of Theorem 7.7, the infinitesimal operators can be interpreted as both **vectors** or **matrices**.

One of the important facts to know is that in $\mathcal{L}$ there is a composition of two elements of $\mathcal{L}$, other than the real linear combination, having properties similar to those of the cross product, namely the composition given by the following

Definition. A **commutator** of two $n \times n$ matrices J and K is a matrix given by

$$[J, K] = JK - KJ \tag{7.24}$$

Theorem 7.8 *If J and K are infinitesimal operators of a continuous matrix group, then so is $[J, K]$.*

Proof. Let two arbitrary families $A(t), B(t)$ with the infinitesimal operators J, K be given. Then[3]

$$\left.\begin{array}{ll} A(t) = E + Jt + Wt^2 + \ldots, & A(t)^{-1} = E - Jt + Yt^2 + \ldots \\ B(t) = E + Kt + Xt^2 + \ldots, & B(t)^{-1} = E - Kt + Zt^2 + \ldots \end{array}\right\} \tag{7.25}$$

The second order terms in t were designated by unknown matrices. In these series as well as in those that follow, the remainders indicated by dots start with third order terms in t.
From

$$A(t) \cdot A(t)^{-1} = E + t^2(W + Y - J^2) + \ldots$$

follows

$$W + Y = J^2, \text{ and similarly, } X + Z = K^2 \tag{7.26}$$

We now form

$$\left.\begin{array}{ll} A(t)B(t) & = E + (J + K)t + (X + W + JK)t^2 \ldots \\ A(t)^{-1}B(t)^{-1} & = E - (J + K)t + (Y + Z + JK)t^2 \ldots \end{array}\right\} \tag{7.27}$$

On multiplying the two expansions, we obtain

$$ABA^{-1}B^{-1} = E - (J^2 + JK + KJ + K^2)t^2 + (2JK + X + Z + W + Y)t^2 + \ldots$$

which by (7.26) is

$$ABA^{-1}B^{-1} = E + (JK - KJ)t^2 + \ldots = E + [J, K]t^2 + \ldots \tag{7.28}$$

In choosing the parameter for this family to be $\tau = t^2$, we see that its infinitesimal operator is $[J, K]$. □

---

[3] That the family $A(t)^{-1}$ has the infinitesimal operator $-J$ follows from differentiating the relation $A(t) \cdot A(t)^{-1} = E$.

For three arbitrary infinitesimal operators H, J, K of a continuous matrix group, definition (7.24) readily implies the following properties:

bilinearity: $[aH + bJ, K] = a[H, K] + b[J, K]$ $a, b \in \mathbf{R}$ (7.29)

skew-symmetry: $[J, K] = -[K, J]$ (7.30)

Jacobi identity: $[[H, J], K] + [[J, K], H] + [[K, H], J] = 0$ (7.31)

Note the cyclic permutation in (7.31).

If, without relating to a group G, a real vector space $\mathcal{L}$ of finite dimension d is such that in it a composition $[\ .\ ,\ .\ ]$ obeying the axioms (7.29), (7.30), and (7.31) is defined, then $\mathcal{L}$ is said to be an **abstract Lie algebra**.

Thus the infinitesimal operators of a continuous matrix group G form a Lie algebra. For the groups in Table (4.3) the following is true due to Theorems (7.1) through (7.5):

Group G	The Lie algebra of G consists of all $n \times n$ matrices that are
GL(n,**C**), GL(n,**R**)	arbitrary, complex or real
SL(n,**C**), SL(n,**R**)	complex or real, with trace = 0
U(n)	skew-hermitian, i.e., $a_{ik} + \overline{a_{ki}} = 0$
SU(n)	skew-hermitian, with trace = 0
O(n), SO(n)	skew-symmetric, i.e., $a_{ik} + a_{ki} = 0$ and real

$$(7.32)$$

Theorem 7.8 can be verified by means of the various examples of Table (7.32) above: For instance, in the group U(n) the commutator of two skew-hermitian matrices is again skew-hermitian.

If $J_1, J_2, \ldots, J_d$ is a basis in $\mathcal{L}$, then

$$[J_i, J_k] = \sum_{\ell=1}^{d} c_{ik}^{\ell} J_\ell \qquad c_{ik}^{\ell} \in \mathbf{R} \tag{7.33}$$

The reason is that $[J_i, J_k]$ is again in $\mathcal{L}$ and, therefore, a linear combination of the basis vectors with certain real coefficients c_{ik}^{ℓ}, the **structure constants**. By (7.29) and (7.30) they completely determine every commutator $[J, K]$ in $\mathcal{L}$.

In the special case of an abelian continuous matrix group G, Equation (7.28) implies

$$[J, K] = 0 \qquad \forall J, K \in \mathcal{L} \tag{7.34}$$

In view of this statement, a **Lie algebra $\mathcal{L}$ is said to be abelian** if (7.34) holds.

An example of an abstract Lie algebra is the 3-dimensional real space $\mathbf{R}^3$ of vectors $J, K, \ldots$ with the cross product as the composition, i.e.,

$$[J, K] = J \times K$$

As an exercise, verify the properties (7.29), (7.30), (7.31).

If one chooses the orthogonal, positively oriented unit vectors J_1, J_2, J_3 as a basis, then the structure relations (7.33) are simply

$$[J_1, J_2] = J_3, \qquad [J_2, J_3] = J_1, \qquad [J_3, J_1] = J_2 \qquad (7.35)$$

Let $J = u_1 J_1 + u_2 J_2 + u_3 J_3$ be an arbitrary vector of $\mathbf{R}^3$. Now we construct the correspondence

$$J = \begin{bmatrix} u_1 \\ u_2 \\ u_3 \end{bmatrix} \longrightarrow J' = \begin{bmatrix} 0 & -u_3 & u_2 \\ u_3 & 0 & -u_1 \\ -u_2 & u_1 & 0 \end{bmatrix} \qquad (7.36)$$

Let further $K = v_1 J_1 + v_2 J_2 + v_3 J_3$ be another vector of $\mathbf{R}^3$ and K' its image. Then the correspondence (7.36) satisifies not only $(J + K) \rightarrow (J' + K')$ and $rJ \rightarrow rJ'$ (r real), but also

$$[J, K]' = [J', K'] = J'K' - K'J' \qquad (7.37)$$

Indeed, by (7.36) the right-most term is

$$J'K' - K'J' = \begin{bmatrix} 0 & -(u_1 v_2 - u_2 v_1) & (u_3 v_1 - u_1 v_3) \\ (u_1 v_2 - u_2 v_1) & 0 & -(u_2 v_3 - u_3 v_2) \\ -(u_3 v_1 - u_1 v_3) & (u_2 v_3 - u_3 v_2) & 0 \end{bmatrix}$$

This is precisely the matrix which by (7.36) is associated with the vector product $[J, K]$. Thus the correspondence (7.36) between vectors and matrices is one-to-one, its images are all of the skew-symmetric real matrices, and by (7.37), it preserves the structure relations (7.35). Hence this correspondence is a **Lie algebra isomorphism**. In particular, the basis vectors $J_1 = (1, 0, 0)^T$, $J_2 = (0, 1, 0)^T$, $J_3 = (0, 0, 1)^T$ of $\mathbf{R}^3$ are associated with the following matrices:

$$J_1' = \begin{bmatrix} 0 & 0 & 0 \\ 0 & 0 & -1 \\ 0 & 1 & 0 \end{bmatrix}, \quad J_2' = \begin{bmatrix} 0 & 0 & 1 \\ 0 & 0 & 0 \\ -1 & 0 & 0 \end{bmatrix}, \quad J_3' = \begin{bmatrix} 0 & -1 & 0 \\ 1 & 0 & 0 \\ 0 & 0 & 0 \end{bmatrix} \qquad (7.38)$$

They form a basis in the Lie algebra of all real skew-symmetric matrices J' of (7.36).

Equations (7.36) and (7.37) imply structure relations isomorphic with (7.35):

$$[J_1', J_2'] = J_3', \quad [J_2', J_3'] = J_1', \quad [J_3', J_1'] = J_2'$$

Not surprisingly, the Lie algebra of SU(2), consisting of all matrices of the form (7.13), possesses a basis that also obeys the relations (7.35); namely,

$$J_1 = \frac{i}{2} \begin{bmatrix} 0 & 1 \\ 1 & 0 \end{bmatrix}, \quad J_2 = -\frac{i}{2} \begin{bmatrix} 0 & -i \\ i & 0 \end{bmatrix}, \quad J_3 = \frac{i}{2} \begin{bmatrix} 1 & 0 \\ 0 & -1 \end{bmatrix} \qquad (7.39)$$

For (4.41) says that the two groups SO(3) and SU(2) are isomorphic with each other in neighborhoods of the two identity matrices. The three matrices (7.39), without the coefficients $i/2$ and $-i/2$, are called the **Pauli spin** matrices[4].

We summarize our results as follows:

Theorem 7.9 *The Lie algebras of* SO(3), SU(2), *and the Lie algebra of the real 3-dimensional vector space* $\mathbf{R}^3$ *with the cross product as the composition law are all isomorphic with one another.*

Finally, one more remark regarding the *dimension of the Lie algebra $\mathcal{L}$ of a continuous matrix group G: It equals the number f of free real parameters of G.* This is easily verified through the groups of Table (4.3) or (7.32). For example, the skew-hermitian matrices with trace zero have

$$2\underbrace{[(n-1)+(n-2)+\ldots+1]}_{\substack{\text{off-diagonal elements} \\ \text{(complex)}}}+\underbrace{n-1}_{\substack{\text{diagonal elements} \\ \text{(purely imaginary, trace 0)}}}=n^2-1$$

real parameters. But this is just the dimension of the Lie algebra of SU(n).

7.3 Representation of Lie Algebras

Earlier in Section 1.4, a realization of a group by linear transformations was called a representation of this group.

Now a realization of a Lie algebra $\mathcal{L}$ by linear transformations shall be called a representation of $\mathcal{L}$. Take (7.36) as an example. More precisely, let $\mathcal{L}$ be an arbitrary Lie algebra with the typical elements J and K. By a **representation** θ **of** $\mathcal{L}$ we mean a map

$$J \rightarrow J'$$

assigning to each element J of $\mathcal{L}$ a linear transformation J' of an n-dimensional vector space V (an $n \times n$ matrix, on introducing a basis in V) such that the following holds:

$$(J+K)' = J'+K' \quad \text{and} \quad (rJ)' = rJ', r \in \mathbf{R} \qquad (7.40)$$
$$[J,K]' = [J',K'] = J'K'-K'J' \qquad (7.41)$$

V is called the **representation space** and n the **dimension** of θ.

Remarks.

(a) θ is completely determined, once the linear transformations J'_k associated with a basis J_k of $\mathcal{L}$ are known.

(b) The dimension of the Lie algebra $\mathcal{L}$ is not related with the dimension of θ.

---

[4]Pauli, Wolfgang (1900-1958). Theoretical physicist. Studied in Munich and other places. From 1929 on professor at ETH, Zürich. Pioneered in the development of quantum mechanics (Pauli exclusion principle). Nobel prize in 1945.

(c) θ describes a homomorphism with respect to the structure of the Lie algebra $\mathcal{L}$.

(d) The J' may be singular.

The following theorem is fundamental:

Theorem 7.10 *Let G be a continuous group of $n \times n$ matrices A and $\mathcal{L}$ be its Lie algebra with the typical element J. Then to each differentiable[5] representation $\vartheta : A \to A'$ of G with representation space V we can construct an associated representation θ of $\mathcal{L}$ with the same representation space V in the following manner:*

$$\theta : J = \left.\frac{dA(t)}{dt}\right|_{t=0} \longrightarrow J' = \left.\frac{dA'(t)}{dt}\right|_{t=0} \tag{7.42}$$

Here $A(t)$ is an arbitrary one-parameter family of G and $A'(t)$ is its representing one-parameter family of linear transformations.

Proof. Properties (7.40), (7.41) are to be satisfied by the correspondence given in (7.42). Let $A(t), B(t)$ be any two one-parameter families in G with the infinitesimal operators J, K and $A'(t), B'(t)$ be the representing families with the infinitesimal operators J', K'.

The representation property of ϑ yields

$$A(t)B(t)A(t)^{-1}B(t)^{-1} \xrightarrow{\vartheta} A'(t)B'(t)A'(t)^{-1}B'(t)^{-1} \tag{7.43}$$

From the derivation of Theorem 7.8 we know that the one-parameter family on the left of (7.43) with the new parameter $\tau = t^2$ has the infinitesimal operator $[J, K]$, and correspondingly, the infinitesimal operator on the right hand side of (7.43) is $[J', K']$. Hence by (7.42), property (7.41) is satisfied.

Similarly, (7.40) is proven from the derivation of Theorem 7.7. $\square$

A conclusion from the above is that the infinitesimal operators of the representing group also form a Lie algebra.

Let us now illustrate Theorem 7.10 through the group $G = \mathrm{SU}(2)$. Its Lie algebra consists of all matrices J of the form (7.13). Let now $A(t)$ be a one-parameter family associated with J. Then

$$A(t) = E + \begin{bmatrix} \alpha & \beta \\ -\bar{\beta} & -\alpha \end{bmatrix} t + \ldots, \qquad A(t)^{-1} = E + \begin{bmatrix} -\alpha & -\beta \\ \bar{\beta} & \alpha \end{bmatrix} t + \ldots \tag{7.44}$$

Now we construct the representations θ_ℓ of the Lie algebra of $\mathrm{SU}(2)$ associated with the unitary representations $\vartheta_\ell : A(t) \to A'(t)$ of $\mathrm{SU}(2)$ described

---

[5]ϑ is differentiable means that the set of representing matrices $A'(t)$ of each one-parameter family $A(t)$ in G is again a one-parameter family (and hence differentiable).

in Section 4.3. According to the transplantation rule (4.47), the basis functions (4.83) are transformed by ϑ_ℓ as follows:

$$X_\lambda = \frac{x_1^{\ell+\lambda} x_2^{\ell-\lambda}}{\sqrt{(\ell+\lambda)!(\ell-\lambda)!}} \xrightarrow{A'(t)} X_\lambda' = \frac{x_1'^{\ell+\lambda} x_2'^{\ell-\lambda}}{\sqrt{(\ell+\lambda)!(\ell-\lambda)!}} \qquad \lambda = \ell, \ell-1, \ldots, -\ell$$

$$\text{where} \quad \begin{bmatrix} x_1' \\ x_2' \end{bmatrix} = A(t)^{-1} \begin{bmatrix} x_1 \\ x_2 \end{bmatrix} = \begin{bmatrix} x_1 - t(\alpha x_1 + \beta x_2) + \ldots \\ x_2 + t(\bar\beta x_1 + \alpha x_2) + \ldots \end{bmatrix}$$

$$(7.45)$$

For convenience, we set $k = \ell + \lambda$, $j = \ell - \lambda$ and $N = \sqrt{(\ell+\lambda)!(\ell-\lambda)!}$. Hence by (7.45), we have

$$X_\lambda' = \frac{1}{N}[x_1 - t(\alpha x_1 + \beta x_2) + \ldots]^k [x_2 + t(\bar\beta x_1 + \alpha x_2) + \ldots]^j$$

$$= \frac{1}{N}[x_1^k - tkx_1^{k-1}(\alpha x_1 + \beta x_2) + t^2 \ldots][x_2^j + tjx_2^{j-1}(\bar\beta x_1 + \alpha x_2) + t^2 \ldots]$$

$$= \frac{1}{N}[x_1^k x_2^j - tkx_1^{k-1} x_2^j(\alpha x_1 + \beta x_2) + tjx_1^k x_2^{j-1}(\bar\beta x_1 + \alpha x_2) + t^2 ..] \quad (7.46)$$

Differentiating (7.46) with respect to t at $t = 0$ yields the function

$$\begin{aligned} W_\lambda &= \frac{1}{N}(-k\alpha x_1^k x_2^j - ki\beta x_1^{k-1} x_2^{j+1} + j\bar\beta x_1^{k+1} x_2^{j-1} + j\alpha x_1^k x_2^j) \\ &= \alpha(j-k)\frac{x_1^k x_2^j}{\sqrt{k!j!}} - \beta\sqrt{k}\sqrt{j+1}\cdot\frac{x_1^{k-1} x_2^{j+1}}{\sqrt{(k-1)!(j+1)!}} \\ &\quad +\bar\beta\sqrt{j}\sqrt{k+1}\cdot\frac{x_1^{k+1} x_2^{j-1}}{\sqrt{(k+1)!(j-1)!}} \end{aligned}$$

From
$$\sqrt{k}\sqrt{j+1} = \sqrt{\ell(\ell+1) - \lambda(\lambda-1)}$$
and
$$\sqrt{j}\sqrt{k+1} = \sqrt{\ell(\ell+1) - \lambda(\lambda+1)}$$
we obtain

$$X_\lambda \xrightarrow{J'} W_\lambda = \bar\beta\sqrt{\ell(\ell+1) - \lambda(\lambda+1)}X_{\lambda+1}$$
$$-\alpha 2\lambda X_\lambda - \beta\sqrt{\ell(\ell+1) - \lambda(\lambda-1)}X_{\lambda-1} \quad (7.47)$$

In particular, for the basis (7.39), Equation (7.13) implies

$$J_3(\alpha = \frac{i}{2}, \beta = 0): \quad X_\lambda \xrightarrow{J_3'} W_\lambda = \frac{i}{2}(-2\lambda)X_\lambda = \frac{\lambda}{i}X_\lambda$$

$$J_2(\alpha = 0, \beta = -\frac{1}{2}): \quad X_\lambda \xrightarrow{J_2'} W_\lambda = \frac{1}{2}[-\sqrt{\ell(\ell+1) - \lambda(\lambda+1)}X_{\lambda+1}$$
$$+\sqrt{\ell(\ell+1) - \lambda(\lambda-1)}X_{\lambda-1}]$$

$$J_1(\alpha = 0, \beta = \frac{i}{2}): \quad X_\lambda \xrightarrow{J_1'} W_\lambda = -\frac{i}{2}[\sqrt{\ell(\ell+1) - \lambda(\lambda+1)}X_{\lambda+1}$$
$$+\sqrt{\ell(\ell+1) - \lambda(\lambda-1)}X_{\lambda-1}]$$

Hence relative to the basis X_λ of the representation space, we obtain the following representing $(2\ell + 1) \times (2\ell + 1)$ matrices:

$$J_1' = \frac{1}{2i} \begin{bmatrix} 0 & d_\ell & & & O \\ d_\ell & 0 & d_{\ell-1} & & \\ & d_{\ell-1} & \ddots & \ddots & \\ & & \ddots & \ddots & d_{-\ell+1} \\ O & & & d_{-\ell+1} & 0 \end{bmatrix}$$

$$J_2' = \frac{1}{2} \begin{bmatrix} 0 & -d_\ell & & & O \\ d_\ell & 0 & -d_{\ell-1} & & \\ & d_{\ell-1} & \ddots & \ddots & \\ & & \ddots & \ddots & -d_{-\ell+1} \\ O & & & d_{-\ell+1} & 0 \end{bmatrix} \qquad (7.48)$$

$$J_3' = \frac{1}{i} \begin{bmatrix} \ell & & & O \\ & \ell-1 & & \\ & & \ddots & \\ O & & & -\ell \end{bmatrix}$$

where $d_\lambda = \sqrt{\ell(\ell+1) - \lambda(\lambda-1)}$, $\lambda = \ell, \ell-1, \ldots, -\ell+1$.
All three matrices are skew-hermitian with vanishing trace. They span a 3-dimensional **Lie subalgebra** of the Lie algebra of $SU(2\ell+1)$.

The representing matrices of J_1, J_2, J_3 for θ_1 and $\theta_{1/2}$ are

$$\left. \begin{array}{l} J_1' = \frac{1}{2i} \begin{bmatrix} 0 & \sqrt{2} & 0 \\ \sqrt{2} & 0 & \sqrt{2} \\ 0 & \sqrt{2} & 0 \end{bmatrix} \quad J_2' = \frac{1}{2} \begin{bmatrix} 0 & -\sqrt{2} & 0 \\ \sqrt{2} & 0 & -\sqrt{2} \\ 0 & \sqrt{2} & 0 \end{bmatrix} \\[2em] J_3' = \frac{1}{i} \begin{bmatrix} 1 & 0 & 0 \\ 0 & 0 & 0 \\ 0 & 0 & -1 \end{bmatrix} \end{array} \right\} \qquad (7.49)$$

$$J_1' = \frac{1}{2i} \begin{bmatrix} 0 & 1 \\ 1 & 0 \end{bmatrix} \quad J_2' = \frac{1}{2} \begin{bmatrix} 0 & -1 \\ 1 & 0 \end{bmatrix} \quad J_3' = \frac{1}{i} \begin{bmatrix} 1/2 & 0 \\ 0 & -1/2 \end{bmatrix} \qquad (7.50)$$

Thus (7.50) is the representation of the matrices J by their complex conjugates. Now the converse of Theorem 7.10 also holds:

Theorem 7.11 *Let $\mathcal{L}$ be the Lie algebra of some continuous matrix group G and $\theta : J \to J'$ be a representation of $\mathcal{L}$ by matrices J'. Then there exists a representation $\vartheta : A \to A'$ of G (defined at least in a neighborhood of the identity matrix of G) by matrices A' such that*

$$J = \left. \frac{dA(t)}{dt} \right|_{t=0} \quad \text{implies} \quad J' = \left. \frac{dA'(t)}{dt} \right|_{t=0}$$

The construction formula for this representation is

$$e^{tJ} \longrightarrow e^{tJ'} \qquad (7.51)$$

Outline of proof. The proof is based on the BCH-formula[6], which evaluates the product $e^{tJ} \cdot e^{tK}$ for two matrices J, K of a Lie algebra. The formula is

$$e^{tJ} \cdot e^{tK} = e^{tH(t)} \qquad (7.52)$$

where

$$H(t) = (J + K) + \frac{1}{2}[J, K]t + \frac{1}{12}\{[[J, K], K] + [[K, J], J]\}t^2 + \ldots \quad (7.53)$$

Equation (7.53) can be deduced from (7.52) by expanding the exponential series on the left and by writing $H(t)$ as a power series in t with unknown coefficients (= matrices) which are then determined by equating like terms. We recommend that the reader work this out as an exercise at least for the first two terms in (7.53).

One of the fundamental things to know is that *the coefficients of all higher order terms with* $t^3, t^4, \ldots$ *of the series* (7.53) *are also linear combinations of repeated commutators of* J *and* K. As $H(t)$ converges at least in a neighborhood of $t = 0$, this implies that for small $|t|$ the matrices $H(t)$ also belong to the Lie algebra $\mathcal{L}$.

Now let a representation θ of $\mathcal{L}$ be given, for example, by

$$J \to J', \qquad K \to K', \qquad H \to H' \qquad (7.54)$$

where H is the matrix described in (7.53).

As in (7.51) we construct the correspondence

$$A = e^{tJ} \to A' = e^{tJ'}, \;\; B = e^{tK} \to B' = e^{tK'}, \;\; C = e^{tH} \to C' = e^{tH'} \quad (7.55)$$

We have

$$A \cdot B = C \qquad (7.56)$$

On the other hand, by the BCH-formula (7.52), (7.53),

$$A'B' = e^{tM} \text{ where } M = J' + K' + \frac{1}{2}[J', K']t + \ldots \qquad (7.57)$$

The representation properties (7.40), (7.41) of θ guarantee that

$$M = \{J + K + \frac{1}{2}[J, K]t + \ldots\}' = H' \qquad (7.58)$$

Here we used the fact that all terms in the BCH-formula are repeated commutators and hence belong to $\mathcal{L}$. Thus $A'B' = C'$, whence $(AB)' = A'B'$.

To complete the proof we use the fact that every matrix in a sufficiently small neighborhood of the identity matrix can be expressed by the exponential map and that the map $U \to e^U$ is continuous (for all U in a neighborhood of the null matrix (Chevalley, 1946)).   □

Now we state a theorem important for determining invariant subspaces:

---

[6]**Baker-Campbell-Hausdorff-formula.** For references see (Magnus, Karrass, Solitar, 1966, Ch. 5), which also treats convergence problems.

Theorem 7.12 *Let ϑ be a differentiable representation of a connected continuous matrix group G and θ be the corresponding representation of the Lie algebra $\mathcal{L}$ of G defined by (7.42). Then the invariant subspaces of the two representations are identical.*

Sketch of proof.

(a) That an invariant subspace of ϑ is also an invariant subspace of θ is an immediate consequence of (7.42), because in a suitable coordinate system possible zero blocks of the representing matrices are preserved under differentiation.

(b) It remains to show that an invariant subspace of θ is also an invariant subspace of ϑ. To this end, we construct the representation (7.51) from θ via the exponential map. Moreover, the following statement is true:

> For a connected group, the set of image matrices $e^{J't}$ consists of all representing matrices of ϑ in a neighborhood U of the identity matrix, and every representing matrix can be written as a finite product of matrices of U. $\left.\right\} (\alpha)$

If in an appropriate coordinate system the J' contain zero blocks, then so do the $e^{J't}$ and all of the representing matrices of ϑ. $\qquad\square$

Remarks.

1. Theorem 7.12, together with Example 5.11, implies that the representation θ_ℓ of the Lie algebra of SU(2) in (7.48) is irreducible.

2. Since the representing matrices (7.48) of θ_ℓ are skew-hermitian and have trace zero, it follows from (7.16), (7.19) and statement (α) that the corresponding representations ϑ_ℓ of SU(2), which in Section 4.4 were constructed via monomials (4.83), are described by **unitary** matrices.

3. Connectedness in Theorem 7.12 is essential, as is exemplified by the following representation ϑ of the group O(2):

$$\begin{bmatrix} \cos\varphi & -\sin\varphi \\ \sin\varphi & \cos\varphi \end{bmatrix} \xrightarrow{\vartheta} \begin{bmatrix} e^{i\varphi} & 0 \\ 0 & e^{-i\varphi} \end{bmatrix}$$

$$\begin{bmatrix} \cos\mu & \sin\mu \\ \sin\mu & -\cos\mu \end{bmatrix} \xrightarrow{\vartheta} \begin{bmatrix} 0 & e^{i\mu} \\ e^{-i\mu} & 0 \end{bmatrix}$$

This representation, unlike the associated representation θ, is irreducible.

An important result in this section obviously is that the representation theory of a connected continuous matrix group can be reduced to that of its Lie algebra, and vice versa. The representation theory of Lie algebras can largely be developed without making use of the concept of continuous matrix groups.

Finally, we show an **application of Schur's lemma** to Lie algebras. We consider an arbitrary representation θ of the Lie algebra of SO(3) or SU(2). For the basis (7.38), here denoted by J_1, J_2, J_3, the following structure relations are valid:

$$[J_1, J_2] = J_3, \qquad [J_2, J_3] = J_1, \qquad [J_3, J_1] = J_2 \qquad (7.59)$$

By (7.40), (7.41) the same relations hold for the image matrices J_k' of θ:

$$[J_1', J_2'] = J_3', \qquad [J_2', J_3'] = J_1', \qquad [J_3', J_1'] = J_2' \qquad (7.60)$$

Now we construct the matrix

$$C = J_1'^2 + J_2'^2 + J_3'^2 \qquad (7.61)$$

The following holds:

$$CJ_k' = J_k'C \qquad k = 1, 2, 3 \qquad (7.62)$$

We prove this only for $k = 1$: The relations (7.60) imply

$$J_2'^2 J_1' = J_2'(J_2'J_1') = J_2'(J_1'J_2' - J_3') = J_2'J_1'J_2' - J_2'J_3'$$

Similarly,

$$J_3'^2 J_1' = J_3'J_1'J_3' + J_3'J_2'$$
$$J_1'J_2'^2 = J_2'J_1'J_2' + J_3'J_2'$$
$$J_1'J_3'^2 = J_3'J_1'J_3' - J_2'J_3'$$

All four of these equations imply $(J_2'^2 + J_3'^2) \cdot J_1' - J_1'(J_2'^2 + J_3'^2) = 0$, thus proving $CJ_1' = J_1'C$. In fact, C commutes with any representing matrix $J' = u_1J_1' + u_2J_2' + u_3J_3'$ of θ.

More generally, one defines the following:

Let $\theta : J \to J'$ be the representation of an arbitrary Lie algebra and let C be a linear operator in the representation space commuting with each of the J'. Then C is said to be a **Casimir operator**.

In general, if the representation θ of a d-dimensional Lie algebra $\mathcal{L}$ is irreducible, i.e., if there is no proper subspace that is simultaneously invariant under the representing linear transformations $J_1', J_2', \ldots, J_d'$ of the basis $J_1, J_2, \ldots, J_d$ of $\mathcal{L}$, then C is a multiple of the identity. The reason is that Theorem 2.2 is not only valid for multiplicative matrix groups, but also more generally for **irreducible matrix systems**. (A remark concerning this was appended to the end of the proof of Schur's lemma in Section 2.1.)

We now determine the eigenvalue λ in our example (7.61) for the case where θ equals the representation θ_ℓ described in (7.48). Equations (7.48) and (7.61) imply that the matrix entry of C at the upper left-hand corner is

$$-\frac{1}{2}\ell - \frac{1}{2}\ell - \ell^2 = -\ell(\ell+1) \qquad (7.63)$$

This is the eigenvalue λ.

7.4 Representations of SU(2) and SO(3)

By virtue of Theorem 7.12, the analysis of a representation ϑ of a connected continuous matrix group can be extended to that of the associated representation θ of its Lie algebra. The same holds for the Cartan methods (see Sections 4.3, 4.4, 4.5, and Chapter 9). As an example we may quote the Clebsch-Gordan coefficients in Section 4.4. Then something similar to that done to finite groups in Section 5.2 is achieved. We demonstrate this through the group SU(2) and thereby automatically include the treatment of SO(3). Its Lie algebra is a 3-dimensional real vector space spanned by the vectors (7.39). They satisfy the structure relations

$$[J_1, J_2] = J_3, \qquad [J_2, J_3] = J_1, \qquad [J_3, J_1] = J_2 \qquad (7.64)$$

Now we again consider the unitary irreducible representation ϑ_ℓ described in Section 4.3. The associated representation θ_ℓ of the Lie algebra is

$$J_k \to J_k' \qquad k = 1, 2, 3 \qquad (7.65)$$

with the matrices J_k' of (7.48).
The matrix

$$L' = i \cdot J_3' \qquad (7.66)$$

has the diagonal elements

$$\ell, \ell - 1, \ldots, -\ell \qquad (7.67)$$

which coincide with the **weights** of ϑ_ℓ (see end of Section 4.3). Thus the weights are identical with the eigenvalues of L'.
If we denote by

$$e_\ell, e_{\ell-1}, \ldots, e_{-\ell} \qquad (7.68)$$

the basis vectors of the representation space (which at the same time are eigenvectors of L'), then e_λ belongs to the weight λ and

$$L'e_\lambda = \lambda \cdot e_\lambda \qquad (7.69)$$

We construct the so-called **lowering operator**[7]

$$L'_- = i \cdot J_1' + J_2' \qquad (7.70)$$

---

[7]Note that the lowering operator (7.70) does not belong to the Lie algebra, which by Theorem 7.7 consists of only the real linear combinations of the J_k'.

whose matrix by (7.48) has nonzero entries only in the lower minor diagonal. More explicitly,

$$L'_-e_\lambda = \begin{cases} d_\lambda \cdot e_{\lambda-1} & \text{if } \lambda \neq -\ell \\ 0 & \text{otherwise} \end{cases} \tag{7.71}$$

with d_λ from (7.48).

By successively applying L'_- to e_λ **one generates the sequence of vectors** (7.68) from left to right, i.e., by decreasing weight subscripts λ.
    Similarly,

$$L'_+ = i \cdot J'_1 - J'_2 \tag{7.72}$$

is the **raising operator**, because

$$L'_+e_\lambda = \begin{cases} d_{\lambda+1} \cdot e_{\lambda+1} & \text{if } \lambda \neq \ell \\ 0 & \text{otherwise} \end{cases} \tag{7.73}$$

Let an arbitrary representation

$$\vartheta = \bigoplus_j c_j\vartheta_j \quad \text{or} \quad \theta = \bigoplus_j c_j\theta_j \tag{7.74}$$

of SU(2) (SO(3)) or the Lie algebra $\mathcal{L}$ of SU(2) (SO(3)) be given with the irreducible inequivalent constituents ϑ_j or θ_j and the multiplicities $c_j \in \{1, 2, \ldots\}$.

The above considerations result in the following

Algorithm. Generating symmetry adapted basis vectors of a representation ϑ of SU(2) (SO(3)) or of a representation θ of the Lie algebra $\mathcal{L}$ of SU(2) (SO(3)).

   1. *Let*
$$J'_1, J'_2, J'_3$$
   be the representing matrices corresponding to a basis J_1, J_2, J_3 of the Lie algebra $\mathcal{L}$.

   2. *J'_3 is diagonalizable and assumed diagonal in the following. Then $L' = iJ'_3$ is diagonal with the weights of θ in the diagonal. To each basis vector of the representation space V of θ there corresponds a weight. However, a given weight λ may correspond to several basis vectors; the latter spanning the eigenspace of L' associated with the eigenvalue λ.*

   3. *Utilizing the weights of θ and the algorithm at the end of Section 4.3 (Cartan's Method), one can determine the irreducible representations appearing in θ and their multiplicities. Let θ_ℓ be such a constituent (ℓ need not be the dominant weight of θ).*

4. *Determine an eigenvector v of $L' = iJ'_3$ corresponding to the eigenvalue ℓ and satisfying*

$$L'_+ v = 0 \qquad where \qquad L'_+ = i \cdot J'_1 - J'_2 \qquad (7.75)$$

The vector v already belongs to an irreducible subspace transformed after the manner of θ_ℓ.

5. *The remaining basis vectors of the irreducible subspace are obtained by successively applying the lowering operator $L'_- = iJ'_1 + J'_2$ to the initial vector v.*

6. *If θ_ℓ appears in θ repeatedly ($c_\ell > 1$), then a total of c_ℓ linearly independent solution vectors v of (7.75) are to be determined. This is possible since the solution space is exactly c_ℓ-dimensional. Each of these c_ℓ solution vectors is an initial vector for the lowering operator, so that one correctly obtains c_ℓ invariant irreducible subspaces, each of which is transformed after the manner of θ_ℓ.*

7. *Steps 4, 5, 6 are to be performed for each of the θ_j appearing in θ.*

Remarks.

1. If ℓ is the dominant weight of θ, then all vectors of the eigenspace of L' corresponding to this eigenvalue ℓ satisfy (7.75).

2. This method also applies to SU(n) and other connected continuous groups (see Section 9.3).

3. This algorithm is comparable to the algorithm "Generating symmetry adapted basis vectors" in Section 5.2. Whatever is afforded by the lowering operator here is achieved by the basis generators $P_{1\mu}^{(\ell)}$ there. Furthermore, solving (7.75) corresponds to Step I in (5.44).

Why are the basis vectors constructed above symmetry adapted? The c_ℓ small boxes inside each of the matrices L'_+, L'_-, L', which correspond to the c_ℓ irreducible invariant subspaces of ϑ_ℓ, are identical: for $L'_-(L'_+)$ they consist of a single lower (upper) minor diagonal[8] given by (7.71) or (7.73), respectively.

A small box inside L' consists of the diagonal $\ell, \ell - 1, \ldots, -\ell$. Because

$$J'_1 = \frac{1}{2i}(L_- + L'_+), \qquad J'_2 = \frac{1}{2}(L_- - L'_+), \qquad J'_3 = \frac{1}{i}L'$$

the representing matrices of the basis J_1, J_2, J_3 of the Lie algebra of SU(2) (SO(3)) also have c_ℓ identical small boxes describing the transformation of the irreducible invariant subspaces of θ_ℓ. The same is true for the matrices $e^{tJ'_k}(k = 1, 2, 3)$. Consequently, by Theorem 7.12, this shows that the basis obtained is symmetry adapted with respect to both the representation of the Lie algebra of SU(2) (SO(3)) and the corresponding representation ϑ of the group SU(2) (SO(3)).

---

[8] All other elements are zero.

7.4.1 Infinitesimal Aspects of the Kronecker Product

We consider the Kronecker product[9] $\vartheta_1 \otimes \vartheta_2$ of the two representations ϑ_1, ϑ_2 of a continuous matrix group G. Let m and n be the dimensions of ϑ_1 and ϑ_2. Let further $D_1(s) = (d^{(1)}_{\lambda\nu}(s))$, $D_2(s) = (d^{(2)}_{\mu\kappa}(s))$ be the representing matrices of the group element s relative to the set of basis vectors $e^{(1)}_i, e^{(2)}_j$ in the two representation spaces. Then the transformations are given by

$$e^{(1)}_i \to \sum_{\lambda=1}^m d^{(1)}_{\lambda i}(s) e^{(1)}_\lambda, \qquad e^{(2)}_k \to \sum_{\mu=1}^n d^{(2)}_{\mu k}(s) e^{(2)}_\mu \qquad (7.76)$$

The Kronecker product transforms the $m \cdot n$ basis vectors $(e^{(1)}_i \cdot e^{(2)}_k)$ as follows:

$$(e^{(1)}_i \cdot e^{(2)}_k) \to \sum_{\lambda,\mu} d^{(1)}_{\lambda i}(s) d^{(2)}_{\mu k}(s) (e^{(1)}_\lambda \cdot e^{(2)}_\mu) \qquad (7.77)$$

Now we consider the infinitesimal operators. In other words, we let the group element s vary in a one-parameter family $s(t)$, where $s(0)$ is assigned to the corresponding identity matrix. With

$$U = (u_{\lambda\nu}) = \left. \frac{dD_1(s(t))}{dt} \right|_{t=0}, \qquad V = (v_{\mu\kappa}) = \left. \frac{dD_2(s(t))}{dt} \right|_{t=0}$$

the two representing infinitesimal operators of the transformations (7.76) for the element $J = \left. \frac{ds(t)}{dt} \right|_{t=0}$ of the Lie algebra of G are

$$e^{(1)}_i \to e^{(1)'}_i = \sum_{\lambda=1}^m u_{\lambda i} e^{(1)}_\lambda, \qquad e^{(2)}_k \to e^{(2)'}_k = \sum_{\mu=1}^n v_{\mu k} e^{(2)}_\mu \qquad (7.78)$$

Because

$$\left. \frac{d}{dt}(d^{(1)}_{\lambda i} d^{(2)}_{\mu k}) \right|_{t=0} = u_{\lambda i} \delta_{\mu k} + \delta_{\lambda i} v_{\mu k}$$

(δ denotes the Kronecker symbol), the representing infinitesimal operator J' of the Kronecker product for the element J is

$$(e^{(1)}_i \cdot e^{(2)}_k) \to (e^{(1)}_i \cdot e^{(2)}_k)' = \sum_{\lambda,\mu} (u_{\lambda i} \delta_{\mu k} + \delta_{\lambda i} v_{\mu k})(e^{(1)}_\lambda \cdot e^{(2)}_\mu) \qquad (7.79)$$

On combining this with (7.78), one finally finds the equation

$$(e^{(1)}_i \cdot e^{(2)}_k)' = (e^{(1)'}_i \cdot e^{(2)}_k) + (e^{(1)}_i \cdot e^{(2)'}_k) \qquad (7.80)$$

which may be conveniently remembered as a generalization of the product rule of differentiation. In (7.79) and (7.80), notice that the infinitesimal

---

[9] See Section 1.10.

operators of the Kronecker product can be described by the infinitesimal operators of each of the individual representations.

If we choose the order of the basis vectors $(e_i^{(1)} \cdot e_k^{(2)})$ in a way such that the pairs of subscripts (i, k) are arranged as follows

$$(1,1), (1,2), \ldots, (1,n); (2,1), (2,2), \ldots, (2,n); \ldots, (m,n)$$

then $(7.79)^{10}$ implies

$$J' = U \otimes E_n \oplus E_m \otimes V = \begin{bmatrix} u_{11}E_n + V & u_{12}E_n & \cdots & u_{1m}E_n \\ u_{21}E_n & u_{22}E_n + V & & \vdots \\ \vdots & u_{32}E_n & & \vdots \\ \vdots & \vdots & & u_{m-1m}E_n \\ u_{m1}E_n & u_{m2}E_n & \cdots & u_{mm}E_n + V \end{bmatrix}$$
(7.81)

7.4.2 Clebsch-Gordan Coefficients

Our algorithm can be employed to compute the Clebsch-Gordan coefficients. We take the complete reduction of the Kronecker product $\vartheta = \vartheta_1 \otimes \vartheta_{1/2}$ of SU(2) (SO(3)) as an example. The subspace transformed after the manner of $\vartheta_{1/2}$ was already found at the end of Section 4.4.2, from Tables (4.86) and (4.87). Retaining the notation used there, the basis vectors in the representation space of ϑ_1 and $\vartheta_{1/2}$ are the monomials

$$X_1 = \frac{x_1^2}{\sqrt{2}}, \quad X_0 = x_1 x_2, \quad X_{-1} = \frac{x_2^2}{\sqrt{2}}, \quad Y_{1/2} = y_1, \quad Y_{-1/2} = y_2$$

The representing matrices of the basis J_1, J_2, J_3 of the Lie algebra of SU(2) for ϑ_1 and $\vartheta_{1/2}$ are explicitly available in (7.49) and (7.50). (Here they play the roles of U and V).

We now perform the algorithm for $\vartheta_1 \otimes \vartheta_{1/2}$ step by step:

1. If the basis vectors for the Kronecker product are chosen in the order

$$X_1 Y_{1/2}, \quad X_1 Y_{-1/2}, \quad X_0 Y_{1/2}, \quad X_0 Y_{-1/2}, \quad X_{-1} Y_{1/2}, \quad X_{-1} Y_{-1/2} \quad (7.82)$$

then by (7.81) and with $w = \sqrt{2}$ the corresponding representing matrices J_1', J_2', J_3' of the Kronecker product are

$$J_1' = \frac{1}{2i} \begin{bmatrix} & 1 & w & & & \\ 1 & & & w & & \\ w & & & 1 & w & \\ & w & 1 & & & w \\ & & w & & & 1 \\ & & & w & 1 & \end{bmatrix} \quad J_2' = \frac{1}{2} \begin{bmatrix} & -1 & -w & & & \\ 1 & & & -w & & \\ w & & & -1 & -w & \\ & w & 1 & & & -w \\ & & w & & & -1 \\ & & & w & 1 & \end{bmatrix} \quad J_3' = \frac{1}{i} \begin{bmatrix} \frac{3}{2} & & & & & \\ & \frac{1}{2} & & & & \\ & & \frac{1}{2} & & & \\ & & & -\frac{1}{2} & & \\ & & & & -\frac{1}{2} & \\ & & & & & -\frac{3}{2} \end{bmatrix}$$
(7.83)

---

^{10}Equation (7.81) can also be obtained from the matrices described in (1.65): $J' = \frac{dD(s(t))}{dt}\Big|_{t=0}$.

All elements not written are zeros.

2. J_3' is already diagonal.

3. Cartan's method yields $\vartheta = \vartheta_{3/2} \oplus \vartheta_{1/2}$. We limit our discussion to finding the subspace that remains invariant under ϑ and corresponds to $\vartheta_{1/2}$.

4. From J_3' in (7.83), we see that every eigenvector of L' corresponding to the eigenvalue $1/2$ must be a linear combination of the second and the third basis functions of (7.82); i.e.,

$$v = \alpha(X_0 Y_{1/2}) + \beta(X_1 Y_{-1/2}) \tag{7.84}$$

Equation (7.83) also furnishes the lowering and the raising operators

$$
L_+' = iJ_1' - J_2' =
\begin{bmatrix}
1 & \sqrt{2} & & \\
 & \sqrt{2} & & \\
 & & 1 & \sqrt{2} \\
 & & & \sqrt{2} \\
 & & & 1
\end{bmatrix}
\qquad
L_-' = iJ_1' + J_2' =
\begin{bmatrix}
1 & & & \\
\sqrt{2} & & & \\
 & \sqrt{2} & 1 & \\
 & & \sqrt{2} & \\
 & & \sqrt{2} & 1
\end{bmatrix}
\tag{7.85}
$$

Thus by (7.84) and (7.85), Equation (7.75) reads

$$
\begin{aligned}
L_+' v &= \alpha L_+'(X_0 Y_{1/2}) + \beta L_+'(X_1 Y_{-1/2}) \\
&= \alpha\sqrt{2}X_1 Y_{1/2} + \beta X_1 Y_{1/2} = 0
\end{aligned}
$$

Hence if, for example, $\alpha = -1$, $\beta = \sqrt{2}$, then with the notation of (4.88), Equation (7.84) becomes

$$v = -X_0 Y_{1/2} + \sqrt{2}X_1 Y_{-1/2} = Z_{1/2} \tag{7.86}$$

5. By (7.85), the lowering operator L_-' transforms (7.86) into

$$L_-'(Z_{1/2}) = X_0 Y_{-1/2} - \sqrt{2}X_{-1}Y_{1/2} = Z_{-1/2} \tag{7.87}$$

which is consistent with the two results at the end of Section 4.4.2.

7.5 Examples from Quantum Mechanics

In this book we limit ourselves to considering examples of **particles in a central force field**. For more detailed treatments we refer to (Wigner, 1964), (Weyl, 1931), (van der Waerden, 1974).

7.5.1 Energy and Angular Momentum

Let a particle with mass μ have the position vector x and be exposed to the potential $v(x)$. The possible energy values E of the particle at the stationary state $\psi(x)$ are the eigenvalues of the Hamiltonian

$$H = -\frac{\hbar^2}{2\mu}\left(\frac{\partial^2}{\partial x_1^2} + \frac{\partial^2}{\partial x_2^2} + \frac{\partial^2}{\partial x_3^2}\right) + v(x) = -\frac{\hbar^2}{2\mu}\Delta + v \qquad (7.88)$$

Here Δ is the Laplacian and $\hbar = h/2\pi$ (h is Planck's constant). Hence the eigenvalue problem is given by

$$H\psi = E\psi \qquad (7.89)$$

together with some boundary conditions for ψ. If v varies only with the distance from the origin, then in spherical coordinates r, θ, φ, the Schrödinger equation (7.89) can be treated by separation of variables

$$\psi = f(r)Y_\ell(\theta, \varphi) \qquad (7.90)$$

where Y_ℓ is some spherical surface function of ℓth degree (Section 4.4). The Laplacian in spherical coordinates is

$$\Delta = \frac{1}{r^2}\left[\frac{\partial}{\partial r}\left(r^2\frac{\partial}{\partial r}\right) + \Lambda\right] \qquad (7.91)$$

where Λ does not depend on r.

With (7.88), (7.90), and (4.99), Equation (7.89) implies

$$H\psi = -\frac{\hbar^2}{2\mu}\frac{1}{r^2}\left[\frac{d}{dr}\left(r^2\frac{df}{dr}\right) - \ell(\ell+1)f\right]Y_\ell + v(r)fY_\ell = EfY_\ell$$

which reduces to the **radial equation**

$$-\frac{\hbar^2}{2\mu}\frac{1}{r^2}\left[\frac{d}{dr}\left(r^2\frac{df}{dr}\right) - \ell(\ell+1)f\right] + v(r)f = Ef \qquad (7.92)$$

with certain boundary conditions for f.

To every fixed number $\ell \in \{0, 1, 2, \ldots\}$—where ℓ is the **azimuthal or orbital quantum number**—the separation of variables (7.90) yields an eigenvalue problem (7.92) in the form of an ordinary differential equation for $f(r)$ and hence a spectrum of energy values denoted by $E_{n\ell}$ ($n = 0, 1, 2, \ldots$). *Each of these values $E_{n\ell}$ occurs at least $(2\ell+1)$ times*, because if $f(r)$ satisfies (7.92), then the $(2\ell+1)$ linearly independent states[11] $\psi = f(r)Y_\ell^m$, where $m = -\ell, -\ell+1, \ldots, \ell$, satisfy the Schrödinger equation (7.89).

Theorem 4.5 says that under the proper rotations, the spherical surface functions Y_ℓ are transformed according to the representation ϑ_ℓ of SO(3),

---

[11]See (4.74).

and so are the wave functions ψ_ℓ associated with the orbital quantum number ℓ (known as the ℓ state). We shall see that the matrices

$$L_k = i\hbar J'_k \qquad k = 1, 2, 3 \tag{7.93}$$

are the operators of the three angular momentum components of the particle (relative to the basis Y_ℓ^m) in the ℓ state. Here the J'_k are the infinitesimal operators (7.48). The J'_k are skew-hermitian, hence the L_k are **hermitian**; i.e., $L_k^T = \overline{L_k}$.

By the reasoning employed for the Casimir operator C at the end of Section 7.3, the square of the magnitude $L^2 = L_1^2 + L_2^2 + L_3^2$ of the angular momentum is a multiple of the identity matrix. This together with (7.48), (7.63), (7.93) implies: *For a particle in the central field in an ℓ state the third component of the angular momentum has the possible values $\hbar\ell$, $\hbar(\ell - 1)$, $\ldots$, $-\hbar\ell$, and the square of the magnitude $L^2 = L_1^2 + L_2^2 + L_3^2$ has the value $\hbar^2 \ell(\ell + 1)$.*

Both L^2 and L_3 are diagonal matrices, and therefore, they commute with each other; i.e., $L^2 L_3 = L_3 L^2$. Physically, this means that the two quantities — **magnitude and third component** of the angular momentum — are **simultaneously and sharply measurable**.

The above arguments were valid for integral quantum numbers ℓ. But the representations ϑ_ℓ and hence the matrices J'_k and L_k are also defined for half integral ℓ. If in particular $\ell = 1/2$, then from (7.50) and (7.93) we obtain the components of the angular momentum

$$L_1 = \frac{\hbar}{2} \begin{bmatrix} 0 & 1 \\ 1 & 0 \end{bmatrix} \qquad L_2 = \frac{\hbar}{2} \begin{bmatrix} 0 & -i \\ i & 0 \end{bmatrix} \qquad L_3 = \frac{\hbar}{2} \begin{bmatrix} 1 & 0 \\ 0 & -1 \end{bmatrix} \tag{7.94}$$

They describe the **spin angular momentum** of electrons and similar particles.

7.5.2   Spin-Orbit Coupling

Preparatory to the spin-orbit coupling discussion, we consider two particles in the same central field and described by the Hamiltonian operators H_1 and H_2. Let further ψ_1, E_1 and ψ_2, E_2 denote the respective stationary states and energy values of the two particles:

$$H_1 \psi_1 = E_1 \psi_1, \qquad H_2 \psi_2 = E_2 \psi_2$$

If there is no interaction between the two systems (which we shall assume), then the Schrödinger equation for the combined system consisting of the two particles is

$$H_1 \psi + H_2 \psi = E\psi \tag{7.95}$$

The eigenfunctions ψ of the entire system are the products $\psi_1 \psi_2$ and the eigenvalues E are the sums $E_1 + E_2$, because

$$H_1(\psi_1 \psi_2) + H_2(\psi_1 \psi_2) = \psi_2 H_1 \psi_1 + \psi_1 H_2 \psi_2 = (E_1 + E_2)(\psi_1 \psi_2) \tag{7.96}$$

If ψ_1 is an ℓ state, then this eigenfunction of H_1 transforms under the proper spherical rotations after the manner of the representation ϑ_ℓ. The angular momentum is determined up to a factor $i\hbar$ by the three matrices (7.48). Similarly, let ψ_2 be an m state of the other particle. Therefore, *the eigenfunction $\psi_1\psi_2$ of the total system transforms after the manner of the Kronecker product $\vartheta_\ell \otimes \vartheta_m$*, and the total angular momentum (again up to the factor $i\hbar$) is given by the three infinitesimal operators obtained via the rule (7.80), which itself *constitutes the law of addition of angular momenta in quantum mechanics*.

This law also applies to systems consisting not of two particles, but of one particle and its spin.

As an example, let us consider an electron in the p state[12] with its orbit ($\ell = 1$) as a first, and its spin ($m = 1/2$) as a second system. This involves $\vartheta_1 \otimes \vartheta_{1/2}$. Because

$$\vartheta_1 \otimes \vartheta_{1/2} = \vartheta_{3/2} \oplus \vartheta_{1/2} \qquad (7.97)$$

the representing matrices relative to an appropriate coordinate system decompose into a 4×4 and a 2×2 block. When passing to the infinitesimal operators, this block form is preserved.

More precisely, the representing matrices of the basis vectors J_1, J_2, J_3 of the Lie algebra of $SO(3)$ are composed of the two matrices (7.48) corresponding to $\ell = 3/2$ and $\ell = 1/2$.

In this adjusted coordinate system, by (7.61), (7.62), (7.63), (7.93), and Theorem 2.7, the square of the magnitude $L^2 = L_1^2 + L_2^2 + L_3^2$ of the angular momentum is a diagonal matrix with the following entries:

$$\hbar^2 \cdot \frac{3}{2} \cdot \frac{5}{2} = \tfrac{15}{4} \cdot \hbar^2 \qquad \text{(4 times)} \qquad (7.98)$$

$$\text{and } \hbar^2 \cdot \frac{1}{2} \cdot \frac{3}{2} = \tfrac{3}{4} \cdot \hbar^2 \qquad \text{(2 times)} \qquad (7.99)$$

They are the eigenvalues of the square of the magnitude L^2 of the total angular momentum for the spin-orbit coupling.

The eigenspace corresponding to (7.99) is spanned by the two vectors $Z_{1/2}$ and $Z_{-1/2}$ found in (7.86), (7.87). Thus the Clebsch-Gordan coefficients furnish the eigenvectors of L^2.

Finally, let us mention that conversely, from the eigenvalues (7.98), (7.99), we can find the eigenvectors of L^2 and hence the Clebsch-Gordan coefficients by solving linear equations. The reason is that the matrix

$$L^2 = -\hbar^2(J_1'^2 + J_2'^2 + J_3'^2)$$

where the J_k' are as given in (7.83), commutes with the representing matrices of $\vartheta_1 \otimes \vartheta_{1/2}$ in the original coordinate system, so that Theorem 2.1 applies.

Remark. If a "small" interaction is present, then the computations done

---

[12] Terminology used in physics.

here serve as a basis for the treatment by perturbation theory. For example, a 4-fold eigenvalue may split into 4 possibly distinct, closely spaced values. In Section 3.4, we gave an outline of the perturbation method in the case of boundary value problems.

PROBLEMS

Problem 7.1 To find the one-parameter group $A(t)$ corresponding to an infinitesimal operator J (where J is an $n \times n$ matrix), one uses an alternative interpretation of the exponential map $A(t) = e^{tJ}$; namely, $A(t)$ is the solution of the system of linear differential equations $dA/dt = J \cdot A$ with constant coefficients under the initial conditions $A(0) = E$. Hence there is a total of n^2 equations. Treat

$$J = \begin{bmatrix} 1 & 1 \\ 0 & 1 \end{bmatrix}$$

under this aspect, and verify that the $A(t)$ form a group.

Problem 7.2 The exponential series

$$A(t) = e^{tJ} = E + tJ + \frac{t^2}{2!}J^2 + \frac{t^3}{3!}J^3 + \cdots \tag{7.100}$$

used to construct the one-parameter group can often be evaluated via the **Cayley-Hamilton equation**. This equation says that each matrix J satisfies its own characteristic equation (replace the powers λ^k in the determinant of $J - \lambda E$ by the powers J^k). Thus for example, matrix J in Problem 7.1 satisfies the equation $J^2 - 2J + E = 0$.

Calculate e^{tJ} for the general skew-symmetric 3×3 matrix

$$\begin{bmatrix} 0 & -u_3 & u_2 \\ u_3 & 0 & -u_1 \\ -u_2 & u_1 & 0 \end{bmatrix} \qquad u_i \in \mathbf{R} \tag{7.101}$$

(Euler's rotation formula).

Problem 7.3 A $2n \times 2n$ matrix A (real or complex) is said to be **symplectic** if

$$A^T S A = S \text{ where } S = \begin{bmatrix} 0 & E_n \\ -E_n & 0 \end{bmatrix} \tag{7.102}$$

E_n is the $n \times n$ identity matrix.

(a) Show that the symplectic matrices form a group.
(b) Which are the infinitesimal operators of this group?
(c) Verify that the infinitesimal operators form a Lie algebra.

Problem 7.4 The Lie algebra of SU(2) consists of skew-hermitian matrices

$$\begin{bmatrix} ir & s+it \\ -s+it & -ir \end{bmatrix} \qquad r, s, t \in \mathbf{R} \tag{7.103}$$

(a)   Show that the matrices

$$\begin{bmatrix} 0 & -r & s & -t \\ r & 0 & t & s \\ -s & -t & 0 & r \\ t & -s & -r & 0 \end{bmatrix}$$

form a representation θ of the Lie algebra of SU(2).

(b)   Find the irreducible representations appearing in θ by use of the weights method.

(c)   Find the symmetry adapted basis vectors using the algorithm in Section 7.4.

Problem 7.5 Let G be a continuous group of $n \times n$ matrices $E, A, \ldots$ and $\mathcal{L}$ be the associated Lie algebra, with the matrices $J_1, J_2, \ldots, J_d$ forming a basis of $\mathcal{L}$.

(a)   Let A be a fixed matrix in G. Verify that the transformations assigned to A given by

$$J \to J' = AJA^{-1} \quad \text{where} \quad J = \sum_{i=1}^{d} \lambda_i J_i, \quad \lambda_i \in \mathbf{C} \tag{7.104}$$

describe a d-dimensional representation ϑ of G. The latter is called the **adjoint representation** of G associated with θ. Here $\mathcal{L}$ is the real subspace of the representation space of ϑ.

(b)   Furthermore, show that the **adjoint representation** θ of $\mathcal{L}$ associated with ϑ assigns to each fixed matrix $K \in \mathcal{L}$ the linear transformation

$$J \to J' = [K, J] \quad \text{where} \quad J = \sum_{i=1}^{d} \lambda_i J_i, \quad \lambda_i \in \mathbf{C} \tag{7.105}$$

In addition, verify the representation property for (7.105).

Problem 7.6 Verify that (7.105) describes a representation also for arbitrary abstract Lie algebras $\mathcal{L}$, namely the **adjoint** representation of $\mathcal{L}$.

Problem 7.7 Find the adjoint representation of the Lie algebra of SU(2), whose elements are described in (7.103).

Problem 7.8 Let $\mathcal{L}$ be a Lie algebra defined by the structure relations

$$[J_k, J_\ell] = \sum_{\lambda} c_{k\ell}^{\lambda} J_\lambda \tag{7.106}$$

with the property that by (7.105) its adjoint representation $J_k \to D(J_k)$ be an isomorphism, thus **faithful**.

Verify that the exponential map applied to the representing matrices $D(J_k)$ is a device to obtain an associated continuous matrix group whose Lie algebra is isomorphic with $\mathcal{L}$.

Problem 7.9 Carry out the construction described in Problem 7.8 for a 3-dimensional Lie algebra spanned by J_1, J_2, J_3 with the structure relations

$$[J_1, J_2] = J_3, \qquad [J_2, J_3] = J_1, \qquad [J_3, J_1] = J_2 \qquad (7.107)$$

Problem 7.10 (A selection rule in quantum mechanics) Verify the following: The Hamiltonian operator (7.88) with $v = v(r)$ of the electron of a hydrogen atom has the energy eigenfunctions

$$\psi_\ell = f(r) \cdot Y_\ell(\theta, \varphi)$$

Here ℓ is the orbital quantum number (angular momentum). The radiation of the atom (due to a weak external electric field), which is observed as a spectrum, is caused by the electron jumping from one state ψ_ℓ into another state.

In the simplest case, where the electric field in x_3-direction is constant in both space and time, the mathematical formulation of this perturbation problem (so-called **Stark effect**) amounts to expanding the functions $x_3\psi_\ell$ in terms of spherical functions: The electron can leap from the ψ_ℓ state into only those states that occur in this expansion. Hence the solution to Problem 4.4 implies that only transitions into states with the orbital quantum numbers $(\ell + 1)$ or $(\ell - 1)$ are possible.

Problem 7.11 Determine by aid of the raising operator the pointwise invariant subspace of the representation $\vartheta_{1/2} \otimes \vartheta_{1/2} = \vartheta_1 \oplus \vartheta_0$ of SU(2).

Problem 7.12 (Addition of angular momenta)   The Kronecker product $\theta_\ell \otimes \theta_m$, where $m, \ell \in \{\, 0, 1/2, 1, 3/2, 2, 5/2, 3, \ldots \,\}$, of the Lie algebra $\mathcal{L}$ of SU(2) is responsible for the addition of two arbitrary angular momenta in a central field. Determine, along with their multiplicities, the values of the third component L_3 and those of the square of the magnitude L^2 of the total angular momentum L.

Chapter 8

Applications to Solid State Physics

The purpose of the present chapter is to give a brief description of some of the notions of crystallography. More detailed treatment of the mathematical theories is available in several texts of the literature (Streitwolf, 1971), (Janssen, 1973), (Bradley, Cracknell, 1972).

8.1 Lattices

Consider the real n-dimensional vector space $\mathbf{R}^n$ with given basis vectors e_1, e_2, ..., e_n, not necessarily orthogonal. Then the position vector y of a point in this space is given by

$$y = y_1 e_1 + y_2 e_2 + \ldots + y_n e_n \qquad y_k \in \mathbf{R} \qquad (8.1)$$

A lattice Γ is defined as the set of all points y with integral coordinates y_k, i.e.,

$$y = y_1 e_1 + y_2 e_2 + \ldots + y_n e_n \qquad y_k \text{ integers} \qquad (8.2)$$

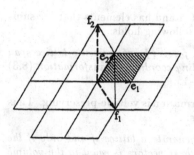

In practical applications, we are interested in the 3-dimensional case ($n = 3$).

The position vector y of a lattice point is called a **lattice vector** with the **lattice coordinates** y_k. The parallelepiped spanned by the basis vectors $e_1, e_2, \ldots, e_n$ is called a **basic cell** or a **primitive cell** of the lattice. In the 2-dimensional lattice ($n = 2$) of Fig. 8.1, it is the shaded parallelogram.

Fig. 8.1

Now let $f_1, f_2, \ldots, f_n$ be n linearly independent lattice vectors. They are of the form

$$f_j = \sum_{k=1}^{n} a_{kj} e_k \qquad a_{kj} \text{ integers} \tag{8.3}$$

The f_j again span a lattice with a typical lattice vector

$$z = \sum_{k=1}^{n} z_j f_j = \sum_{j,k} (a_{kj} z_j) e_k \tag{8.4}$$

Thus each lattice point of the form (8.4) is also a lattice point of the form (8.2) with coordinates

$$y_k = \sum_{k=1}^{n} a_{kj} z_j \qquad k = 1, 2, \ldots, n \tag{8.5}$$

Under what conditions does the converse hold? That is, when is a point of the e-lattice also a point of the f-lattice? Or in still other words, how must the basis (8.3) look like in order for the two lattices to be identical? To answer this question, we introduce the column vectors $y = (y_1, y_2, \ldots, y_n)^T$, $z = (z_1, z_2, \ldots, z_n)^T$, and the matrix $A = (a_{kj})$. Then (8.5) reads

$$y = Az, \quad \text{i.e.,} \quad z = A^{-1}y \qquad A \text{ integral} \tag{8.6}$$

The converse holds if the coordinate transformation (8.6) produces an integral z-column for each integral y-column. This is possible if and only if all elements of A^{-1} are integral, as is seen on substituting the basis vectors $e_1, e_2, \ldots, e_n$ for y. Then $\det(A^{-1}) = 1/\det(A)$ is an integer likewise. Since $\det(A)$ is already an integer, this implies that $\det(A) = \pm 1$. If, conversely, $\det(A) = \pm 1$ and A is an integral matrix, then so is A^{-1}. This is a direct consequence of the equation

$$A^{-1} = \frac{1}{\det(A)} \cdot \mathrm{adj}(A)$$

where $\mathrm{adj}(A)$ is the so-called adjoint[1] of A and has elements that are sums of products of elements of A. Thus the following holds:

Theorem 8.1 *The n lattice vectors $f_1, f_2, \ldots, f_n$ of a given e-lattice span the e-lattice if and only if the corresponding coordinate transformation (8.3) is integral and its determinant is 1 or -1.*

As $\det(A) = \pm 1$, the coordinate transformation is volume-preserving. Thus Theorem 8.1 can be reworded as follows:

Theorem 8.2 *n given lattice vectors generate a lattice if and only if the volume of the parallelepiped formed by these vectors is equal to the volume of a basic cell.*

---

[1] See linear algebra.

In this case, they form a new basic cell. Fig. 8.1 shows such a new basic parallelogram, where

$$f_1 = e_1 - e_2, \quad f_2 = -e_1 + 2e_2, \quad A = \begin{bmatrix} 1 & -1 \\ -1 & 2 \end{bmatrix}, \quad \det(A) = 1$$

$$A^{-1} = \begin{bmatrix} 2 & 1 \\ 1 & 1 \end{bmatrix}, \quad e_1 = 2f_1 + f_2, \quad e_2 = f_1 + f_2$$

8.1.1 Reciprocal Lattice

The concept of the reciprocal lattice is not absolutely necessary for calculations in solid state physics, but convenient for illustrating certain situations[2]. Let again $e_1, e_2, \ldots, e_n$ span an n-dimensional lattice $\Gamma \subset \mathbf{R}^n$. We denote by (x, y) the inner product of two vectors in $\mathbf{R}^n$ in the Euclidean sense. Now we construct vectors $\epsilon_1, \epsilon_2, \ldots, \epsilon_n$ such that

$$(\epsilon_i, e_k) = \delta_{ik} \qquad i, k \in \{1, 2, \ldots, n\} \tag{8.7}$$

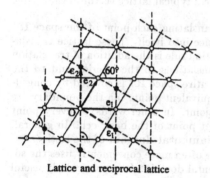

Lattice and reciprocal lattice

Fig. 8.2

Hence ϵ_i is perpendicular to all vectors e_k with $k \neq i$, and its inner product with e_i is 1. The vectors $\epsilon_1, \epsilon_2, \ldots, \epsilon_n$ span a lattice Γ^{-1}, the so-called **reciprocal lattice** of Γ. In Fig. 8.2 ($n = 2$) let e_1, e_2 be vectors of unit length, forming an angle of 60°. Both Γ and the reciprocal lattice Γ^{-1} (dashed lines) are hexagonal; they are carried into themselves by rotations through 60°.

Now let us proceed to the general theory: Let B be a proper or improper rotation of the space $\mathbf{R}^n$ about the origin and B^{-1} be its inverse. The inner product is invariant under B, thus the following holds:

Theorem 8.3 *The bilinear term* $\sum_{i=1}^{n} q_i y_i$, *where* $q_i, y_i \in \mathbf{R}$, *can be regarded as the inner product of a vector* $y = \sum_{i=1}^{n} y_i e_i$ *and a vector* $q = \sum_{i=1}^{n} q_i \epsilon_i$. *Here the e_i and ϵ_i are the basis vectors relative to the lattice Γ and its reciprocal lattice Γ^{-1}, respectively. This inner product satisfies the relation*

$$(q, \mathrm{B}^{-1} y) = (\mathrm{B} q, y) \tag{8.8}$$

---

[2] E.g., group theoretic classification of energy bands in a crystal and related calculations (see Section 8.5); investigations of so-called Fermi-surfaces of metals, X-ray interferences from crystals.

Thus the action of B^{-1} *on* y *can be replaced by the action of the rotation* B *on* q.

The bilinear term $\sum\limits_{i=1}^{n} q_i y_i$ can be written as $q^T y$. Here q and y are one-column matrices with the respective components q_i and y_i. Relative to the basis e_i, let B^{-1} be described by the matrix B^{-1}. The transformed bilinear term then is

$$q^T (B^{-1} y) = (q^T B^{-1}) y \qquad (8.9)$$

Equation (8.9) shows that the reciprocal lattice is not used for calculations.

At this stage, the difference between the notions of the coordinate-independent B and the coordinate-dependent B should be noted.

8.1.2   Brillouin Zone, Stabilizer

Let again $y = \sum\limits_{k=1}^{n} y_k e_k$ (y_k integers) be a typical lattice vector of the lattice $\Gamma \subset \mathbf{R}^n$ spanned by the basis e_k. Translating each point of the space $\mathbf{R}^n$ by the lattice vector y maps the lattice into itself. These lattice translations form an abelian group (composition rule is the addition of translation vectors). If we apply all lattice translations to the interior and to the boundary of the basic cell, then the entire space $\mathbf{R}^n$ is covered by congruent parallelepipeds. If, in addition, equivalent points[3] on the boundary are identified, then this cover is even disjoint. In other words, to every point of $\mathbf{R}^n$ there exists a unique equivalent point of the basic cell. In general, such a domain of $\mathbf{R}^n$ is called a **fundamental domain** of the lattice.

Rather than using a basic cell, one often more conveniently uses the so-called **Brillouin zone** as a fundamental domain. It is particularly useful when working with the reciprocal lattice. In this case the construction law is as follows: Let O be a fixed lattice point and $P \neq O$ be some general lattice point of Γ^{-1}. Let further π_P be the perpendicular bisecting plane of the line segment OP and finally, let H_P be the half space bounded by π_P and containing O. The Brillouin zone of O is the intersection of all half spaces H_P, where equivalent points on the boundary are identified.

It remains to show that the Brillouin zone indeed is a fundamental domain of the reciprocal lattice. By construction, the boundaryless Brillouin zone consists of exactly the set of points of $\mathbf{R}^n$ whose distances to the point O are less than the distance to any other lattice point. These sets relative to each of the other lattice points form a partition of the whole space $\mathbf{R}^n$ and are disjoint up to boundary faces. Since the lattice is carried into itself by each lattice translation, the zones are congruent with one another. Observing that equivalent points on the boundary are identified, this completes the proof.

---

[3]**Two points** of $\mathbf{R}^n$ are said to be **equivalent** if and only if there exists a lattice translation carrying one point into the other.

The advantage of using the Brillouin zone is that every proper or improper rotation of the lattice carrying the lattice into itself (with fixed point O) also carries the Brillouin zone into itself. In short, *the Brillouin zone possesses the symmetry of the lattice.*

Moreover, two points of the interior of a Brillouin zone are inequivalent.

Remarks.

1. In the case of plane lattices, the words space, face, and edge must be replaced by plane, edge, and point.

2. The definition of the Brillouin zone readily implies that, for its construction, only the lattice points P in the vicinity of O are relevant.

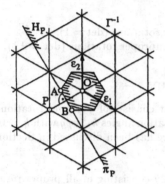

Fig. 8.3 shows an example of a Brillouin zone for the reciprocal lattice ϵ_1, ϵ_2 of the hexagonal lattice e_1, e_2 of Fig. 8.2. Notice that the set of all inequivalent points on the boundary is a connected sequence of edges consisting of 3 edges but only 2 vertices (e.g., A and B).

Fig. 8.3

Let us now continue our general considerations: Let G be an arbitrary group of proper or improper rotations about O leaving our lattice Γ (and hence also Γ^{-1})[4] invariant (in the lattice of Fig. 8.3, the group G may be one of $C_2, C_3, C_6, D_2, D_3, D_6$; see Section 1.9). Furthermore, let P be a point of the Brillouin zone. We define a **small group** or the **stabilizer of P with respect to the group G** to be the subgroup of all those elements of G that carry P into a point P', where P' is equivalent to P with respect to the reciprocal lattice (i.e., $\overrightarrow{PP'}$ is a lattice vector of Γ^{-1}). Thus the small group depends on the choice of the point P. To illustrate this, we again consider the reciprocal lattice Γ^{-1} of Fig. 8.3 and choose, for example, $G = D_6$. Next, for P we successively select the points O, R, S, T, U, V, W of Fig. 8.4. Their stabilizers then are

$$D_6, D_3, D_2, D_1, D_1, D_1, C_1 = \text{ trivial group}$$

---

[4]See Problem 8.10.

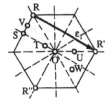

Fig. 8.4

Another example is $G = D_3$ with the three reflection axes through the vertices of the hexagon. Here the stabilizers of, say, the points R and V are the groups D_3 and D_1. However, if we choose $G = D_3$ with the reflection axes through the midpoints of the edges of the hexagon, then we obtain for R and V the stabilizers C_3 and C_1.

8.2   Point Groups and Representations

In this section we shall give a survey of all finite 3-dimensional point groups, i.e., subgroups of the orthogonal group $O(3)$ of $\mathbf{R}^3$, and their representations.

We distinguish the following groups:
(+)-groups consisting exclusively of proper rotations (det $= 1$);
($\pm$)-groups consisting of both proper and improper rotations (det $= \pm 1$).

8.2.1   The List of All (+)-Groups

1. **The cyclic group C_g** of order g: It consists of g proper rotations about some fixed axis through the rotation angles $k \cdot 2\pi/g$ with $k = 0$, $1, 2, \ldots, g - 1$. Its abstract structure has been discussed in Section 1.9.

2. **The dihedral group D_h** of order $2h$ consisting of all proper rotations carrying a plane regular h-sided polygon in space onto itself: namely, the h proper rotations about an axis perpendicular to the plane of the h-sided polygon forming a subgroup C_h together with h proper $180°$ rotations about the symmetry axes in the plane of the h-sided polygon.

    If in discussing the transformations just described we confine ourselves to the plane, then we obtain the group D_h as defined in Section 1.9: The $180°$ rotations in $\mathbf{R}^3$ become axial reflections and the rotations about an axis perpendicular to the plane of the h-sided polygon become rotations about the center of the h-sided polygon. The **spatial and the plane dihedral groups are isomorphic**[5] to each other in the abstract sense. Hence the similarity classes can be translated from the solution to Problem 5.2.

3. **The tetrahedral group T** of order 12 consists of the proper rotations carrying a regular tetrahedron onto itself: This group was treated in Example 5.6. Its similarity classes are listed in Table (5.112).

---

[5]This is the reason why they are denoted the same.

4. The **octahedral group O** of order 24 consists of the proper rotations carrying a cube onto itself. This group was discussed in Example 5.4. Its similarity classes are listed in Table (5.111).

5. The **icosahedral group I** consists of the 60 proper rotations carrying a regular icosahedron onto itself. This group is isomorphic to the alternating group A_5, as can be proven by inscribing 5 cubes.

We state without proof that this is the complete list of **all** finite 3-dimensional (+)-groups.

8.2.2   Representations and Characters of (+)-Groups

1. For the abelian group C_g, they are found in (1.52).

2. The irreducible representations of D_h are listed in (1.59) and (1.60), where the element there denoted by s must here be regarded as the 180° rotation. The character table was found in Problem 5.2.

3. For the tetrahedral group T one first finds the unit representation ϑ_1^1 and the 3-dimensional representation ϑ_1^3 obtained by describing the tetrahedral rotations by 3×3 matrices in some coordinate system. They enable us to fill the first and fourth rows of Table (8.10). The irreducibility criterion of Theorem 5.10 shows that ϑ_1^3 is irreducible. Equation (5.165) implies that there are exactly two more 1-dimensional representations, ϑ_2^1 and ϑ_3^1. But a 120° rotation must necessarily be represented by a cube root of unity. The entries 1 in the last column of χ_2^1, χ_3^1 result from the orthogonality with respect to the first row (compare (5.141)).

Class Cardinality	Identity 1	+120° rot. 4	−120° rot. 4	180° rot. 3
χ_1^1	1	1	1	1
χ_2^1	1	ω	ω^{-1}	1
χ_3^1	1	ω^{-1}	ω	1
χ_1^3	3	0	0	-1

Tetrahedral group T $\omega = e^{i2\pi/3}$

$$(8.10)$$

4. Every transformation of the octahedral group **permutes the four interior diagonals of the cube.** It is readily seen that all permutations of these four objects occur. This implies that the group O is isomorphic to the symmetric group S_4. The irreducible representations are listed in Table (5.77), their characters in (5.138). Here the

isomorphism for similarity classes translates as follows:

S_4	O
(.)(.)(.)(.)	identity
(...)(.)	120° rotations about interior diagonals
(....)	90° rotations about centroids of faces
(..)(..)	180° rotations about centroids of faces
(..)(.)(.)	180° rotations about midpoints of edges

(8.11)

5. The irreducible representations of the icosahedral group I are 1-, 3-, 4-, or 5-dimensional[6].

8.2.3   (±)-Groups of the First Kind

Among the improper transformations, the reflection Z about the origin (also called the **inversion**) plays a special role, as it commutes with all orthogonal transformations. Furthermore,

$$Z^2 = \text{identity} \qquad (8.12)$$

Claim. *If G is a (+)-group of order g and s is some typical element of G, then the $2g$ elements s and Zs $(s \in G)$ form a (±)-group.*

Proof. The following multiplication rules hold:

$$s_1(Zs_2) = Z(s_1s_2) \qquad (8.13)$$

$$(Zs_1)(Zs_2) = Z^2(s_1s_2) = s_1s_2 \qquad (8.14)$$

Thus the set of the $2g$ elements is closed under multiplication. Moreover, there is a neutral element, namely, the one of G (see (8.13) where $s_1 = e$). There is also an inverse $(Zs)^{-1} = Zs^{-1}$.

The associative law is an immediate consequence of the fact that s and Zs are transformations (compare Section 1.1). □

The group obtained is denoted by $G + ZG$ and is called a (±)-**group of the first kind**[7].

Hence the following holds:

Theorem 8.4 *If G is a (+)-group and Z the reflection about the origin, then $G + ZG$ is a (±)-group of the first kind.*

Z commutes with all elements of $G + ZG$. Consequently, the thus constructed group of order $2g$ has exactly twice as many similarity classes as has G. In fact, from the similarity class K of G we obtain the two similarity

---

[6]A complete list of irreducible representations of I can be found in (Fässler, Schwarzenbach, 1979).

[7]$G + ZG$ is isomorphic to the direct product $G \times \{e, Z\}$. Thus G is a normal subgroup of index 2.

classes[8] K and ZK of $G + ZG$. Hence by Theorem 5.14 there are twice as many irreducible inequivalent representations of $G + ZG$ as there are of G. Clearly, they are constructed as follows:

Theorem 8.5 *To each irreducible n-dimensional representation of a (+)-group G one can construct two irreducible inequivalent representations of $G + ZG \cong G \times \{e, Z\}$ by the following correspondence:*

$$Z \to E_n \qquad or \qquad Z \to -E_n \qquad (8.15)$$

E_n *is the $n \times n$ identity matrix.*

Inequivalence follows from the characters being different for the element Z. Correspondingly, one obtains the following simple character table for such groups:

Example (Character table for $D_3 + ZD_3$)

Class	e	120°	180°	Z	$Z(120°)$	$Z(180°)$
Cardinality	1	2	3	1	2	3
χ_1^1	1	1	1	1	1	1
χ_2^1	1	1	-1	1	1	-1
χ_1^2	2	-1	0	2	-1	0
$\widetilde{\chi}_1^1$	1	1	1	-1	-1	-1
$\widetilde{\chi}_2^1$	1	1	-1	-1	-1	1
$\widetilde{\chi}_1^2$	2	-1	0	-2	1	0

(8.16)

The character table for D_3 in the upper left corner appears four times; in the lower right corner, however, it appears with reversed signs.

8.2.4 (±)-Groups of the Second Kind

By definition, (±)-groups of the second kind are the *groups that do not contain the inversion Z*. They can be constructed in the following manner: Let G be a (+)-group of even order $g = 2h$ and containing a subgroup H of order h. We denote by u the elements of H and by a the remaining elements of G.

Claim. *The 2h elements u and Za $(u \in H, a \in G \setminus H)$ form a (±)-group.*

Proof. We first show that the set of $2h$ elements described above is closed under multiplication. By commutativity of Z, we have

$$(Za)u = Z(au) \qquad u(Za) = Z(ua) \qquad (8.17)$$

The element au is not in H. For, if $au = u_1 \in H$, then $a = u_1 u^{-1} \in H$, contradicting the hypothesis. Nor is ua an element of H. Furthermore,

$$(Za_1)(Za_2) = Z^2(a_1 a_2) = a_1 a_2 \qquad (8.18)$$

---

[8] ZK means the set $\{Z \cdot s \mid s \in K\}$.

As $a_1^{-1} \notin H$ and with the considerations made for (8.17), this implies that $a_1^{-1} u \in (G \setminus H)$. Thus two distinct elements u yield distinct elements of $G \setminus H$. If u ranges over the subgroup H, then $a_1^{-1} u$, where a_1^{-1} is fixed, ranges over the complete set $G \setminus H$, since H and $G \setminus H$ have the same cardinality. Hence there exists an element $u^* \in H$ such that $a_1^{-1} u^* = a_2$, therefore, $u^* = a_1 a_2 \in H$. We thus showed that multiplication does not lead outside this set.

Associativity, existence of the neutral element and of the inverse also hold. □

The group thus constructed is denoted by (H, G). This group has the same order as G, namely, $g = 2h$.

Hence the following holds:

Theorem 8.6 *If G is a (+)-group of even order $2h$ and H is a subgroup of G of order h, then (H, G) is a (±)-group of the second kind. This group is composed of the elements of H and the composites of elements of $G \setminus H$ with the inversion Z.*

Theorem 8.7 *The groups (H, G) and G are isomorphic.*

Proof. Find a bijective correspondence $\varphi: (H, G) \to G$ with

$$\varphi(\alpha) \cdot \varphi(\beta) = \varphi(\alpha\beta) \qquad \forall \alpha, \beta \in (H, G)$$

By (8.17) and (8.18), this is achieved by the correspondence

$$u \to u \qquad Za \to a \tag{8.19}$$

□

8.2.5   Review

We state without proof that 8.2.3 and 8.2.4 list **all** finite 3-dimensional (±)-groups. The following table collects them together with all of the (+)-groups:

(+)-groups		C_n	D_n	T	O	I
(±)-groups	1st kind	$C_n + ZC_n$	$D_n + ZD_n$	T+ZT	O +ZO	I +ZI
$n=1,2,3,\ldots$	2nd kind	(C_n, C_{2n})	(C_n, D_n)	(D_n, D_{2n})	(T, O)	

The finite 3-dimensional point groups

$$(8.20)$$

(T, O) is the group of all proper and improper rotations carrying a regular tetrahedron into itself. It is isomorphic to the symmetric group S_4 (permutations of the vertices of the tetrahedron) and has already been used in Example 5.3 (molecular oscillation).

$O + ZO$ is the group of all proper and improper rotations carrying a cube into itself; $T + ZT$ is a (±)-subgroup of $O + ZO$.

We leave it to the reader to describe the remaining groups.

8.3 The 32 Crystal Classes

A **crystallographic point group** or **crystal class** K is a point group carrying a 3-dimensional lattice into itself, with the origin being a lattice point. An example is the group $C_1 + Z$ consisting of the identity and the reflection in a lattice point. It, in fact, admits every lattice of this group. If we describe the elements of a crystal class relative to a lattice basis e_1, e_2, e_3 (which is the same as saying that these three vectors span a basic cell), then they are described by integral 3×3 matrices denoted by A. Reason: The columns of A contain the components of the images of the lattice vectors, which themselves describe lattice vectors. Moreover, the proper and the improper rotations preserve volumes. Thus the following holds:

Theorem 8.8 *Relative to a lattice basis, congruence mappings of a lattice (i.e., isometries taking a lattice onto itself) leaving a lattice point fixed are described by integral 3×3 matrices A with $\det(A) = \pm 1$.*

Congruence mappings with a fixed point are either proper rotations about some rotation axis through some rotation angle φ or improper rotations that are composites of proper rotations with reflections in a plane perpendicular to the rotation axis. Hence relative to an orthonormal basis, where the first basis vector is in direction of the rotation axis, they are described by matrices of the following form:

$$\begin{bmatrix} \pm 1 & 0 & 0 \\ 0 & \cos\varphi & -\sin\varphi \\ 0 & \sin\varphi & \cos\varphi \end{bmatrix} \qquad \begin{array}{l} +1 \quad \text{for proper rotations} \\ -1 \quad \text{for improper rotations} \end{array} \qquad (8.21)$$

The trace is $\pm 1 + 2\cos\varphi$.

When passing to lattice coordinates, i.e., to the matrix A, this trace remains unchanged. Since A is integral, this implies that

$$\text{tr}(A) = \pm 1 + 2\cos\varphi \quad \text{is an integer} \qquad (8.22)$$

As $2\cos\varphi$ is an integer and $|\cos\varphi| \le 1$, the angle φ can take on only the values

$$0°, \pm 60°, \pm 90°, \pm 120°, 180° \qquad (8.23)$$

This can be phrased as follows:

Theorem 8.9 *A lattice admits no other than 1-, 2-, 3-, 4- and 6-fold rotation axes.*

This condition, referring to both the proper and the improper rotations, reduces the list (8.20) to $5+5+1+1 = 12$ groups in each of the first two rows and to $3+5+3+1=12$ groups in the last row. But since C_2 and D_1 are isomorphic groups (consisting of the identity and a 180° rotation), this implies that

$$D_1 \cong C_2; \quad D_1 + ZD_1 \cong C_2 + ZC_2;$$
$$(C_1, C_2) \cong (C_1, D_1); \quad (D_1, D_2) \cong (C_2, D_2)$$

The list (8.20) is then reduced to precisely $2 \cdot 12 + 12 - 4 = 32$ different groups. Here is an opportunity to remark that to each of these groups there exist invariant lattices in nature. Hence there are precisely 32 different crystal classes. They are listed in the second column of Table (8.24). The third column shows the notations used by **Schoenflies**, the founder of this theory. The fourth column contains the new international nomenclature[9].

The classification of the crystal classes[10] by so-called **crystal systems** (first column in Table (8.24)) will not be discussed in detail here. We just mention that in every crystal system there is one crystal class (the so-called **holohedry** or **holohedral crystal class**, framed in Table (8.24)) to be singled out. It contains any other crystal class of the same system as a subgroup and is defined as follows: If Γ is the lattice of the crystal, then the holohedral crystal class consists of **all** proper and improper rotations leaving a lattice point, taken as the origin, fixed and carrying Γ into itself. Since every lattice admits the reflection Z about the origin, the holohedral crystal class is a crystallographic point group of the first kind.

8.4   Symmetries and the Ritz Method

Let V be a finite- or infinite-dimensional Hilbert space with the inner product $\langle f, g \rangle$ for arbitrary vectors $f, g \in V$. (In applications, they are often functions). Let further a linear operator $H : V \to V$ be given.

We now select an n-dimensional subspace $V^n (n < \infty)$ of V, preferably spanned by an orthonormal basis $f_1, f_2, \ldots, f_n$. Then

$$\langle f_j, f_k \rangle = \delta_{jk} \qquad j, k = 1, 2, \ldots, n \tag{8.25}$$

For each vector f of the Hilbert space we define

$$Pf = \sum_{j=1}^{n} \langle f, f_j \rangle f_j \tag{8.26}$$

Bilinearity of the inner product and (8.25), (8.26) imply

$$\langle Pf, f_k \rangle = \langle f, f_k \rangle \tag{8.27}$$

Furthermore,

$$\langle f - Pf, f_k \rangle = 0 \qquad k = 1, 2, \ldots, n \tag{8.28}$$

By (8.26), the vector Pf is in V^n and by (8.28), the "error vector" $f - Pf$ is orthogonal to V^n. Hence Pf is the **orthogonal projection** of f onto the subspace V^n. Now let f in particular be a vector of V^n. The image Hf is, in general, not in V^n, but the projection PHf always is. The linear operator $M = PH$ restricted to V^n is called the **Ritz operator**[11].

---

[9] Also called the Hermann-Mauguin symbols.

[10] See (Burzlaff, Thiele, 1977).

[11] Ritz, Walter (1878-1909). In addition to the approximation method described here, he also developed the combination principle of spectroscopy named after him.

Table 8.24: Crystal classes

Crystal systems with number of crystal classes	Crystal classes with various nomenclatures		
	abstract	Schoenflies	international
triclinic 2	C_1 =identity	C_1	1
	$C_1 + Z$	$C_i = S_2$	$\bar{1}$
monoclinic 3	C_2	C_2	2
	$C_2 + ZC_2$	C_{2h}	2/m
	(C_1, C_2)	$C_s = C_{1h}$	m
orthorhombic 3	D_2	$D_2 = V$	2 2 2
	$D_2 + ZD_2$	$D_{2h} = V_h$	m m m
	(C_2, D_2)	C_{2v}	m m 2
rhombohedral 5	C_3	C_3	3
	D_3	D_3	3 2
	$C_3 + ZC_3$	$C_{3i} = S_6$	$\bar{3}$
	$D_3 + ZD_3$	D_{3d}	$\bar{3}$ m
	(C_3, D_3)	C_{3v}	3 m
tetragonal 7	C_4	C_4	4
	D_4	D_4	4 2 2
	$C_4 + ZC_4$	C_{4h}	4/m
	$D_4 + ZD_4$	D_{4h}	4/m m m
	(C_2, C_4)	S_4	$\bar{4}$
	(C_4, D_4)	C_{4v}	4 m m
	(D_2, D_4)	$D_{2d} = V_d$	$\bar{4}$ 2 m
hexagonal 7	C_6	C_6	6
	D_6	D_6	6 2 2
	$C_6 + ZC_6$	C_{6h}	6/m
	$D_6 + ZD_6$	D_{6h}	6/m m m
	(C_3, C_6)	C_{3h}	$\bar{6}$
	(C_6, D_6)	C_{6v}	6 m m
	(D_3, D_6)	D_{3h}	$\bar{6}$ m 2
cubic 5	T	T	2 3
	O	O	4 3 2
	$T + ZT$	T_h	m 3
	$O + ZO$	O_h	m 3 m
	(T, O)	T_d	$\bar{4}$ 3 m

Z = inversion, m = plane of mirror
S_i is not to be mistaken for the symmetric group.

The Ritz method consists in approximating the given operator H by M as closely as possible[12]. Of course, the quality of the approximation depends on the choice of V^n. For example, the eigenvalues of M will approximate those of H acceptably only if the corresponding eigenvectors of H are "in the proximity" of V^n.

By (8.26), (8.27) the image of f_k under M has the components

$$m_{jk} = \langle M f_k, f_j \rangle = \langle P H f_k, f_j \rangle = \langle H f_k, f_j \rangle \qquad j, k = 1, 2, \ldots, n \quad (8.29)$$

Summary of the Ritz method. After choosing a finite orthonormal set f_1, f_2, ..., f_n spanning the Ritz space $V^n \subset V$, we replace a given linear operator H of the Hilbert space V by the so-called **Ritz operator** M of V^n. The **Ritz matrix** describing M with respect to the above-mentioned basis has the elements

$$m_{jk} = \langle H f_k, f_j \rangle \qquad j, k = 1, 2, \ldots, n \quad (8.30)$$

Now let G be an arbitrary group and ϑ a representation of G on the Hilbert space V. The representation ϑ may have been obtained by transplanting functions (see (4.47)). We shall denote the representing linear transformations by $D(s), s \in G$. In dealing with symmetries, we shall make use of

Theorem 8.10 *Let G be an arbitrary group, V a finite- or an infinite-dimensional Hilbert space with the inner product $\langle f, g \rangle$, and let ϑ be a representation of G defined on V such that it satisfies the following*

Assumptions:

1. *ϑ leaves the desired Ritz space V^n with the basis vectors f_1, f_2, ..., f_n invariant, i.e.,*

$$D(s) f_i \in V^n \qquad \forall s \in G, \qquad i = 1, 2, \ldots, n \quad (8.31)$$

2. *ϑ transforms every vector $f \in V$ orthogonal to V^n into another such vector, i.e.,*

$$\langle f, f_i \rangle = 0 \quad \forall i \quad \Rightarrow \quad \langle D(s) f, f_i \rangle = 0 \quad \forall i, \forall s \in G \quad (8.32)$$

3. *ϑ commutes with the linear operator H, i.e.,*

$$H D(s) = D(s) H \qquad \forall s \in G \quad (8.33)$$

Claim: *The Ritz operator M commutes with the representation ϑ restricted to the Ritz space V^n, i.e.,*

$$M D(s) f = D(s) M f \qquad \forall s \in G, \quad \forall f \in V^n \quad (8.34)$$

---

[12]Alternative methods (e.g., Trefftz, balance calculations with collocation) are given in (Schwarzenbach, 1979).

If, in addition, G is a compact Lie group, then the assumptions for the fundamental theorem of Section 2.3 and for the algorithm of Section 5.2 are satisfied.

Remark. Assumption 2 of Theorem 8.10 can be replaced by the requirement that ϑ be **unitary**. In this case, by (8.33), the Ritz matrix (8.30) has the property

$$m_{jk} = \langle HD(s)f_k, D(s)f_j \rangle \qquad \forall s \in G \qquad (8.35)$$

Hence in the inner product (8.30), we may replace f_k and f_j by their images under arbitrary transformations of ϑ.

Proof of Theorem 8.10. We decompose $f \in V$ into two components $g \in V^n$ and h orthogonal to V^n. Hence

$$g = Pf, \qquad h = f - Pf, \qquad f = g + h \qquad (8.36)$$

To simplify notations we put $D = D(s)$. Since $Pg = g$ and $Ph = 0$, we have

$$DP(g+h) = DPg + DPh = DPg = Dg \qquad (8.37)$$

Assumptions 1 and 2 further imply

$$PD(g+h) = PDg + PDh = PDg = Dg \qquad (8.38)$$

Combining (8.36), (8.37), (8.38), we obtain

$$DPf = PDf \qquad \forall f \in V, \quad \forall s \in G \qquad (8.39)$$

By (8.39), we obtain

$$DMg = DPHg = PDHg \qquad (8.40)$$

On the other hand, by (8.33)

$$MDg = PHDg = PDHg \qquad (8.41)$$

Thus

$$MDg = DMg \qquad \forall g \in V^n \qquad (8.42)$$

$\square$

8.5 Examples of Applications

Here we consider eigenvalue problems in which H is the quantum mechanical Hamiltonian operator from Section 7.5; i.e.,

$$H\psi = E\psi \qquad (8.43)$$

or written out explicitly,

$$-\frac{\hbar^2}{2m}\Delta\psi(y) + v(y)\psi(y) = E\psi(y) \qquad (8.44)$$

Here y is the position vector in $\mathbf{R}^3$, ψ a wave function, that is, a vector in the Hilbert space of complex-valued functions satisfying certain boundary conditions; v is a given potential, E an energy eigenvalue, Δ the Laplacian operator[13], $2\pi\hbar = h$ Planck's constant, and m the mass of the electron.

Example 8.1 We consider a plane crystal characterized by a square lattice with cartesian coordinates y_1, y_2 and the point group K equal to the plane dihedral group D_4, which leaves the lattice invariant[14]. We assume the following for (8.44):

1. $v(y)$ is lattice periodic, i.e.,

$$v(y + n) = v(y) \tag{8.45}$$

   where n is an arbitrary lattice vector, that is, a vector with integral components.

2. $v(y)$ admits the point group D_4, i.e.,

$$v(y) = v(Ry) \qquad \forall R \in D_4 \tag{8.46}$$

3. $\psi(y)$ is lattice periodic, i.e.,

$$\psi(y + n) = \psi(y) \tag{8.47}$$

   Thus the wave function $\psi(y)$ need only be determined on the unit square.

We now work in the Hilbert space of complex-valued lattice periodic functions with the inner product

$$\langle f, g \rangle = \int_0^1 \int_0^1 f(y)\overline{g(y)} \, dy_1 \, dy_2 \tag{8.48}$$

The bar denotes complex conjugation. As elements of the Ritz space V^n we choose linear combinations of lattice periodic functions of the form

$$f_q(y) = e^{2\pi i(q_1 y_1 + q_2 y_2)} = e^{2\pi i(q,y)} = e(q, y) \qquad q_1, q_2 \text{ integers} \tag{8.49}$$

(The first and the last terms were introduced for notational purposes.) In our cartesian coordinate system, q may be thought of as a 2-dimensional vector with integral components q_1 and q_2, thus as a lattice vector. The term (q, y) is the inner product of the plane. We say q is a **numbering lattice vector**[15]. The following orthogonality relations hold:

$$\langle f_p, f_q \rangle = \delta_{pq} = \left\{ \begin{array}{ll} 1 & \text{if } p = q \\ 0 & \text{otherwise} \end{array} \right. \tag{8.50}$$

---

[13]In non-orthogonal coordinates y_i, the Laplacian Δ is not just the sum of the second partial derivatives. See, e.g., (Madelung, 1964).

[14]In this case, K is identical with the holohedral crystal class of the lattice.

[15]If the lattice is not square, then by Theorem 8.3 the vector q must be regarded as one of the reciprocal lattice. In our example, we have the special case $\Gamma = \Gamma^{-1}$.

If one selects a finite number of pairwise distinct lattice vectors q, then the corresponding functions form an orthonormal basis of the Ritz space.

The set of such q-vectors drawn in the lattice is called the **vector star** associated with the Ritz space. Relative to some tacitly specified numbering of the q-vectors, the Ritz matrix (8.30) is

$$m_{pq} = \langle H f_q, f_p \rangle \qquad (8.51)$$

where p and q independently range through the vector star. The matrix (m_{pq}) describes the Ritz operator M relative to the basis (8.49). More explicitly, with (8.44), (8.48), (8.49), this reads as follows:

$$m_{pq} = \frac{h^2}{2m} |q|^2 \delta_{pq} + \int_0^1 \int_0^1 v(y) e(q - p, y) \, dy_1 \, dy_2 \qquad (8.52)$$

By transplanting a function f of the Hilbert space in the sense of (4.47), we now construct a representation ϑ of the dihedral group D_4:

$$f(y) \xrightarrow{\vartheta} f'(y) = f(R^{-1}y) \qquad R \in D_4 \qquad (8.53)$$

Indeed, the function $f'(y)$ is again lattice periodic, since the lattice admits the group D_4. The representation ϑ is **unitary**, because the inner product (8.48) is invariant under the transplantation (8.53). Since the Laplace operator commutes with the proper and the improper rotations of D_4 and since by (8.46) the potential $v(y)$ admits the dihedral group D_4, Assumption 3 of Theorem 8.10 is satisfied. Now we need to check Assumption 1 (Assumption 2 is satisfied, because ϑ is unitary); i.e., every basis function $f_q(y)$ of the Ritz space V^n is to be mapped by ϑ into a function of the Ritz space. For this purpose we consider the transplantation (8.53) of the functions (8.49):

$$f_q(y) \xrightarrow{\vartheta} f'_q(y) = f_q(R^{-1}y) = e(q, R^{-1}y) = e(Rq, y) \qquad (8.54)$$

The last step in (8.54) follows from Theorem 8.3.
An immediate consequence of the relation

$$e(Rq, y) = f_{Rq}(y) \qquad (8.55)$$

is that Assumption 1 of Theorem 8.3 is satisfied if and only if the *vector star of the Ritz space is invariant under the group D_4*.

If this requirement is satisfied, then our problem can be replaced by a finite-dimensional approximation, and Theorem 8.10 applies.

Here we point out the fact that the group D_4 permutes the q-vectors. Thus by (8.54), (8.55), the representation ϑ permutes the functions $f_q(y)$. This fact considerably simplifies the computation of the Ritz matrix (8.30), in this case as much as in many other cases. Namely, if such a permutation carries the vectors p, q into the vectors p', q', then by (8.35)

$$m_{pq} = m_{p'q'} \qquad (8.56)$$

As to the choice of the vector star, we notice that the Ritz method approximates eigenoscillations with low frequencies better than those with high frequencies. But the frequency of a function (8.49) is essentially the length $|q|$ of the numbering vector. Hence to form a star, one should take the shortest possible vectors.

Before we proceed to arbitrarily large stars, we choose as a simple example the star shown in Fig. 8.5 consisting of 5 vectors (zero vector included) labeled 1, 2, 3, 4, 5.

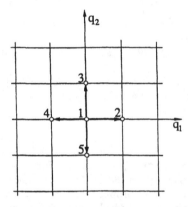

The functions (8.49) associated with these q-vectors span a 5-dimensional Ritz space V^5. The representation of the group D_4 restricted to V^5 permutes the functions (8.49) associated with the five q-vectors, hence leaves the Ritz space spanned by them unchanged. Therefore, we denote this representation by ϑ_{per}.

Fig. 8.5

For example, according to Fig. 8.5, the 180° rotation about the origin results in the permutation

$$\begin{array}{ccccc} 1 & 2 & 3 & 4 & 5 \\ 1 & 4 & 5 & 2 & 3 \end{array}$$

of the five q-vectors. Hence the corresponding representing permutation matrix is

$$\begin{bmatrix} 1 & 0 & 0 & 0 & 0 \\ 0 & 0 & 0 & 1 & 0 \\ 0 & 0 & 0 & 0 & 1 \\ 0 & 1 & 0 & 0 & 0 \\ 0 & 0 & 1 & 0 & 0 \end{bmatrix} \tag{8.57}$$

In order to calculate the multiplicities of the irreducible constituents of ϑ_{per}, we need the character table (5.149) but also the character χ_{per} of our representation ϑ_{per}. To determine χ_{per}, we recall that the trace of a permutation matrix equals the number of fixed elements under this permutation (see (5.137)). Table (8.58) reproduces the character table (5.149). χ_{per} is shown in the last row, and the multiplicities of the various irreducible representations, calculated from (5.148), are given in the last column. σ_x, σ_y denote the reflections in the horizontal and the vertical lines; σ_+, σ_- denote the reflections in the diagonals (see Fig. 5.2).

Class	e	$\pm 90°$ rot.	$180°$ rot.	σ_x, σ_y	σ_+, σ_-	Multi-
Number of elements	1	2	1	2	2	plicity
$\vartheta_1^1, \ \chi_1^1$	1	1	1	1	1	2
$\vartheta_2^1, \ \chi_2^1$	1	1	1	-1	-1	0
$\vartheta_3^1, \ \chi_3^1$	1	-1	1	1	-1	1
$\vartheta_4^1, \ \chi_4^1$	1	-1	1	-1	1	0
$\vartheta_1^2, \ \chi_1^2$	2	0	-2	0	0	1
$\vartheta_{\text{per}}, \ \chi_{\text{per}}$	5	1	1	3	1	

$$(8.58)$$

Thus the representation ϑ_{per} reduces completely to

$$\vartheta_{\text{per}} = 2\vartheta_1^1 \oplus \vartheta_3^1 \oplus \vartheta_1^2 \tag{8.59}$$

By Theorem 2.7 and because M commutes with ϑ_{per} (Theorem 8.10), the vectors of the irreducible subspaces of ϑ_3^1 and ϑ_1^2 are **eigenvectors** of the Ritz operator M, as their multiplicities equal 1. The two 1-dimensional subspaces, transformed according to the unit representation ϑ_1^1, span a 2-dimensional subspace invariant under M (see fundamental theorem of Section 2.3).

Now we consider the constituent ϑ_1^2. This representation is 2-dimensional and we construct[16] the corresponding irreducible subspace via the projector (5.159). Here the projector is

$$P = \frac{1}{4} \sum_{s \in D_4} \overline{\chi_1^2(s)} D_{\text{per}}(s) \tag{8.60}$$

where the $D_{\text{per}}(s)$ are the representing permutation matrices. Since χ_1^2 in Table (8.58) has nonzero values only for the identity e and the $180°$ rotation, only the identity matrix E and (8.57) appear in the sum. For the second column of (8.60) one finds, up to a factor $1/2$, the eigenvector

$$w = \begin{bmatrix} 0 \\ 1 \\ 0 \\ -1 \\ 0 \end{bmatrix} \tag{8.61}$$

According to Fig. 8.5 and definition (8.49) the function $e^{2\pi i y_1}$ belongs to the q-vector no. 2, because $q_1 = 1$, $q_2 = 0$. Similarly, the function $e^{-2\pi i y_1}$ belongs to the q-vector no. 4 and hence the wave function

$$\psi(y) = e^{2\pi i y_1} - e^{-2\pi i y_1} = 2i \cdot \sin 2\pi y_1 \tag{8.62}$$

or $\sin 2\pi y_1$, for short, belongs to (8.61). This is the eigenfunction of the Ritz operator and a rough approximation to an eigenfunction of the Hamiltonian

---

[16] Instead of constructing the projector, one could also directly use Table (3.7) and pass to the real equation (3.13).

operator. It is worthwhile to mention that this function was found **without quantitative knowledge** of the behavior **of the potential function** $v(y)$.

The Ritz matrix M is real, because the imaginary part of (8.52) vanishes due to

$$\int_{-1/2}^{1/2}\int_{-1/2}^{1/2} v(y)\sin[2\pi(n,y)]\,dy_1\,dy_2 = 0 \quad \text{where} \quad n = q - p \qquad (8.63)$$

Here we used the periodicity properties when changing the integration limits. Equation (8.63) is valid, because v admits the group D_4, which in particular implies that $v(y) = v(-y)$, but also because $\sin[2\pi(n,y)] = -\sin[2\pi(n,-y)]$.

As M is real, Equation (8.52) implies

$$m_{pq} = m_{qp} \qquad (8.64)$$

Furthermore, Equation (8.56) helps to calculate the eigenvalue corresponding to (8.62). If, for example, we use the 90° rotation with the corresponding permutation of the numbering lattice vectors (Fig. 8.5)

$$\Big\downarrow \begin{array}{ccccc} 1 & 2 & 3 & 4 & 5 \\ 1 & 3 & 4 & 5 & 2 \end{array} \qquad (8.65)$$

then (8.56) furnishes the relations

$$\left. \begin{array}{llll} m_{12} = & m_{13} = m_{14} = m_{15} & = \alpha \\ m_{23} = & m_{34} = m_{45} = m_{52} & = \beta \\ m_{24} = & m_{35} = m_{42} = m_{53} & = \gamma \\ m_{22} = & m_{33} = m_{44} = m_{55} & = \delta \\ m_{11} = & & = \epsilon \end{array} \right\} \qquad (8.66)$$

Hence the Ritz matrix has the form

$$M = \begin{bmatrix} \epsilon & \alpha & \alpha & \alpha & \alpha \\ \alpha & \delta & \beta & \gamma & \beta \\ \alpha & \beta & \delta & \beta & \gamma \\ \alpha & \gamma & \beta & \delta & \beta \\ \alpha & \beta & \gamma & \beta & \delta \end{bmatrix} \qquad (8.67)$$

For illustration, we calculate $\gamma = m_{24}$ by Equation (8.52). In this case $p = (1,0)$, $q = (-1,0)$, $q - p = (-2,0)$; thus

$$\gamma = \int_0^1\int_0^1 v(y)e^{-4\pi i y_1}\,dy_1\,dy_2 \qquad (8.68)$$

Furthermore, the diagonal entries are equal to the averages

$$\delta = \frac{h^2}{2m} + \int_0^1\int_0^1 v(y)\,dy_1\,dy_2 \quad \text{and} \quad \epsilon = \delta - \frac{h^2}{2m} \qquad (8.69)$$

The Ritz matrix (8.67) applied to the vector w yields $(\delta - \gamma) \cdot w$. Therefore, the eigenvalue is

$$\delta - \gamma = \frac{h^2}{2m} + \int_0^1 \int_0^1 v(y)(1 - e^{-4\pi i y_1})\, dy_1\, dy_2 \qquad (8.70)$$

Since ϑ_1^2 is 2-dimensional, this eigenvalue appears at least twice.

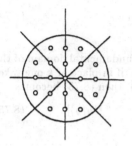

Finally, we treat by means of the theory of Section 3.1 a vector star having arbitrary size and admitting the group D_4. In view of the statement after (8.56), we shall choose all lattice vectors within a circle of arbitrary radius r as a vector star (see Fig. 8.6).

Fig. 8.6

In the following, we adopt the definitions and notations of Section 3.1. By labeling the n vectors of the vector star, we obtain, as in the above example, the corresponding representation ϑ_{per} of $n \times n$ permutation matrices. The representation space decomposes into 8-, 4-, and 1-dimensional subspaces, depending on whether the corresponding orbit consists of 8 or 4 points or of the origin alone.

Let i be the number of points[17] of the vector star in the interior of the fundamental domain and h and k be the number of points on the vertical or the 45° line, respectively. Then, in the order of the head row of Table (8.58), the character χ_{per} of ϑ_{per} is

$$\chi_{\mathrm{per}} = 8i + 4(h+k) + 1, \quad 1, \quad 1, \quad 2h+1, \quad 2k+1 \qquad (8.71)$$

On calculating the multiplicities from Table (8.58) and Equation (5.148), one obtains

$$\vartheta_{\mathrm{per}} = (i+h+k+1)\vartheta_1^1 + i \cdot \vartheta_2^1 + (i+h)\vartheta_3^1 + (i+k)\vartheta_4^1 + (2i+h+k)\vartheta_1^2 \quad (8.72)$$

This result can also be directly obtained from Table (3.11).

All symmetry adapted basis vectors are linear combinations of functions of the form (8.49), where q-vectors belong to one and only one orbit of the star and the coefficients are tabulated in (3.7). By the fundamental theorem of Section 2.3, the Ritz operator decomposes for $i \approx n$ into two identical problems of dimension $\approx n/4$ and four different problems, each of dimension $\approx n/8$.

Example 8.2 (Bloch waves) Let us revisit Example 8.1 (cartesian coordinate system) and retain the requirements (8.45) and (8.46). However,

---

[17] Compare definition (3.2).

we shall abandon condition (8.47) that the wave function $\psi(y)$ be periodic. Instead, we let ψ have the form of a **Bloch wave**; namely,

$$\psi(y) = e^{2\pi i(k,y)} u(y) \tag{8.73}$$

where k is an arbitrary fixed vector[18] of the plane and $u(y)$ a lattice periodic function.

We shall work in the Hilbert space of Bloch waves with the inner product

$$\langle \psi, \varphi \rangle = \int_0^1 \int_0^1 \psi(y)\overline{\varphi(y)}\, dy_1\, dy_2 \tag{8.74}$$

which does not depend on k.

In choosing k, one can confine oneself to the fundamental domain of the reciprocal lattice Γ^{-1} defined in Section 8.1. For, if in place of k one takes the vector $k + q$ (q being a lattice vector of Γ^{-1}), then (8.73) becomes

$$\psi'(y) = e^{2\pi i(k+q,y)} u(y) = e^{2\pi i(k,y)}[e^{2\pi i(q,y)} u(y)] \tag{8.75}$$

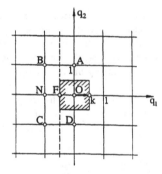

Since the expression within the right bracket is lattice periodic, ψ' is also a function of our Hilbert space.

As a fundamental domain, we choose the Brillouin zone of Γ^{-1}. In Fig. 8.7 it is the shaded domain.

The Bloch waves have the following weak periodicity behavior:

Fig. 8.7

$$\psi(y + n) = e^{2\pi i(k,n)} \psi(y) \qquad n \text{ is a lattice vector of } \Gamma \tag{8.76}$$

For the Ritz approach, we choose, in accordance with (8.49), the special Bloch waves

$$f_q(y) = e^{2\pi i(k,y)} e^{2\pi i(q,y)} = e(k + q, y) \tag{8.77}$$

q is again called the numbering lattice vector of the function in question. If one selects a finite number of pairwise distinct lattice vectors q, then the corresponding functions f_q form an orthonormal set, because (8.74) and (8.77) imply that

$$\langle f_p, f_q \rangle = \delta_{pq} = \begin{cases} 1 & \text{if } p = q \\ 0 & \text{otherwise} \end{cases} \tag{8.78}$$

---

[18] To be regarded as a vector of the reciprocal lattice Γ^{-1} (Theorem 8.3). Here we have the special case $\Gamma = \Gamma^{-1}$.

We construct a representation ϑ of D_4 by transplanting the functions f of our Hilbert space in the sense of (4.47):

$$f(y) \xrightarrow{\vartheta} f'(y) = f(R^{-1}y) \qquad R \in D_4 \tag{8.79}$$

After these preparations, we are now in the position to satisfy the assumptions of Theorem 8.10.

The representation (8.79) applied to the basis functions (8.77) gives

$$f_q(y) \to f_q(R^{-1}y) = e(k + q, R^{-1}y) = e(Rk + Rq, y)$$

Here we used Theorem 8.3. On multiplying by $e(k, y) \cdot e(-k, y) = 1$, we obtain

$$f_q(y) \to e(k, y) \cdot e(Rk - k + Rq, y) = f_{Rk-k+Rq}(y) \tag{8.80}$$

In order that $f_{(Rk-k+Rq)}$ be another Ritz function (8.77), $Rk - k + Rq$ must be a reciprocal lattice vector. But since the lattice admits the dihedral group D_4, the vector Rq is again a reciprocal lattice vector. Hence $Rk - k$ likewise must be such a reciprocal lattice vector. This is true if and only if R *is an operation of the small group of the point associated with* k relative to the group D_4. This requires that we restrict D_4 to a subgroup, namely, to the stabilizer of k (Section 8.1.2).

In Fig. 8.7 we selected as an example the vector $k = (1/2, 0)$. The stabilizer of k is the Klein four group D_2 consisting of the identity, the rotation through $180°$, and the reflections in the coordinate axes.

In order that Assumption 1 of Theorem 8.10 be satisfied

$$q' = Rq + (Rk - k) \tag{8.81}$$

must again be a **numbering lattice vector** of the Ritz basis. Moreover, it must be such a one *for every R of the small group of k*. By rewriting (8.81) in the form

$$(q' - (-k)) = R(q - (-k)) \tag{8.82}$$

it becomes apparent that R carries the vector connecting the points $(-k)$ and q into a vector of the same length connecting $(-k)$ and q'. Hence the point F associated with the vector $(-k)$ is a fixed point of the mapping $q \to q'$. *The operations (8.82) form the small group of k translated into the fixed point F.*

In our example of Fig. 8.7 this translated group consists of the identity, the $180°$ rotation about the fixed point F, and the reflections in the vertical and horizontal lines through F.

In order that Assumption 2 of Theorem 8.10 be satisfied, the figure consisting of the endpoints of the **lattice vectors** of the vector star must admit the translated small group. A simple example illustrating this is the set of the points A, B, C, D, N, O in Fig. 8.7 (F is not included, as it is not a lattice point). Here again the representation ϑ is unitary, because the inner product (8.74) is invariant under the transplantation by orthogonal transformations. Thus Assumption 2 of Theorem 8.10 is satisfied.

Since Assumption 3 is satisfied by the same reason as that given in the discussion after (8.53), we may apply Theorem 8.10 to our representation of the small group of k. We continue the procedure (finding the Ritz matrix, reading off the symmetry adapted basis vectors from Table (3.7), and passing to real vectors (3.13)) in a way similar to that in Example 8.1, though for arbitrarily large vector stars of arbitrary crystals with an arbitrary fixed vector k.

Remark. In the example above, we regarded the small group as a subgroup of the plane dihedral group D_4. The relations between the representations of a group and the representations of one of its subgroups are the objects of study treated by the so-called *induced representations*. They are particularly useful when in the example the wave vector k is changed and hence several subgroups are involved (Janssen, 1973, Ch. 4).

8.6   Crystallographic Space Groups

A length-preserving mapping of $\mathbf{R}^n$ $(n = 2, 3)$ onto itself is an **isometry**. For an arbitrary position vector $x \in \mathbf{R}^n$, it is described by

$$x \to x' = Rx + t \tag{8.83}$$

or, in the compact matrix notation (shown below for $n = 3$),

$$
\begin{bmatrix} x'_1 \\ x'_2 \\ x'_3 \\ 1 \end{bmatrix}
=
\begin{bmatrix}
r_{11} & r_{12} & r_{13} & t_1 \\
r_{21} & r_{22} & r_{23} & t_2 \\
r_{31} & r_{32} & r_{33} & t_3 \\
0 & 0 & 0 & 1
\end{bmatrix}
\cdot
\begin{bmatrix} x_1 \\ x_2 \\ x_3 \\ 1 \end{bmatrix}
$$

Here $R = (r_{ij})$ describes a proper or improper rotation, and t is a fixed vector. R is the **rotational part** and t the **translative part** of the isometry (8.83), which we denote, for short, by (R/t).

A group of isometries of $\mathbf{R}^n$ is called an n-dimensional **space group** G. If we first apply (R_1/t_1) and then (R_2/t_2) to $x \in \mathbf{R}^n$, then

$$x \to R_2(R_1 x + t_1) + t_2 = R_2 R_1 x + R_2 t_1 + t_2 \tag{8.84}$$

Thus the composition law is given by

$$(R_2/t_2)(R_1/t_1) = (R_2 R_1 / R_2 t_1 + t_2) \tag{8.85}$$

Furthermore, the inverse of (R/t) is

$$(R/t)^{-1} = (R^{-1}/ - R^{-1}t) \tag{8.86}$$

All pure translations (identity/t) form an abelian subgroup T of G.

An n-dimensional space group G with the property that the subgroup T is an n-dimensional lattice Γ is called an n-dimensional **crystallographic**

space group. The rotational parts also form a group by (8.85), (8.86). An important feature of crystallographic space groups is the fact that *the rotational parts carry lattice vectors n of Γ into other such vectors*. For, if n is an arbitrary lattice vector and (R/t) an arbitrary element of a crystallographic space group G, then by

$$(R/t)(\mathrm{id}/n)(R/t)^{-1} = (R/Rn + t)(R^{-1}/ - R^{-1}t) = (\mathrm{id}/Rn) \qquad (8.87)$$

it follows that $(\mathrm{id}/Rn) \in T$. Furthermore, by definition of G, it follows that Rn is a lattice vector of Γ.

Hence the group of all rotational parts of G is a **crystallographic point group**, which in accordance with Section 8.3 is denoted by K. This group **characterizes the crystal classes of a crystal** and hence, in the 3-dimensional case, is identical with one of the groups listed in Table (8.24).

If K is a subgroup of G (which is the same as saying that the translative parts of the elements of G belong to the subgroup $T \subset G$), then G is said to be a **symmorphic**[19] space group.

Example 8.3 (Non-symmorphic crystallographic space group)

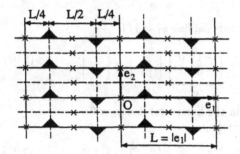

Let the crystal structure be visualized as the infinitely extended triangular structure of Fig. 8.8. We consider a plane crystallographic space group of isometries carrying this structure into itself.

Fig. 8.8

We now claim that

(a) the translations $n_1 e_1 + n_2 e_2$, $(n_1, n_2$ integers) of the rectangular lattice,

(b) the 180° rotations about the points marked with a ×,

(c) the reflections about the dashed vertical axes, together with

(d) the compositions of a rotation (b) with a reflection (c) (called *glide reflections* in horizontal axes through the rotation poles × with the translational parts $\pm\frac{1}{2}e_1, \pm\frac{3}{2}e_1, \pm\frac{5}{2}e_1, \ldots$)

form a crystallographic space group G.

In the following, the isometries of G are always described by the coordinates y_1, y_2 relative to the basis e_1, e_2.

---

[19] If G is symmorphic, i.e., $K \subset G$, then K and T as subgroups generate the group G.

Let us consider the elements listed in (c): A reflection in the axis

$$y_1 = \frac{1}{4} + \frac{k}{2} \qquad k = \dots, -2, -1, 0, 1, 2, \dots$$

means

$$y_1' = y_1 + 2(\frac{1}{4} + \frac{k}{2} - y_1) = k + \frac{1}{2} - y_1 \qquad y_2' = y_2$$

That is,

$$\begin{bmatrix} y_1' \\ y_2' \end{bmatrix} = \begin{bmatrix} -1 & 0 \\ 0 & 1 \end{bmatrix} \begin{bmatrix} y_1 \\ y_2 \end{bmatrix} + \begin{bmatrix} k + \frac{1}{2} \\ 0 \end{bmatrix} \qquad (8.88)$$

The rotational part describes a vertical reflection V in the y_2-axis, which indeed carries the lattice into itself. For later use, we record the following elements of G:

$$s_{180°} = 180° \text{ rotation about the origin O:}$$

$$\begin{bmatrix} y_1' \\ y_2' \end{bmatrix} = \begin{bmatrix} -y_1 \\ -y_2 \end{bmatrix} = \underbrace{\begin{bmatrix} -1 & 0 \\ 0 & -1 \end{bmatrix}}_{180° \text{ rot.}} \begin{bmatrix} y_1 \\ y_2 \end{bmatrix} \qquad (8.89)$$

s_V = reflection in the vertical line $y_1 = 1/4$:

$$\begin{bmatrix} y_1' \\ y_2' \end{bmatrix} = \begin{bmatrix} \frac{1}{2} - y_1 \\ y_2 \end{bmatrix} = \underbrace{\begin{bmatrix} -1 & 0 \\ 0 & 1 \end{bmatrix}}_{V} \begin{bmatrix} y_1 \\ y_2 \end{bmatrix} + \begin{bmatrix} \frac{1}{2} \\ 0 \end{bmatrix} \qquad (8.90)$$

s_H = glide reflection in the horizontal line $y_1 = 0$
with translational part $\frac{1}{2}e_1$:

$$\begin{bmatrix} y_1' \\ y_2' \end{bmatrix} = \begin{bmatrix} \frac{1}{2} + y_1 \\ -y_2 \end{bmatrix} = \underbrace{\begin{bmatrix} 1 & 0 \\ 0 & -1 \end{bmatrix}}_{H} \begin{bmatrix} y_1 \\ y_2 \end{bmatrix} + \begin{bmatrix} \frac{1}{2} \\ 0 \end{bmatrix} \qquad (8.91)$$

We leave it to the reader to verify by elementary geometry, as in (8.88), that each element of G differs from the four elements identity, $s_{180°}$, s_V, s_H only by an integral translative part.

Thus

the crystal class K = the plane dihedral group D_2

and consists of the two reflections V, H in the coordinate axes, the $180°$ rotation about the origin O, and the identity[20]. G is non-symmorphic; because, for example, V does not belong to G.

In analogy to Example 8.1, we attempt to simplify the eigenvalue problem (8.44) of our non-symmorphic crystal by group theory methods where the following requirements must be met:

---

[20] In this example, K is identical with the holohedral class of the lattice.

(a) The potential v admits the space group G; that is,

$$v(y) = v((R/t)y) \qquad \forall\, (R/t) \in G \qquad\qquad (8.92)$$

(b) The wave function ψ is lattice periodic; that is,

$$\psi(y) = \psi(y+n) \qquad \forall \text{ integers } n \qquad\qquad (8.93)$$

In the Hilbert space of lattice periodic functions with the inner product

$$\langle f, g \rangle = \int_0^1 \int_0^1 f(y)\overline{g(y)}\, dy_1\, dy_2 \qquad\qquad (8.94)$$

we select as an orthonormal basis the functions

$$e(q, y) = e^{2\pi i (q, y)} = e^{2\pi i (q_1 y_1 + q_2 y_2)} \qquad\qquad (8.95)$$

Here $q = (q_1, q_2)$, with integers q_1, q_2, is a vector of the reciprocal lattice.

Now we construct a representation $(R/t) \to D(R/t)$ of G by transplantation[21] according to (4.47). By Theorem 8.3 and Equation (8.86), the basis vectors (8.95) transform as follows:

$$\left. \begin{aligned} D(R/t) : e(q, y) \to e(q, (R/t)^{-1}y) &= e(q, R^{-1}y - R^{-1}t) \\ &= e(q, -R^{-1}t) \cdot e(q, R^{-1}y) \\ &= e(Rq, -t) \cdot e(Rq, y) \end{aligned} \right\} \qquad (8.96)$$

Since R leaves the lattice Γ invariant and thus also the reciprocal lattice Γ^{-1} (which consists of vertically positioned rectangles)[22], $Rq = q'$ is again a lattice vector of Γ^{-1}. The representation thus sends every function of (8.95), up to a factor of absolute value 1, into another such function.

Since all elements of G differ from the elements (8.89), (8.90), (8.91), and from the identity $(E/0)$ only by an integral translative part (see (8.88), for example), there exist at most 4 distinct representing transformations, namely,

$$D(s_{180°}), \quad D(s_V), \quad D(s_H), \quad E \qquad\qquad (8.97)$$

because *by periodicity in t, the scalar factor $e(Rq, -t)$ is insensitive against integral changes of t.*

Hence by (8.85), the transformations (8.97) are composed of the corresponding elements of the crystal class $D_2 = \{180° \text{ rotation}, V, H, \text{identity}\}$. *Therefore, they can be regarded as a representation of the group D_2.*

Assumption 1 of Theorem 8.10 is satisfied upon choosing a vector star in the reciprocal lattice invariant under the group D_2. Then (8.96) describes a finite-dimensional *representation ϑ by pseudo-permutation matrices:* In each row and each column there exists a unique nonzero element of absolute value 1.

---

[21] The definition of transplantation in Section 4.3 was given only for linear homogeneous transformations, but certainly also applies to the inhomogeneous case.
[22] See Problem 8.10.

Assumption 2 is also satisfied, because the representation is unitary. Finally, the operator (8.44) satisfies Assumption 3, because (8.92) holds, but also because the Δ-operator commutes with any of the isometries.

Therefore, the finite-dimensional representation (8.97) of the group D_2 can be completely reduced by the algorithm "Generating symmetry adapted basis vectors" (Section 5.2). By employing the fundamental theorem of Section 2.3, the eigenvalue problem of the Ritz operator M (8.51) is simplified by group theory methods.

Finally, let us mention that we can always pass from the crystallographic space group G to the corresponding (finite) crystallographic point group K, once (8.92), (8.93) are satisfied[23].

Remark. For more general problems (e.g., Bloch waves in a non-symmorphic crystal) the representation of the crystallographic space group can no longer be replaced by a representation of a finite group in the way described above.

PROBLEMS

Problem 8.1 For the hexagonal lattice in Fig. 8.2 find the matrices describing the holohedral crystal class of the lattice relative to the basis e_1, e_2 (group D_6).

Problem 8.2 A spatial lattice Γ is spanned by the three edges of a regular tetrahedron that meet at the vertex O. Construct the reciprocal lattice Γ^{-1}.

Problem 8.3 Give two generating elements for each of the groups T and O (see (5.111), (5.112)). Also find matrices of their irreducible representations.

Problem 8.4 Which are the character tables of $C_3 + ZC_3$ and (C_3, D_3)?

Problem 8.5 Verify without group theory methods that in Example 8.1 the function $\sin 2\pi y_1$ is an eigenfunction in the Ritz sense. To this end, apply the Hamiltonian operator H of (8.43) and project the result to the selected Ritz space. Observe (8.45), (8.46), (8.47).

Problem 8.6 Find the remaining eigenvalues and eigenfunctions of the Ritz matrix (8.67) of Example 8.1.

Problem 8.7 Treat Example 8.1 with the vector star drawn in the figure on the right.
Hint. The symmetry adapted basis vectors are tabulated in Table (3.7). Group theory methods reduce the eigenvalue problem to solving quadratic equations and a cubic equation (compare (3.11)).

Problem 8.8 Find the eigenvalues and eigenvectors of Example 8.2 for the vector star A, B, C, D, N, O (Fig. 8.7).

---

[23] This involves the isomorphism $G/T \cong K$, where G/T is the factor group of G modulo T.

Problem 8.9 Let a hexagonal lattice be given according to Fig. 8.2 and spanned by e_1, e_2. Solve the eigenvalue problem

$$-\frac{\hbar^2}{2m}\Delta\psi(y) + v(y)\psi(y) = E\psi(y)$$

Assume that the potential v is lattice periodic and that it admits the point group D_6 of the lattice. ψ is required to be lattice periodic. In the reciprocal lattice choose the vector star as in Fig. 8.2:

$$q_1 = \text{zero vector}, \quad q_2 = \epsilon_1, \quad q_3 = \epsilon_2, \quad q_4 = -\epsilon_1 - \epsilon_2$$

(Caution: This star admits only the dihedral group D_3.)

Problem 8.10 Show that if a point group K leaves a lattice Γ invariant, then it also leaves the reciprocal lattice Γ^{-1} invariant.

Problem 8.11 Treat Example 8.3 with the vector star consisting of the zero vector, the vectors $\pm\epsilon_1$ and $\pm\epsilon_2$ of the reciprocal lattice (compare Fig. 8.8).
Hint. Use the algorithm "Generating symmetry adapted basis vectors".

Problem 8.12 (The group of all proper isometries of the plane $\mathbf{R}^2$)
With respect to a cartesian coordinate system the group of all proper isometries of the plane $\mathbf{R}^2$ are the transformations

$$(R/t) : x \rightarrow x' = Rt + x \quad \text{where } R \in SO(2),$$
$$t \text{ is an arbitrary translation vector}$$

Here x, x' range through the points of the plane.

(a) Show by graphical means: Each transformation (R/t) with $R \neq E$ is a rotation about a well-defined point of the plane through the same oriented rotation angle φ as appearing in the rotation described by R about O.
  Hint. Regard each of R and t as a combination of two axial reflections and show the existence of a fixed point $x_0 = Rx_0 + t =$ center of rotation.

(b) Describe all similarity classes of this group by aid of (5.110).

Chapter 9

Unitary and Orthogonal Groups

Here we discuss the representation theory of the compact Lie groups $U(n)$, $SU(n)$, $SO(n)$[1]. We shall describe algorithms to determine the weights of a given representation ϑ in the sense of Cartan's method (thus generalizing Sections 4.3 and 4.4). This also allows us to determine the irreducible representations constituting ϑ. We shall frequently omit the proofs, referring to the literature instead.

9.1 The Groups $U(n)$ and $SU(n)$

We first consider the unitary group $U(n)$ of all $n \times n$ matrices A with $A^T \bar{A} = E$ (see (4.3)). The case $SU(2)$ was discussed in detail in Chapter 4. Hence we shall work with $n = 3$, noting that dealing with higher dimensions n does not require any additional care.

In Sections 4.3.3 and 4.5.2 we pointed out that the representation theory of continuous groups G is based on an appropriately chosen abelian subgroup T of G. For $G = U(3)$, the abelian subgroup T consists of all unitary 3×3 diagonal matrices. The reader can readily verify that T possesses the following **maximality property**: There are no other matrices of $SU(3)$ that commute with every matrix of T. The main diagonal of each matrix of T consists of complex numbers of absolute value 1, thus can be written in the form

$$e^{2\pi i x_1}, \qquad e^{2\pi i x_2}, \qquad e^{2\pi i x_3} \qquad \text{where } x_i \in \mathbf{R} \qquad (9.1)$$

Here the x_i are determined up to an additive integer. If x_1, x_2, x_3 are regarded as cartesian coordinates of space (see Fig. 9.1), then two points in space represent the same element of T provided they are sent into each

---

[1]Evidently, this automatically includes the groups $O(n)$. See (Miller, 1972, Ch. 9).

other by a translation of the cubic lattice Γ (Section 8.1) spanned by the unit vectors e_1, e_2, e_3. In the unit cube ($0 \leq x_i \leq 1$) spanned by these vectors there exists to each element (9.1) of T a unique point x, provided that equivalent points of Γ lying in opposite faces of the cube are identified. This makes the cube into a 3-dimensional torus. Therefore, T is also called the **maximal torus of U(3)**.

9.1.1　Similarity Classes, Diagram

To each unitary matrix A there exists a unitary matrix U such that UAU^{-1} is diagonal and hence an element of T. In other words, T contains at least one element from each similarity class of $U(3)$. The character[2] of a representation of $U(3)$ can thus be regarded as a function on T and hence as a periodic function in the space of Fig. 9.1, the period in all three coordinate directions being 1.

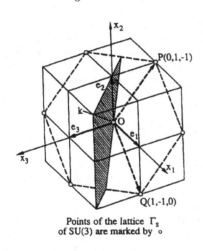

Two elements x, x' of T may lie in the same similarity class; namely, if and only if the coordinates x_1', x_2', x_3' are the images of a permutation of the coordinates x_1, x_2, x_3. For, when transforming a matrix to diagonal form, the diagonal elements are determined up to their order only. In interpreting geometrically the 6 permutations of the coordinates x_1, x_2, x_3 in Fig. 9.1, we obtain, apart from the identity, three reflections in diagonal planes passing through the interior diagonal k of the cube (one of which is shaded in Fig. 9.1) together with two rotations about k through angles $\pm 120°$.

Points of the lattice Γ_s of SU(3) are marked by ○

Fig. 9.1

They constitute the point group $(C_3, D_3) = C_{3v} = 3m$ (see (8.24)) shown in Fig.9.2.

　　Thus two points x, x' that are carried into each other by a transformation of the group (C_3, D_3) describe elements of T which belong to the same similarity class of $U(3)$. The point group (C_3, D_3) is called the **Weyl[3] group** ψ associated with the continuous group $U(3)$.

---

[3]Weyl, Hermann (1885-1955). Held professorships at ETH, Zürich from 1913, in Göttingen from 1930, and in Princeton from 1933. Worked in all fields of mathematics and particularly in representation theory of continuous groups. Collected works published by Springer, 1968.

The translations of the lattice Γ together with ψ generate a discontinuous symmorphic space group[4] denoted by Ω. We may summarize this as follows: Two points $x, x' \in \mathbf{R}$ describe similar elements in U(3) if and only if they are mapped into each other by a transformation of Ω of $\mathbf{R}^3$. This already indicates a remarkable *relation between the continuous group* U(3) *and the discontinuous space group* Ω.

The region $x_1 \geq x_2 \geq x_3$ is a **fundamental domain** of the corresponding Weyl group $\psi = (C_3, D_3)$; that is, to every point $x \in \mathbf{R}^3$ of this domain there exists a unique point which is equivalent with respect to (C_3, D_3). This is because the three numbers x_k can be permuted into an arrangement by ascending order.

Similar statements hold for the group SU(3) of unitary matrices with det $A = +1$. For the subgroup T_S of all diagonal matrices of SU(3) the condition that $x_1 + x_2 + x_3$ be an integer must be added to (9.1). Since the x_i are determined only up to additive integers, we may put, for example,

$$x_1 + x_2 + x_3 = 0 \tag{9.2}$$

Equation (9.2) describes a plane perpendicular to k spanned by the hexagon in Fig. 9.1. For SU(3), the cubic lattice is replaced by a plane hexagonal lattice Γ_S consisting of all lattice points of Γ lying in this hexagon plane. The lattice Γ_S is spanned by, say, the two vectors with coordinates $(1, -1, 0)$ and $(0, 1, -1)$. As in the case of U(3), two points equivalent with respect to the lattice Γ_S represent the same element of T_S.

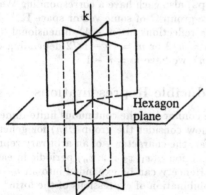

The point group (C_3, D_3) described above leaves our plane unchanged and furnishes in it the plane dihedral group D_3 as the Weyl group ψ for SU(3). This group can be generated by reflections in the lines of intersection of the three planes of mirror with the hexagon plane (see Figs. 9.1, 9.2, 9.3).

Fig. 9.2

As in the preceding, the lattice translations of Γ_S and the group D_3 generate the corresponding discrete and now 2-dimensional space group Ω of SU(3). The group Ω can also be generated by reflections about the three families of parallel equidistant reflection axes that are plotted in Fig. 9.3. For, the composition of two reflections about neighboring parallel lines perpendicular to OP or OQ yields the primitive lattice vector $\overrightarrow{OP}$ or $\overrightarrow{OQ}$, respectively.

---

[4]See Section 8.5.

The group Ω of SU(3) is also called the **kaleidoscope group**; since, if we think of three mirrors placed on each side of one of the small triangles of Fig. 9.3 and perpendicular to the plane of the paper, then we obtain a familiar toy known as the kaleidoscope.

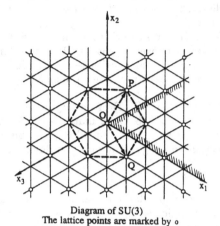

The figure of the three families of reflection axes is called the **diagram** of the special unitary group SU(3). A fundamental domain of the Weyl group D_3 is bounded by two adjacent reflection half axes (see Section 3.1) and is shaded in Fig. 9.3.

Diagram of SU(3)
The lattice points are marked by o

Fig. 9.3

Let us remark generally that aside from U(n), SU(n), and SO(n), the **symplectic** and the 5 **exceptional groups** (see (Stiefel, 1942)) (this list includes the so-called **Cartan groups**) also each have a corresponding Weyl group ψ and a discontinuous space group Ω of some vector space $\mathbf{R}^m$. In these cases, Ω is generated by the reflections in $(m-1)$-dimensional hyperplanes. For SO(n), we have $m = n/2$ or $m = (n-1)/2$ depending on whether n is even or odd; for SU(n), we have $m = n - 1$.

9.1.2 Characters of Irreducible Representations

In this chapter our interest will be confined to the continuous finite-dimensional representations. We shall now consider the group U(n) for general n. From the statements made so far, the character χ of an arbitrary representation ϑ of U(n) clearly is a function $\chi(x_1, x_2, \ldots, x_n)$ periodic in each of its variables x_k with period 1. Hence χ can be expanded into an n-fold Fourier series, i.e., into a linear combination of expressions of the form[5]

$$e^{2\pi i(q_1 x_1 + q_2 x_2 + \ldots + q_n x_n)} = e(q_1 x_1 + q_2 x_2 + \ldots + q_n x_n) = e(q, x) \qquad (9.3)$$

Here the q_k are integers, q is the vector with components q_k, and (q, x) is meant to suggest the inner product of the fixed vector q with the position vector x. Moreover, $\chi(x_1, x_2, \ldots, x_n)$ must be a function **automorphic** with respect to the corresponding Weyl group ψ; that is, it must remain invariant under the permutations of the x_k.

---

[5] We adopt the short notation from (8.49).

If ϑ is irreducible, then the Fourier series of χ will break off after a finite number of terms (9.3). Thus the characters of irreducible representations are **automorphic Fourier polynomials**, which can be written out explicitly. To do this, we introduce the notion of the **alternating elementary sum** $S_\ell(x_1, x_2, \ldots, x_n)$, where $\ell = (\ell_1, \ell_2, \ldots, \ell_n)$ is an integral vector with $\ell_1 > \ell_2 > \ldots > \ell_n$, that is, a vector lying in the interior of the fundamental domain of the Weyl group ψ of $U(n)$. If we denote the transformations of ψ by R, then S_ℓ is defined as follows:

$$S_\ell(x_1, x_2, \ldots, x_n) = \sum_{R \in \psi} \pm e(R\ell, x) \qquad (9.4)$$

Here $+$ is used for proper, $-$ for improper rotations R. More explicitly, S_ℓ is given by a complete permutation of the ℓ_k ($+$ for even, and $-$ for odd permutations).

In the special case of $n = 3$, we have

$$S_\ell(x_1, x_2, x_3) = \qquad (9.5)$$
$$\begin{cases} +e(\ell_1 x_1 + \ell_2 x_2 + \ell_3 x_3) + e(\ell_2 x_1 + \ell_3 x_2 + \ell_1 x_3) + e(\ell_3 x_1 + \ell_1 x_2 + \ell_2 x_3) \\ -e(\ell_1 x_1 + \ell_3 x_2 + \ell_2 x_3) - e(\ell_3 x_1 + \ell_2 x_2 + \ell_1 x_3) - e(\ell_2 x_1 + \ell_1 x_2 + \ell_3 x_3) \end{cases}$$

As the name "alternating elementary sum" suggests, the action of $R \in \psi$ on a point x leaves the sum S_ℓ invariant or changes its sign, depending on whether R is a proper or an improper rotation. The reason for excluding ℓ-vectors on the boundary of the fundamental domain $\ell_1 \geq \ell_2 \geq \ldots \geq \ell_n$ is that the sum (9.4) vanishes identically if among the ℓ_k's there is an equality.

In the following, we shall frequently use the particularly simple alternating elementary sum denoted by Δ and corresponding to the vector $\ell = (n - 1, n - 2, \ldots, 1, 0)$. The following holds for $U(n)$:

Theorem 9.1 *Let $\ell = (\ell_1, \ell_2, \ldots, \ell_n)$ be an integral vector with*

$$\ell_1 > \ell_2 > \ldots > \ell_n \qquad (9.6)$$

and

$$S_\ell(x) = \sum_{R \in \psi} \pm e(R\ell, x) \qquad (9.7)$$

be the corresponding alternating elementary sum, where R ranges through all permutations of the integral components ℓ_k ($k = 1, 2, \ldots, n$) of the vector ℓ. Let further Δ be the alternating elementary sum associated with the vector $(n - 1, n - 2, \ldots, 1, 0)$. Then

$$\chi_\ell(x) = \frac{S_\ell(x)}{\Delta} \qquad (9.8)$$

is the character of a general irreducible representation ϑ_ℓ of $U(n)$. Here the characters for different ℓ-vectors are pairwise distinct, hence they belong to inequivalent representations. The list obtained describes all irreducible inequivalent continuous representations of $U(n)$.

For proofs we refer to (Weyl, 1950), (Stiefel, 1944).

Obviously, the character χ_ℓ is an automorphic function since it is the quotient of two alternating elementary sums. One can show that the division in (9.8) has no remainder, hence gives a Fourier polynomial. We shall carry out the division in the next subsection.

Prior to this, let us digress on the problem of **extending these results to the group SU(n)**. This is done in a few words, for the representations ϑ_ℓ in Theorem 9.1 will turn out to remain irreducible when restricted to the subgroup SU(n). Hence all irreducible representations of SU(n) are automatically included in the treatment. Furthermore, we have the following property, useful for **normalization:** The exponential term $e(\ell, x)$ is insensitive against addition of an integral vector $(d, d, \ldots, d)$ to ℓ, because

$$
\begin{aligned}
& e((\ell_1 + d)x_1 + (\ell_2 + d)x_2 + \ldots + (\ell_n + d)x_n) \\
= {} & e(\ell_1 x_1 + \ell_2 x_2 + \ldots + \ell_n x_n) \cdot e(d(x_1 + x_2 + \ldots + x_n)) \\
= {} & e(\ell_1 x_1 + \ell_2 x_2 + \ldots + \ell_n x_n)
\end{aligned}
$$

as, in the diagram plane, $x_1 + x_2 + \ldots + x_n = 0$.

Hence we may confine ourselves to ℓ-vectors with, say, $\ell_n = 0$. This is restated more rigorously in

Theorem 9.2 *If in (9.6) the ℓ-vectors are restricted to*

$$\ell_n = 0 \tag{9.9}$$

then Theorem 9.1 for U(n) applies also to the group SU(n) verbatim if, as in the case of U(n), the transformation R ranges through all permutations of the integral components ℓ_k (k = 1, 2, ..., n) of the vector ℓ.

Remark. The irreducible representations of the groups U(n) and SU(n) can be constructed by means of the so-called **Young tableaux**; see (Miller, 1972, Ch. 9).

9.1.3   Division Algorithm

This method will be explained by aid of the irreducible representation corresponding to the vector $\ell = (5, 2, 0)$ of U(3) or SU(3). The division (9.8) is computed via the exponents of the summands of S_ℓ and Δ, that is, in practice via the integral factors of x_1, x_2, x_3, in this order. The denominator Δ is written in abbreviated form on the following **auxiliary file:** It ranges over all images $R\ell$ of the vector $\ell = (2, 1, 0) = 2\ 1\ 0$ [6] under the action of the group ψ. Here the minus sign on the right indicates the sign of the corresponding summand. We comment on the various steps below

---

[6]In the following, we shall frequently use this simplified vector notation without brackets.

(see Table (9.10)).

$$
\begin{array}{ccc}
2 & 1 & 0 \\
1 & 0 & 2 \\
0 & 2 & 1 \\
2 & 0 & 1- \\
0 & 1 & 2- \\
1 & 2 & 0-
\end{array}
\qquad (9.10)
$$

auxiliary file

Row	What?	Computation of vectors						Number of vectors
Z_1		5 2 0						
Z_2	Δ	2 1 0						
Z_3	$\psi(Z_1 - Z_2)$	3 1 0	1 0 3	0 3 1	3 0 1	0 1 3	1 3 0	6
Z_4	$\Delta \cdot Z_3$	5 2 0 4 3 0−			4 2 1−			
Z_5	$Z_1 - Z_4$	4 3 0		4 2 1				
Z_6	Δ	2 1 0						
Z_7	$\psi(Z_5 - Z_6)$	2 2 0	2 0 2	0 2 2				3
Z_8	$\Delta \cdot Z_7$	4 3 0 4 2 1−						
Z_9	$Z_5 - Z_8$	4 2 1		4 2 1				
Z_{10}	Z_9	4 2 1$_2$						
Z_{11}	Δ	2 1 0						
Z_{12}	$\psi(Z_{10} - Z_{11})$	2 1 1$_2$	1 2 1$_2$ 1 1 2$_2$					$3 \cdot 2 = 6$
Z_{13}	$\Delta \cdot Z_{12}$	4 2 1$_2$						
Z_{14}	$Z_{10} - Z_{13}$	—						total 15

Description of the rows

Z_1 : The vector associated with the irreducible representation.

Z_2 : The vector associated with the denominator function Δ.

Z_3 : Division of the exponential terms amounts to subtracting the vectors. Since the quotient χ_{520} of the division must be an **automorphic function**, there must result 6 vectors from the action of ψ on 3 1 0. They provide the so-called **automorphic elementary sum**

$$
\begin{aligned}
\Sigma \ = \ & e(3x_1 + x_2) + e(x_1 + 3x_3) + e(3x_2 + x_3) \\
& + e(3x_1 + x_3) + e(x_2 + 3x_3) + e(x_1 + 3x_2)
\end{aligned}
\qquad (9.11)
$$

Z_4 : As in the usual division of polynomials the "intermediary multiplication" $\Sigma \cdot \Delta$ is performed by placing the auxiliary file under each vector of Z_3 and by adding (there is a total of 36 additions). *The sign is given by the usual multiplication rules.* The result is an alternating function. Only the vectors lying in the interior of the fundamental domain (9.6), that is, only those satisfying the condition $\ell_1 > \ell_2 > \ell_3$, are listed. For example, 3 1 0 (Z_3) + 1 0 2 (auxiliary file) = 4 1 2 violates this condition; and so does 3 1 0 (Z_3) + 0 2 1 (auxiliary file) =

3 3 1, because the vector 3 3 1 lies on the boundary of the fundamental domain. However, for example, 3 0 1 (Z_3) + 1 2 0− (auxiliary file) = 4 2 1− satisfies the requirement.

Z_5 : Find the remainder of the division by performing the subtraction $Z_1 - Z_4$. Note that the remainder is just the *sum of the two alternating elementary sums* associated with the vectors 4 3 0 and 4 2 1, i.e., it is just $S_{430}(x) + S_{421}(x)$. We define an ordering as follows: A vector $\ell_1 \ell_2 \ell_3$ is said to be **higher** than $\ell'_1 \ell'_2 \ell'_3$ if among the differences $\ell_1 - \ell'_1, \ell_2 - \ell'_2, \ell_3 - \ell'_3$ the first nonzero difference is positive, or if all are zero (identical vectors). A vector from the set of ℓ-vectors is called **dominant** if it is higher than any other vector of the set. *The division method is always iterated by taking the dominant vector of the remainder*, that is, here by taking 4 3 0, as it is higher than 4 2 1.

Z_7 : The vector 2 2 0 lies on the boundary of the fundamental domain and action of ψ produces only two more pairwise distinct vectors. Thus the automorphic function described by them consists of three summands, each of which is counted once only.

Z_9 : Subtraction $Z_5 - Z_8$ yields the vector 4 2 1 twice.

Z_{10} : The two vectors in Z_9 are combined: The subscript 2 indicates that the corresponding alternating elementary sum has to be taken twice.

Now the character χ_{520} is the sum of exponential functions whose exponent vectors are given in the three ψ-rows. The last column of Table (9.10) shows the number of these terms; in total there are 15 of them. Thus the representation is 15-dimensional, because the dimension equals the character value of the unit element, that is, in U(3) of $x_1 = x_2 = x_3 = 0$.
The character itself is

$$\chi_{520} = \left\{ \begin{array}{l} e(3x_1 + x_2) + e(x_1 + 3x_3) + e(3x_2 + x_3) \\ +e(3x_1 + x_3) + e(x_2 + 3x_3) + e(x_1 + 3x_2) \\ +e(2x_1 + 2x_2) + e(2x_1 + 2x_3) + e(2x_2 + 2x_3) \\ +2[e(2x_1 + x_2 + x_3) + e(x_1 + 2x_2 + x_3) + e(x_1 + x_2 + 2x_3)] \end{array} \right. \tag{9.12}$$

A convenient check may be obtained by noting that the dimension d of an arbitrary irreducible representation ϑ_ℓ of U(n) or SU(n) corresponding to the vector $\ell = (\ell_1, \ell_2, \ldots, \ell_n)$ is given by

$$d = \frac{\prod_{i<j}(\ell_i - \ell_j)}{\prod_{i>j}(i - j)} \tag{9.13}$$

Some special **examples** (compare Theorems 9.1 and 9.2):

1. ϑ_{210} is the unit representation of U(3) and SU(3).

2. ϑ_{321} assigns to each group element (9.1) the number $e(x_1 + x_2 + x_3)$, which happens to be the determinant. This holds for every element of U(3). For SU(3) it is again the unit representation, as in the foregoing Example 1.

3. ϑ_{310} has the character $e(x_1) + e(x_2) + e(x_3)$, hence it is the representation of the group U(3) or SU(3) by itself.

4. ϑ_{410} has, by the algorithm (9.10), the character

$$e(2x_1) + e(2x_2) + e(2x_3) + e(x_1 + x_2) + e(x_1 + x_3) + e(x_2 + x_3) \quad (9.14)$$

9.1.4 Weights

Our methods accomplish more than just finding characters. Every representation $\vartheta : s \rightarrow D(s)$ of U(n) or SU(n) is completely reducible, as we are working with compact Lie groups. By statement (α) in the text after (5.197), the representation ϑ is equivalent to one by unitary matrices. Furthermore, Corollary 2.4 applies to the representation ϑ restricted to the torus of the group in question. As all representing matrices of $D(s)$ have eigenvalues of absolute value 1, this implies that *the exponential terms occurring as summands in the character of ϑ* (see Section 9.1.3) *are identical with the diagonal entries of the representing diagonal matrices of the torus.*

As in Chapter 4, the integral vectors $q = (q_1, q_2, \ldots, q_n)$ occurring in these diagonal matrices in the form of $e(q_1 x_1 + q_2 x_2 + \ldots + q_n x_n)$ are called the **weights** of the corresponding representation ϑ of U(n) or SU(n). In other words, *the division algorithm (9.10) provides the weights in the ψ-rows and hence the representing diagonal matrices of ϑ for the elements of the torus* T. To know the diagonal matrices completely, it suffices to find the weights (together with the multiplicities) lying in the fundamental domain

$$q_1 \geq q_2 \geq \ldots \geq q_n \quad (9.15)$$

of the Weyl group ψ. In the algorithm (9.10), the first vectors in the ψ-rows for the computed example are then the vectors

$$3\ 1\ 0, \qquad 2\ 2\ 0, \qquad 2\ 1\ 1_2 \quad (9.16)$$

Permuting the components of the various vectors (9.16) furnishes all of the 15 weights, and so one obtains the diagonal elements that in (9.12) appear as summands.

Remarks.

1. In the case of SU(n), the weight vectors q can be normalized to, say, $q_n = 0$ (see the paragraph preceding Theorem (9.2)).

2. For large values of n, the algorithm just described is too labor-intensive and consumes considerable amount of computer time, because the rapid growth of the order of the Weyl group ψ of U(n) is decisive for the computational complexity. Freudenthal found an algorithm that largely avoids the action of this group; see (Freudenthal, 1954), also see (Flöscher, 1978).

9.1.5   Algorithm for the Complete Reduction

Complete reducibility of the representation ϑ of U(n) or SU(n) implies that $\vartheta = \bigoplus_j c_j \vartheta_j$, where ϑ_j denotes the irreducible inequivalent representation occurring with multiplicity $c_j \geq 1$. The subscripts $j = (j_1, j_2, \ldots, j_n)$ are integral n-dimensional vectors. If χ and χ_j are the characters of ϑ and ϑ_j, then Theorem 9.1 implies that

$$\chi = \sum_j c_j \chi_j = \frac{1}{\Delta} \sum_j c_j S_j$$

where again S_j is the alternating elementary sum associated with the vector j. We rewrite the above equation in the form

$$\chi \cdot \Delta = \sum_j c_j S_j \tag{9.17}$$

The **algorithm for complete reduction** now consists in *computing the alternating function $\chi \cdot \Delta$ by the auxiliary file method and then reading off the reduction directly from Equation (9.17).*

Example. Complete reduction of the product $\vartheta = \vartheta_\ell \otimes \vartheta_m$, where ϑ_ℓ and ϑ_m are two irreducible representations of U(n) or SU(n) corresponding to the vectors $(\ell_1, \ell_2, \ldots, \ell_n)$ and $(m_1, m_2, \ldots, m_n)$.
By (5.133), the character of $\vartheta = \vartheta_\ell \otimes \vartheta_m$ satisfies

$$\chi = \chi_\ell \cdot \chi_m \tag{9.18}$$

As a computational example we select the two representations ϑ_ℓ and ϑ_m of U(3) or SU(3) with $\ell = 3\ 1\ 0$ and $m = 4\ 1\ 0$ (see the Examples 3 and 4 after (9.13)). Their characters are

$$\chi_{310} = e(x_1) + e(x_2) + e(x_3)$$
$$\chi_{410} = e(2x_1) + e(2x_2) + e(2x_3) + e(x_1 + x_2) + e(x_2 + x_3) + e(x_1 + x_3)$$

Consequently, the corresponding weights are

$$
\begin{array}{llllll}
1\ 0\ 0 & 0\ 1\ 0 & 0\ 0\ 1 & & & \tag{9.19}\\
2\ 0\ 0 & 0\ 2\ 0 & 0\ 0\ 2 & 1\ 1\ 0 & 0\ 1\ 1 & 1\ 0\ 1 \qquad (9.20)
\end{array}
$$

The $3 \cdot 6 = 18$ sums, each formed by a vector of the first row and a vector of the second row, yield the weights of ϑ. Among the results we only record those lying in the fundamental domain, namely,

$$3 \ 0 \ 0 \qquad 2 \ 1 \ 0_2 \qquad 1 \ 1 \ 1_3 \tag{9.21}$$

The product $\chi \cdot \Delta$ in (9.17) yields an alternating function with the two vectors 5 1 0 and 4 2 0 in the fundamental domain. Thus the complete reduction is given by

$$\vartheta_{310} \otimes \vartheta_{410} = \vartheta_{510} \oplus \vartheta_{420} \tag{9.22}$$

Checking dimensions via (9.13) yields $3 \cdot 6 = 10 + 8$.

For Kronecker products $\vartheta_\ell \otimes \vartheta_m$, calculations may be simplified as follows: Equation (9.17), together with $\chi = \chi_\ell \cdot \chi_m$ and $\chi_m \cdot \Delta = S_m$, implies

$$\chi_\ell \cdot S_m = \sum_j c_j S_j \tag{9.23}$$

where $\chi_\ell \cdot S_m$ is computed by the auxiliary file method. We take the same example $\ell = 3 \ 1 \ 0$, $m = 4 \ 1 \ 0$ as above:

Weight of ϑ_{310} in the fundamental domain $\ . \ = \quad 1 \ 0 \ 0$

Vector of ϑ_{410} in the interior of the fundamental domain $= \quad 4 \ 1 \ 0$

4	1	0
1	0	4
0	4	1
4	0	1-
0	1	4-
1	4	0-

auxiliary file

Row	What?	Computation of vectors		
Z_1	$\psi(1\,0\,0)$	1 0 0	0 0 1	0 1 0
Z_2	$Z_1 \cdot$ auxiliary file	5 1 0		4 2 0

$$\tag{9.24}$$

Description of the rows

Z_1 : Action of ψ produces all weights of ϑ_{310}.

Z_2 : Multiplication of Z_1 with the auxiliary file. Only the vectors in the interior of the fundamental domain are recorded.

From row Z_2 we see that $\chi_{310} \cdot S_{410} = S_{510} + S_{420}$, which again implies the desired complete reduction (9.22).

Remark. There are other computational methods; for instance, we can perform the complete reduction directly by generalizing Cartan's method (Section 4.3.3), provided we know the characters of the irreducible representations (division method). This involves only automorphic functions.

9.2   The Special Orthogonal Group $SO(n)$

Arguments similar to those of Section 9.1 work here. We first consider the group $SO(5)$ as a representative example. A maximal torus T consists of matrices of the form

$$
\begin{bmatrix}
1 & & 0 \\
 & D(\varphi_1) & \\
0 & & D(\varphi_2)
\end{bmatrix}
\quad \text{with } D(\varphi_i) = \begin{bmatrix} \cos\varphi_i & -\sin\varphi_i \\ \sin\varphi_i & \cos\varphi_i \end{bmatrix} \begin{array}{l} i = 1,2 \\ \varphi_i \in \mathbf{R} \end{array} \Bigg\}
$$

$$(9.25)$$

Two matrices M and M' of T with the boxes 1, $D(\varphi_1)$, $D(\varphi_2)$ and 1, $D(\varphi_1')$, $D(\varphi_2')$, respectively, along the main diagonal are similar with respect to $SO(5)$ if and only if

$$
\left. \begin{array}{l}
\varphi_1' = \pm\varphi_1 \\
\varphi_2' = \pm\varphi_2
\end{array} \right\} \quad \text{or} \quad \left. \begin{array}{l}
\varphi_1' = \pm\varphi_2 \\
\varphi_2' = \pm\varphi_1
\end{array} \right\} \tag{9.26}
$$

where all sign combinations are admitted. The reason is that one easily finds 8 pseudo-permutation matrices P with $PMP^T = M'$; furthermore, similar matrices have the same eigenvalues.

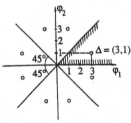

By (9.26), the Weyl group ψ of $SO(5)$ consists of the permutations of the φ_i with arbitrary sign changes. A geometric interpretation in the diagram plane of Fig. 9.4 is as follows: ψ is the plane dihedral group D_4 consisting of the four reflections in the plotted axes and the rotations about the origin through angles $k \cdot 90°$ ($k = 0, 1, 2, 3$).

Fig. 9.4

The fundamental domain $\varphi_1 \geq \varphi_2 \geq 0$ is shaded. As in Section 9.1, an irreducible representation of $SO(5)$ is determined by an integral vector $\ell = (\ell_1, \ell_2)^7$ in the fundamental domain (dimension of the vector ℓ = dimension of the torus). Both the division algorithm and the algorithm for complete reduction remain valid provided we take the vector $(3, 1)$ for Δ.

For example, let us determine **the weights of the irreducible representation ϑ_{53} of $SO(5)$**.

According to Table 9.29, we obtain for ϑ_{53} the weights

$$
\left. \begin{array}{cccccccccc}
2 & 2, & 2 & -2, & -2 & 2, & -2 & -2, & 2 & 0 \\
0 & 2, & -2 & 0, & 0 & -2, & 0 & 0, & 0 & 0
\end{array} \right\} \tag{9.27}
$$

---

[7]In addition, the numbers ℓ_1, ℓ_2 must be either both even or both odd; compare end of this Section.

Compared with Section 9.1, there is a slight modification here: The diagonal entry associated with some weight (q_1, q_2) is $e^{\frac{1}{2}(q_1\varphi_1 + q_2\varphi_2)}$. Hence, by (9.27), the character of ϑ_{53} is

$$\chi_{53} = e^{\frac{1}{2}(2\varphi_1 + 2\varphi_2)} + e^{\frac{1}{2}(2\varphi_1 - 2\varphi_2)} + \ldots + 1 + 1 \tag{9.28}$$
$$= 2[\cos(\varphi_1 + \varphi_2) + \cos(\varphi_1 - \varphi_2) + \cos\varphi_1 + \cos\varphi_2 + 1]$$

3	1
−1	3
−3	−1
1	−3
1	3−
−3	1−
−1	−3−
3	−1−

auxiliary file

Note that a negative sign on the left of a number refers to the vector component in question, while a sign on the right, as before, indicates the sign of the corresponding summand. In Fig. 9.4, the 8 vectors of the auxiliary file are marked by circles. In the terminology of Section 3.1 they form an orbit.

Row	What?	Computation of vectors				Number of vectors
Z_1		5 3				
Z_2	Δ	3 1				
Z_3	$\psi(Z_1 - Z_2)$	2 2	2 −2	−2 2	−2 −2	4
Z_4	$\Delta \cdot Z_3$	$\left\{\begin{array}{l} 5\ 3 \quad 3\ 1- \\ 5\ 1- \end{array}\right.$				
Z_5	$Z_1 - Z_4$	5 1	3 1			
Z_6	Δ	3 1				
Z_7	$\psi(Z_5 - Z_6)$	2 0	0 2	−2 0	0 −2	4
Z_8	$\Delta \cdot Z_7$	5 1	3 1−			
Z_9	$Z_5 - Z_8$		3 1₂			
Z_{10}	Δ	3 1				
Z_{11}	$\psi(Z_9 - Z_{10})$	0 0₂				$2 \cdot 1 = 2$
Z_{12}	$\Delta \cdot Z_{11}$	3 1₂				
Z_{13}	$Z_9 - Z_{12}$	—				total 10

$$\tag{9.29}$$

As another example we consider ϑ_{51}. In the fundamental domain the division algorithm yields the weights

$$2 \quad 0, \qquad 0 \quad 0 \tag{9.30}$$

Consequently, the character is

$$\chi_{51} = 1 + 2(\cos\varphi_1 + \cos\varphi_2) \tag{9.31}$$

But this is the trace of the matrix (9.25), so that ϑ_{51} is a representation of the group SO(5) by itself.

Finally, **a last example:** ϑ_{42} is a 4-dimensional representation with the weights

$$1 \quad 1, \qquad 1 \quad -1, \qquad -1 \quad 1, \qquad -1 \quad -1$$

Thus its character is

$$\chi_{42} = 2 \left[\cos \left(\frac{\varphi_1}{2} + \frac{\varphi_2}{2} \right) + \cos \left(\frac{\varphi_1}{2} - \frac{\varphi_2}{2} \right) \right] = 4 \cos \frac{\varphi_1}{2} \cos \frac{\varphi_2}{2} \qquad (9.32)$$

In the terminology of Section 4.3.3, this is a double-valued representation, since half angles occur as rotation angles.

9.2.1   Weyl Group and Diagram of SO(n)

For even numbers $n = 2\rho$ the torus of SO(n) consists of ρ proper 2×2 rotation matrices with arbitrary rotation angles $\varphi_1, \varphi_2, \ldots, \varphi_\rho \in \mathbf{R}$ arranged along the main diagonal (all other matrix entries are zero). For odd $n = 2\rho + 1$, there is an additional entry 1 in, say, the upper left hand corner (compare special case (9.25)). Hence in both cases the torus is ρ-dimensional.

If $\varphi_1, \varphi_2, \ldots, \varphi_\rho$ are regarded as cartesian coordinates of $\mathbf{R}^\rho$, then the Weyl group is generated by the reflections in the following $(\rho - 1)$-dimensional hyperplanes (subsequently called planes):

Case SO($2\rho + 1$):

1st kind:	$\varphi_i = 0$	$(i = 1, 2, \ldots, \rho)$
2nd kind:	$\varphi_i - \varphi_j = 0$	$(i < j)$
3rd kind:	$\varphi_i + \varphi_j = 0$	$(i < j)$

The reflections in these ρ^2 planes generate the so-called **hyper-octahedral group** as a Weyl group ψ. This group comprises all permutations of the φ_i with arbitrary sign changes, thus has group order $\rho! \cdot 2^\rho$. In geometrical parlance, it consists of all proper and improper rotations that carry the ρ-dimensional cube into itself. A rotation is proper if the permutation and the number of sign changes are either both even or both odd, otherwise the rotation is improper. As a **fundamental domain** we may take

$$\varphi_1 \geq \varphi_2 \geq \ldots \geq \varphi_\rho \geq 0 \qquad (9.33)$$

Case SO(2ρ):

1st kind:	$\varphi_i - \varphi_j = 0$	$(i < j)$
2nd kind:	$\varphi_i + \varphi_j = 0$	$(i < j)$

The reflections in these $\rho(\rho - 1)$ planes generate all permutations of the φ_i with even number of sign changes. Thus the Weyl group has order $\rho! \cdot 2^{\rho-1}$. Geometrically speaking, this Weyl group ψ is the subgroup of the hyper-octahedral group generated by reflections in diagonal planes of the ρ-dimensional cube. A rotation is proper or improper depending on whether the permutation is even or odd. As a **fundamental domain** we may take

$$\varphi_1 \geq \varphi_2 \geq \ldots \geq \varphi_{\rho-1} \geq \varphi_\rho \geq -\varphi_{\rho-1} \qquad (9.34)$$

In the case $\rho = 3$, the Weyl group $\psi \cong (T, O)$ is the full tetrahedral group.

The diagram of the group $SO(n)$ consists of infinitely many equidistant planes parallel to each of the various reflection planes of the Weyl group in question.

9.2.2 Algorithm for the Computation of Weights

The algorithms described in Section 9.1 for complete reduction and for finding the weights of a representation are valid also for the group $SO(n)$ provided one introduces the following modifications:

1. The integral vector $\ell = (\ell_1, \ell_2, \ldots, \ell_\rho)$ of an irreducible representation ϑ_ℓ must conform to the following conditions:

   (α) ℓ must lie in the interior of the fundamental domain of the Weyl group; that is, the following must hold:

   $$\left. \begin{array}{ll} \ell_1 > \ell_2 > \ldots > \ell_\rho > 0 & \text{for } SO(2\rho + 1) \\ \ell_1 > \ell_2 > \ldots > \ell_{\rho-1} > \ell_\rho > -\ell_{\rho-1} & \text{for } SO(2\rho) \end{array} \right\} \quad (9.35)$$

   (β) The integers ℓ_k must be either all even or all odd.

2. To determine Δ, we employ the vector

   $$n = (n_1, n_2, \ldots, n_\rho) = \begin{cases} (2\rho - 1, 2\rho - 3, \ldots, 1) & \text{for } SO(2\rho + 1) \\ (2\rho - 2, 2\rho - 4, \ldots, 0) & \text{for } SO(2\rho) \end{cases} \right\}$$
   $$(9.36)$$

3. Hence action of ψ on the integral vector $\ell = (\ell_1, \ell_2, \ldots, \ell_\rho)$ is equivalent to complete permutation of the ℓ_k with all possible sign changes, where in the case of $SO(2\rho)$ the additional condition that the number of sign changes be even must be satisfied.

4. To each vector $q = (q_1, q_2, \ldots, q_\rho)$ occurring in the ψ-rows there corresponds an entry of the representing diagonal matrices of the torus, namely,

   $$e^{\frac{1}{2}(q_1\varphi_1 + q_2\varphi_2 + \ldots + q_\rho\varphi_\rho)} \quad (9.37)$$

We say q is the **weight vector** of the corresponding representation. Its character is the sum of all diagonal entries (9.37).

Finally, we shall give an equation for checking the dimension d of an arbitrary irreducible representation ϑ_ℓ. For $i, j = 1, 2, \ldots, \rho$ it is

$$d = \frac{\prod_j \ell_j [\prod_{i<j} (\ell_i^2 - \ell_j^2)]}{\prod_j n_j [\prod_{i<j} (n_i^2 - n_j^2)]} \quad \text{in the case of } SO(2\rho + 1) \quad (9.38)$$

$$d = \frac{\prod_{i<j}(\ell_i^2 - \ell_j^2)}{\prod_{i<j}(n_i^2 - n_j^2)} \qquad \text{in the case of SO}(2\rho) \qquad (9.39)$$

9.3   Irreducible Subspaces of Representations of SU(3)

The groups $U(n)$ and $SU(n)$ play an important part in elementary particle physics (Lichtenberg, 1978). As in Section 7.4, our aim shall be to deduce an algorithm which for a given representation ϑ of the group $SU(3)$ generates a complete list of symmetry adapted basis vectors of the invariant irreducible subspaces of ϑ.

The derivation of the algorithm for the group $SU(3)$ can, in principle, be extended to the groups $U(n)$ and $SU(n)$.

In view of Theorem 7.12, the problem of complete reduction of a representation ϑ of the group $SU(3)$ can be reformulated for the corresponding representation θ of its Lie algebra $\mathcal{L}$ (see Remark 1 after (4.9)).

If in the maximal torus T_S of $SU(3)$ described in (9.1) and (9.2) the one-parameter family is given by

$$x_k = \frac{\xi_k}{2\pi}t \qquad \text{with } \xi_1 + \xi_2 + \xi_3 = 0 \qquad \xi_i \in \mathbf{R} \text{ fixed} \qquad (9.40)$$

where $t \in \mathbf{R}$ is the parameter, then the infinitesimal operator, that is, the general element of the 2-dimensional Lie algebra of T_S, is the diagonal matrix with diagonal entries

$$i\xi_1, \ i\xi_2, \ -i(\xi_1 + \xi_2) \qquad \text{where } \xi_1, \xi_2 \in \mathbf{R} \qquad (9.41)$$

As a basis, we choose

$$U_1 = \begin{bmatrix} i & 0 & 0 \\ 0 & 0 & 0 \\ 0 & 0 & -i \end{bmatrix}, \qquad U_2 = \begin{bmatrix} 0 & 0 & 0 \\ 0 & i & 0 \\ 0 & 0 & -i \end{bmatrix} \qquad (9.42)$$

The 8-dimensional Lie algebra $\mathcal{L}$ (compare end of Section 7.2) consists of skew-hermitian matrices of trace 0 (Theorem 7.4). Aside from U_1, U_2, we select as the remaining basis vectors the following matrices:

$$J_1 = \begin{bmatrix} 0 & 0 & 1 \\ 0 & 0 & 0 \\ -1 & 0 & 0 \end{bmatrix}, \quad J_2 = \begin{bmatrix} 0 & 0 & i \\ 0 & 0 & 0 \\ i & 0 & 0 \end{bmatrix}, \quad J_3 = \begin{bmatrix} 0 & 1 & 0 \\ -1 & 0 & 0 \\ 0 & 0 & 0 \end{bmatrix}$$

$$J_4 = \begin{bmatrix} 0 & i & 0 \\ i & 0 & 0 \\ 0 & 0 & 0 \end{bmatrix}, \quad J_5 = \begin{bmatrix} 0 & 0 & 0 \\ 0 & 0 & i \\ 0 & i & 0 \end{bmatrix}, \quad J_6 = \begin{bmatrix} 0 & 0 & 0 \\ 0 & 0 & 1 \\ 0 & -1 & 0 \end{bmatrix} \qquad (9.43)$$

The following structure relations are available:

$[\,\cdot\,,\,\cdot\,]$	J_1	J_2	J_3	J_4	J_5	J_6
U_1	$2J_2$	$-2J_1$	J_4	$-J_3$	$-J_6$	J_5
U_2	J_2	$-J_1$	$-J_4$	J_3	$-2J_6$	$2J_5$

e.g., $[U_1, J_4] = -J_3$

$$(9.44)$$

Let now U be a fixed element of the Lie algebra of the maximal torus T_S (i.e., $U = r_1 U_1 + r_2 U_2$, $r_i \in \mathbf{R}$) and let J be a complex[8] linear combination of the basis vectors (9.42), (9.43). We consider the adjoint representation (7.105) restricted to the elements U of the Lie subalgebra (9.41):

$$J \to J' = [U, J] \qquad J = \sum_{i=1}^{2} \mu_i U_i + \sum_{i=1}^{6} \nu_i J_i \qquad \mu_i, \nu_i \in \mathbf{C} \qquad (9.45)$$

Since this is a representation of an abelian Lie algebra, the representing 8×8 matrices also commute and are simultaneously diagonalizable.

Owing to the simplicity of the structure relations (9.44), one easily finds a diagonalized representation of (9.45); for example, U_1, U_2, together with the new basis vectors

$$\left. \begin{array}{lll} L_{+2,+1} = J_1 - iJ_2, & L_{-2,-1} = J_1 + iJ_2, & L_{+1,-1} = J_3 - iJ_4 \\ L_{-1,+1} = J_3 + iJ_4, & L_{+1,+2} = J_5 + iJ_6, & L_{-1,-2} = J_5 - iJ_6 \end{array} \right\} \qquad (9.46)$$

of the representation space of (9.45).

The notations are adapted to the new commutator rules. Namely, the following is true:

$$[U_1, L_{\lambda,\mu}] = i\lambda L_{\lambda,\mu} \qquad [U_2, L_{\lambda,\mu}] = i\mu L_{\lambda,\mu} \qquad \forall(\lambda, \mu) \qquad (9.47)$$

Now we consider an arbitrary irreducible representation ϑ_ℓ of SU(3) as described in Section 9.1 (ℓ is a 3-dimensional vector) and the corresponding representation θ_ℓ of the Lie algebra $\mathcal{L}$ of SU(3). The representing matrices $U_1', U_2', J_1', J_2', \ldots, J_6'$ of the basis (9.42), (9.43) of $\mathcal{L}$ also satisfy (9.44). Hence by (9.47) and using consistent notation, we obtain

$$[U_1', L_{\lambda,\mu}'] = i\lambda L_{\lambda,\mu}' \qquad [U_2', L_{\lambda,\mu}'] = i\mu L_{\lambda,\mu}' \qquad (9.48)$$

According to Section 9.1.4, the representing matrices of ϑ_ℓ of the torus (9.1), (9.2) are simultaneously diagonal with respect to some suitable basis of the representation space. The diagonal entries are

$$e(q, x) = e^{2\pi i(q_1 x_1 + q_2 x_2 + q_3 x_3)} \qquad (9.49)$$

Here q is an integral vector, namely the **weight** of the corresponding basis vector of a suitably chosen basis (see Section 9.1.4). A weight q is always

---

[8] J is an element of the Lie algebra $\mathcal{L}$ only in the case of real linear combinations.

uniquely determined up to an integral vector (d, d, d) (see discussion preceding (9.9)). We normalize $q_3 = 0$ and set

$$(q_1, q_2, 0) = (q_1, q_2) \tag{9.50}$$

By (9.40), (9.49), (9.50), the diagonals of the representing matrices of θ_ℓ, restricted to the 2-dimensional Lie subalgebra of the torus, are

$$\{i(q_1\xi_1 + q_2\xi_2 + q_3\xi_3)\} = \{i(q_1\xi_1 + q_2\xi_2)\} \tag{9.51}$$

Here (q_1, q_2, q_3) ranges through all weights of θ_ℓ.

In particular, (9.51) implies for the two diagonals of the representing matrices U_1', U_2' of U_1, U_2 (see (9.41),(9.42)) that

$$U_1' = \{iq_1\}, \qquad U_2' = \{iq_2\} \tag{9.52}$$

where q_1 and q_2 range through the first and second components, respectively, of the weights of θ_ℓ. For a basis vector, denoted by b_q, with weight $q = (q_1, q_2)$, Equation (9.52) implies

$$U_1' b_q = iq_1 b_q, \qquad U_2' b_q = iq_2 b_q \tag{9.53}$$

The operators (9.46) now take the roles of the raising and lowering operators of Section 7.4. Namely, by (9.48) and (9.53),

$$\left.\begin{array}{lll} U_1' L_{\lambda,\mu}' b_q = & L_{\lambda,\mu}' U_1' b_q + \lambda i L_{\lambda,\mu}' b_q & = i(q_1 + \lambda) L_{\lambda,\mu}' b_q \\ U_2' L_{\lambda,\mu}' b_q = & \cdots & = i(q_2 + \mu) L_{\lambda,\mu}' b_q \end{array}\right\} \tag{9.54}$$

On comparing (9.54) with (9.53), we obtain the following result:

$$\left.\begin{array}{l} \text{Given an irreducible representation } \vartheta_\ell \text{ of SU(3) with the weight} \\ q = (q_1, q_2). \text{ If } b_q \text{ is a basis vector of the representation space of} \\ \vartheta_\ell, \text{ then the vector } L_{\lambda,\mu}' b_q \text{ has either the weight } (q_1 + \lambda, q_2 + \mu) \\ \text{or is equal to the zero vector.} \end{array}\right\} \tag{9.55}$$

We illustrate this result in the diagram plane of SU(3). For this purpose, we use a hexagonal basis adjusted to the 2-dimensional lattice Γ_S in Fig. 9.3 such that the points P, Q have the respective coordinates $(1, 2)$ and $(1, -1)$. This geometrically takes care of the normalization of the weights with $q_3 = 0$ (compare (9.50)).

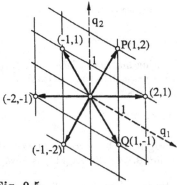

From Fig. 9.5 we see that each of the operators $L_{\lambda,\mu}'$ maps a given lattice point into one of the 6 neighboring lattice points in Γ_S.

To any representation ϑ there corresponds a set of weights (q_1, q_2). These weights, regarded as vectors in the hexagonal coordinate system of Fig. 9.5, constitute the **weight diagram** of ϑ.

Fig. 9.5

If θ denotes the representation of the Lie algebra $\mathcal{L}$ of SU(3) associated with the representation ϑ, let

$$U_1', U_2', J_1', \ldots, J_6' \tag{9.56}$$

be the representing linear transformations of θ. Let further $L_{\lambda,\mu}'$ be the linear combinations of the transformations (9.56) as in (9.46). Then the relations (9.48) are in force, and in addition to statement (9.55) the following holds:

The vectors b_q and $L_{\lambda,\mu}' b_q$ belong to the same invariant irreducible subspace of θ.

We shall first illustrate the procedure for finding irreducible subspaces through the following

Example. $\vartheta = \vartheta_{310} \otimes \vartheta_{310}$ (ϑ_{310} is the representation of SU(3) by itself). In appealing to Section 1.10, we denote by $e_i^{(1)}$ and $e_i^{(2)}$ ($i = 1, 2, 3$) the three basis vectors of the two copies of ϑ_{310} in ϑ. The weights of ϑ_{310} are the vectors $(1, 0, 0)$, $(0, 1, 0)$, $(0, 0, 1)$. Hence the normalized weights (with corresponding basis vectors) are

$$e_1^{(k)} : (1,0) \qquad e_2^{(k)} : (0,1) \qquad e_3^{(k)} : (-1,-1) \qquad k = 1,2 \tag{9.57}$$

The 9 products $e_i^{(1)} \cdot e_j^{(2)}$ form a basis for the representation space of ϑ (Section 1.10). The representing matrices of the torus are diagonal with respect to the basis $e_i^{(1)} \cdot e_j^{(2)}$, and the weights of ϑ are obtained by adding two weights (9.57) each.

$\cdot$ $\nearrow$	$e_1^{(2)}$	$e_2^{(2)}$	$e_3^{(2)}$
$e_1^{(1)}$	(2, 0)	(1, 1)	(0, -1)
$e_2^{(1)}$	(1, 1)	(0, 2)	(-1, 0)
$e_3^{(1)}$	(0,-1)	(-1,0)	(-2, -2)

(9.58)

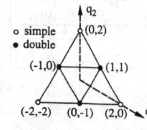

o simple
• double

Fig. 9.6

Fig. 9.6 shows the weight diagram of $\vartheta_{310} \otimes \vartheta_{310}$. The method of complete reduction (9.24) gives

$$\vartheta_{310} \otimes \vartheta_{310} = \vartheta_{410} \oplus \vartheta_{320} \tag{9.59}$$

For an irreducible representation ϑ_ℓ of SU(3), Equation (9.13) yields

$$\text{dimension of } \vartheta_\ell = \frac{1}{2}\ell_1\ell_2(\ell_1 - \ell_2) \qquad \ell = (\ell_1, \ell_2, 0) \tag{9.60}$$

Thus the Kronecker product decomposes into a 3-dimensional and a 6-dimensional irreducible representation. For ϑ_{320}, the division algorithm (9.10) furnishes the normalized weights q:

$$(-1,0) \qquad (0,-1) \qquad (1,1) \tag{9.61}$$

They differ from the weights (9.57) only by their signs. Therefore, by (9.49), $\vartheta_{320} = \overline{\vartheta_{310}}$. Rather than (9.59), physicists use the following convenient notation:

$$3 \otimes 3 = 6 \oplus \bar{3}$$

where representations are denoted by their dimensions. $\bar{3}$ is the complex conjugate representation of 3. This notation, however, has the disadvantage of not always being unique. For example, there are 4 inequivalent irreducible representations of SU(3) of dimension 15 (see Table (9.71)).

Now let us find the invariant subspace of $\bar{3}$. Among the weights (9.61) we pick one, say $(1, 1)$. By virtue of (9.58), the weight $(1, 1)$ of $3 \otimes 3$ is associated with the vectors $e_1^{(1)} \cdot e_2^{(2)}$ and $e_2^{(1)} \cdot e_1^{(2)}$. Thus a vector v belonging to the desired subspace with weight $(1, 1)$ must have the form

$$v = \alpha(e_1^{(1)} \cdot e_2^{(2)}) + \beta(e_2^{(1)} \cdot e_1^{(2)}) \tag{9.62}$$

How do we find the numbers α and β? For example, by (9.55), we have

$$L'_{-1,1}v = 0 \tag{9.63}$$

since the weight $(0, 2)$ does not occur in (9.61) (see Figs. 9.5 and 9.6). So by (7.80), the representing linear transformations J'_k obey, for instance,

$$J'_3(e_1^{(1)} \cdot e_2^{(2)}) = (J_3 e_1^{(1)}) \cdot e_2^{(2)} + e_1^{(1)} \cdot (J_3 e_2^{(2)}) = -e_2^{(1)} \cdot e_2^{(2)} + e_1^{(1)} \cdot e_1^{(2)}$$

with J_3 from (9.43). Thus one obtains the following table:

k =	1	2	3
$J'_k(e_1^{(1)} \cdot e_2^{(2)}) =$	$-e_3^{(1)} \cdot e_2^{(2)}$	$ie_3^{(1)} \cdot e_2^{(2)}$	$e_1^{(1)} \cdot e_2^{(2)} - e_2^{(1)} \cdot e_2^{(2)}$
$J'_k(e_2^{(1)} \cdot e_1^{(2)}) =$	$-e_2^{(1)} \cdot e_3^{(2)}$	$ie_2^{(1)} \cdot e_3^{(2)}$	$e_1^{(1)} \cdot e_1^{(2)} - e_2^{(1)} \cdot e_2^{(2)}$
k =	4	5	6
$J'_k(e_1^{(1)} \cdot e_2^{(2)}) =$	$i(e_1^{(1)} \cdot e_1^{(2)} + e_2^{(1)} \cdot e_2^{(2)})$	$ie_1^{(1)} \cdot e_3^{(2)}$	$-e_1^{(1)} \cdot e_3^{(2)}$
$J'_k(e_2^{(1)} \cdot e_1^{(2)}) =$	$i(e_1^{(1)} \cdot e_1^{(2)} + e_2^{(1)} \cdot e_2^{(2)})$	$ie_3^{(1)} \cdot e_1^{(2)}$	$-e_3^{(1)} \cdot e_1^{(2)}$

$$\tag{9.64}$$

For $L'_{-1,1} = J'_3 + iJ'_4$, Table (9.64) implies

$$L'_{-1,1}(e_1^{(1)} \cdot e_2^{(2)}) = -2e_2^{(1)} \cdot e_2^{(2)}, \qquad L'_{-1,1}(e_2^{(1)} \cdot e_1^{(2)}) = -2e_2^{(1)} \cdot e_2^{(2)}$$

Thus (9.63) reads

$$L'_{-1,1}v = (\alpha + \beta)(-2e_2^{(1)} \cdot e_2^{(2)}) = 0$$

whence
$$v = e_1^{(1)} \cdot e_2^{(2)} - e_2^{(1)} \cdot e_1^{(2)}$$
is a first basis vector of the subspace of $\overline{3}$. By (9.55) and (9.61), the operators $L'_{-2,-1}$ and $L'_{-1,-2}$ generate from this the two remaining basis vectors $L'_{-2,-1}v$ and $L'_{-1,-2}v$ associated with the weights $(-1,0)$ and $(0,-1)$. Up to scalar factors, they are of the form

$$v' = e_2^{(1)} \cdot e_3^{(2)} - e_3^{(1)} \cdot e_2^{(2)}, \qquad v'' = e_3^{(1)} \cdot e_1^{(2)} - e_1^{(1)} \cdot e_3^{(2)} \qquad (9.65)$$

This example also shows that, instead of the operator $L'_{-1,1}$ in (9.63), one could just as well choose one of the operators

$$L'_{2,1}, \quad L'_{1,-1}, \quad L'_{1,2}$$

Then a basis of the invariant subspace associated with $\overline{3}$ could also be given by the three vectors v, $L'_{-2,-1}v = w$, and $L'_{1,-1}w$. The three weights of the small triangle in the middle of Fig. 9.6 correspond to the irreducible representation 3; the six weights of the large triangle correspond to the irreducible representation 6.

Now let us proceed to the general theory: If (q_1, q_2, q_3) is a weight vector of an irreducible representation ϑ_ℓ, then so are any of the vectors obtained by permuting the q_i (Section 9.1). Thus the weight diagram of ϑ_ℓ and hence the weight diagram of each representation ϑ is invariant under the corresponding Weyl group ψ and, therefore, invariant under the 120° rotation and the reflections in coordinate axes.

For an arbitrary n-dimensional representation ϑ_ℓ of SU(3), where $n > 1$, Equation (9.55) implies the following: To each point P of the weight diagram there exists at least a point Q such that the line segment PQ equals the length of an arm of the star of Fig. 9.5.

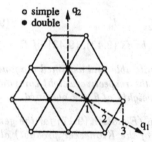

The weight diagram of the irreducible representation ϑ_{520} is drawn in Fig. 9.7. The weights were calculated in (9.10). The weight $(2, 1, 1)$ occurring there corresponds to the normalized weight $(1, 0)$ here.

Fig. 9.7

Finally, Fig. 9.8 shows the weight diagram of the Kronecker product

$$\vartheta_{310} \otimes \vartheta_{310} \otimes \vartheta_{310} = 2\vartheta_{420} \oplus \vartheta_{510} \oplus \vartheta_{210} \qquad (9.66)$$

with the corresponding multiplicities[9].

---

[9]The reader may verify (9.66) and Fig. 9.8. Compare Table (9.71).

In dimension notation, the above equation (9.66) reads

$$3 \otimes 3 \otimes 3 = 8 \oplus 8 \oplus 10 \oplus 1 \qquad (9.67)$$

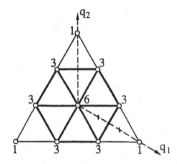

Equation (9.66) tells us that the largest triangle of Fig. 9.8, center included, belongs to ϑ_{510}. The thickly drawn hexagon including the center, which is taken twice, corresponds to each of the two copies of the representation ϑ_{420}. The center corresponds to the unit representation ϑ_{210}.

Fig. 9.8

Now let an arbitrary representation

$$\vartheta = \bigoplus_j c_j \vartheta_j \quad \text{or} \quad \theta = \bigoplus_j c_j \theta_j \quad (j \text{ is a 3-dimensional vector}) \qquad (9.68)$$

of SU(3) or the Lie algebra $\mathcal{L}$ of SU(3) be given, where ϑ_j or θ_j are the respective irreducible inequivalent constituents and $c_j \in \{1, 2, \ldots\}$ their multiplicities. The considerations made earlier in this section, together with the reasoning of Section 7.4.1 applied to our case, imply the following

Algorithm (Generating symmetry adapted basis vectors associated with a representation ϑ of SU(3) or a representation θ of the Lie algebra $\mathcal{L}$ of SU(3))

1. Let

$$U_1', U_2', J_1', J_2', \ldots, J_6' \qquad (9.69)$$

be the representing[10] matrices of the basis (9.42), (9.43) of $\mathcal{L}$.

2. U_1' and U_2' are simultaneously diagonalizable and assumed diagonal in the following. The diagonal contains the respective weight components q_1 and q_2 (with factor i). Thus all weights of θ are obtained.

3. The algorithm for complete reduction (Equation (9.17) for the general case, and (9.23) for the special case of the Kronecker product) allows us to determine, from the weights of θ, the vectors j of the irreducible constituents θ_j occurring in θ along with their multiplicities c_j.

4. In order to determine an irreducible subspace of the representation space V of θ associated with θ_ℓ, we establish the complete list of

---

[10]If ϑ is to be completely reduced, then the corresponding representation θ should be considered.

weights of θ_ℓ by use of the division algorithm (9.10) and select among them the dominant weight q. (See definition given under item Z_5 after (9.11).)

5. *Next we determine all basis vectors of V associated with the weight q and consider the subspace V_q spanned by them.*

6. *Among the 6 operators $L'_{\lambda,\mu}$ corresponding to (9.46) (i.e., $L'_{+2,+1} = J'_1 - iJ'_2, \ldots$) we select those that do not map the weight q into a weight of θ_ℓ (compare Fig. 9.5). These operators are called* **annihilation operators**[11], *the others are* **basis generators**.

7. *Determine a vector v of the subspace V_q such that the various annihilation operators satisfy*

$$L'_{\lambda,\mu}v = L'_{\lambda',\mu'}v = \ldots = 0 \qquad (9.70)$$

(In (9.70) it suffices to consider as many annihilation operators as to obtain a c_ℓ-dimensional solution space for v.) Every such vector v lies in one of the c_ℓ irreducible subspaces that transform like θ_ℓ.

8. *The remaining basis vectors of one of these irreducible subspaces are generated by successively applying an appropriate sequence of basis generators $L'_{\lambda,\mu}$ to some initial vector v. An appropriate sequence can be found easily from the known weights of θ_ℓ.*

9. *If θ_ℓ occurs in θ repeatedly ($c_\ell > 1$), then a total of c_ℓ linearly independent solution vectors v of (9.70) need to be determined. This is possible since the solution space of v for all requirements (9.70) is exactly c_ℓ-dimensional. Each of these c_ℓ solution vectors is an initial vector, so that one correctly finds c_ℓ invariant irreducible subspaces with symmetry adapted bases, each of which transforms like θ_ℓ.*

10. *Do Steps 4 through 9 for each of the θ_ℓ occurring in θ.*

Remarks.

(a) For the special case of Kronecker products (of any number of factors), the requirement of diagonal form in Step 2 is already satisfied. This case is important in both mathematics and physics.

(b) In elementary particle physics the representations 3 and $\overline{3}$ serve to describe the **quarks** and **antiquarks**, respectively. In the language of physics, Equation (9.67) tells us that from 3 quarks (= triplets) one can build

1 baryon decuplet (10)

2 baryon octets ($8 \oplus 8$)

and   1 baryon singlet (1)

---

[11] If q is dominant, then there are at least 3 annihilation operators; namely, $L'_{1,2}$, $L'_{2,1}$, $L'_{1,-1}$.

Thus one can describe proton, neutron, Λ-, Ξ-, Δ-, and Ω- particles.

(c) In dealing with complete reduction problems, one best establishes a weight table of the expected irreducible representations by way of the division algorithm. We confine ourselves to the following list:

Vector ℓ	Dimension	Normalized weights of irreducible representations of SU(3) in the fundamental domain				
(2,1)	1	(0,0)	unit representation			
(3,1)	3	(1,0)	representation by itself			
(4,1)	6	(2,0)	(1,1)			
(4,2)	8	(2,1)	$(0,0)_2$			
(5,1)	10	(3,0)	(2,1)	(0,0)		
(5,2)	15	(3,1)	(2,2)	$(1,0)_2$		
(6,1)	15	(4,0)	(3,1)	(2,2)	(1,0)	
(6,2)	24	(4,1)	(3,2)	$(2,0)_2$	$(1,1)_2$	
(6,3)	27	(4,2)	(3,0)	(3,3)	$(2,1)_2$	$(0,0)_3$

$$(9.71)$$

The normalized vector $\ell = (\ell_1, \ell_2)$ with $\ell_1 > \ell_2 > 0$ characterizes the irreducible representation ϑ_ℓ. Furthermore,

$$\vartheta_{\ell_1, \ell_1 - \ell_2} = \overline{\vartheta_{\ell_1, \ell_2}} \tag{9.72}$$

is the complex conjugate representation of ϑ_ℓ.

PROBLEMS

Problem 9.1 Develop the theory of Section 9.1 step by step for the groups U(2) and SU(2). Show that this produces the known weights of Theorem 4.4 and the character formulas (5.201) and (5.204).

Problem 9.2 Formulate the results of Section 9.2 for the case SO(3). Show that again the weights of Theorem 4.4 and the character formulas (5.201) and (5.204) are obtained. Notice that the weights of Chapter 4 differ from those of Chapter 9 by a factor 2. This is due to an irrelevant normalization chosen for practical reasons (in order to avoid computations with nonintegers in the division algorithm).

Problem 9.3 In the group SO(4), show that

(a) The representation corresponding to $\ell = (4, 0)$ is equivalent to the representation by itself;

(b) The representation corresponding to $\ell = (3, \pm 1)$ assigns to the matrices of SO(4) the matrices of SU(2) (up to equivalence).

Problem 9.4 In the group SO(6), show that the representation corresponding to $\ell = (5, 3, 1)$ maps the elements of SO(6) into those of SU(4). This representation is multiple valued. We state without proof that all matrices of SU(4) are images. The relationship is similar to that between SO(3) and SU(2).

Problem 9.5 In the group SU(4), show that

(a) The representation corresponding to $\ell = (4, 2, 1, 0)$ is the representation by itself;

(b) The representation corresponding to $\ell = (4, 3, 1, 0)$ is equivalent to a representation by matrices of SO(6) (compare Problem 9.4);

(c) The representation corresponding to $\ell = (4, 3, 2, 0)$ assigns to each matrix of SU(4) its complex conjugate.

Problem 9.6 (a) Determine the irreducible constituents of the representation $3 \otimes 3 \otimes \bar{3}$ of SU(3) together with their multiplicities (two quarks and an antiquark).

(b) Determine the irreducible subspaces of the quark model $\bar{3} \otimes 3$ for a meson (quark and antiquark).

Appendix A

Answers to Selected Problems

1.1 Abelian, since the group table is symmetric with respect to the main diagonal. Thus the irreducible invariant subspaces are 1-dimensional, and hence spanned by vectors that are eigenvectors for all representing matrices.

The eigenvalues of $D(a), D(b)$ are ± 1 (2-fold each). Pairwise intersection of the eigenspaces of $D(a)$ with those of $D(b)$ produce the 4 invariant subspaces:

$$\begin{bmatrix} \lambda \\ \lambda \\ \lambda \\ \lambda \end{bmatrix}, \quad \begin{bmatrix} \lambda \\ -\lambda \\ \lambda \\ -\lambda \end{bmatrix}, \quad \begin{bmatrix} \lambda \\ \lambda \\ -\lambda \\ -\lambda \end{bmatrix}, \quad \begin{bmatrix} \lambda \\ -\lambda \\ -\lambda \\ \lambda \end{bmatrix} \qquad \lambda \in \mathbf{C}$$

The irreducible representations are

$$a \to \pm 1, \qquad b \to \pm 1 \qquad \text{(all 4 sign combinations occur)}$$

1.2 (a) 1. ϑ_1^3: Assume ϑ_1^3 is reducible. Since ϑ_1^3 is unitary, there exists a 1-dimensional invariant subspace; that is, there is a vector v which is an eigenvector for all representing matrices $D(s)$ of ϑ_1^3.

Let s_1 be the reflection about the plane $x = y$. Then

$$D(s_1) = \begin{bmatrix} 0 & 1 & 0 \\ 1 & 0 & 0 \\ 0 & 0 & 1 \end{bmatrix}$$

with the eigenvalues $\lambda_1 = 1$ (double), $\lambda_2 = -1$ (simple) and the corresponding eigenvectors

$$\begin{bmatrix} \alpha \\ \alpha \\ \beta \end{bmatrix}, \quad \begin{bmatrix} \gamma \\ -\gamma \\ 0 \end{bmatrix} \qquad \alpha, \beta, \gamma \in \mathbf{C}$$

Let s_2 be the reflection about the plane $x = -z$. Then

$$D(s_2) \begin{bmatrix} \alpha \\ \alpha \\ \beta \end{bmatrix} = \begin{bmatrix} -\beta \\ \alpha \\ -\alpha \end{bmatrix} \qquad D(s_2) \begin{bmatrix} \gamma \\ -\gamma \\ 0 \end{bmatrix} = \begin{bmatrix} 0 \\ -\gamma \\ \gamma \end{bmatrix}$$

For $\alpha = -\beta$, we have $\begin{bmatrix} \alpha \\ \alpha \\ \beta \end{bmatrix}$ as an eigenvector of $D(s_2)$.

Let s_3 be the reflection about the plane $x = z$. Then

$$D(s_3) \begin{bmatrix} -\beta \\ -\beta \\ \beta \end{bmatrix} = \begin{bmatrix} \beta \\ -\beta \\ -\beta \end{bmatrix}$$

That is, for all $\beta \neq 0$, we have that $\begin{bmatrix} -\beta \\ -\beta \\ \beta \end{bmatrix}$ is not an eigenvector of $D(s_3)$, contradicting the hypothesis!

2. ϑ_2^3 is irreducible, since ϑ_1^3 is.

3. $S_4 \longrightarrow S_3 \xrightarrow{\vartheta_2} 2 \times 2$ matrices

   ϑ_2 is the representation described in Example 1.8. It is irreducible by Example 1.10, and so is ϑ_1^2.

(b) Consider the determinants of the matrices of ϑ_1^3, ϑ_2^3.

(c) Consult (1.50): $1^2 + 1^2 + 2^2 + 3^2 + 3^2 = 24 = 4!$

(d)

	ϑ_1^1	ϑ_2^1	ϑ_1^3	ϑ_2^3	ϑ_1^2
$\begin{vmatrix} 1 & 2 & 3 & 4 \\ 2 & 1 & 3 & 4 \end{vmatrix}$	1	-1	U	$-U$	$\begin{bmatrix} 1/2 & \sqrt{3}/2 \\ \sqrt{3}/2 & -1/2 \end{bmatrix}$
$\begin{vmatrix} 1 & 2 & 3 & 4 \\ 2 & 3 & 4 & 1 \end{vmatrix}$	1	-1	V	$-V$	$\begin{bmatrix} -1 & 0 \\ 0 & 1 \end{bmatrix}$

where $U = \begin{bmatrix} 0 & 0 & -1 \\ 0 & 1 & 0 \\ -1 & 0 & 0 \end{bmatrix}$, $V = \begin{bmatrix} -1 & 0 & 0 \\ 0 & 0 & -1 \\ 0 & 1 & 0 \end{bmatrix}$

1.3 $T = \begin{bmatrix} -1/2 & -1/2 \\ \sqrt{3}/2 & -\sqrt{3}/2 \end{bmatrix}$ $D'(s) = T^{-1} D(s) T$

1.4 Consider $D = (d_{ij}^{(1)} D_2)$ in (1.65). Thus $\overline{D}^T = (\overline{d}_{ji}^{(1)} \overline{D}_2^T)$ and $\overline{D}^T D = D \overline{D}^T = E_{mn}$.

1.5 In order to determine the invariant irreducible subspaces, see solution to Problem 1.1. $\{a, c\}$ is the generating set of S_3 (Example 1.6). Invariant subspaces are given by

$$\underbrace{\begin{bmatrix} r \\ 0 \\ 0 \\ r \end{bmatrix} \begin{bmatrix} 0 \\ s \\ -s \\ 0 \end{bmatrix}}_{\text{each 1-dim.}} \underbrace{\begin{bmatrix} t \\ u \\ u \\ -t \end{bmatrix}}_{\text{2-dim.}} \quad r, s, t, u \in \mathbb{C}$$

They are pairwise orthogonal (see solution 1.4).

Complete reduction gives $\vartheta_2 \otimes \vartheta_2 = \vartheta_1 \oplus \vartheta_{\text{alt}} \oplus \vartheta_2$ (see (1.29), (1.51)), where ϑ_1 is the unit representation.

1.6 Invariant inner product $\langle x, y \rangle = \frac{1}{3}[4x_1\bar{y}_1 + 2x_1\bar{y}_2 + 2x_2\bar{y}_1 + 4x_2\bar{y}_2]$ according to (1.77). The property $\langle D(a)x, D(a)y \rangle = \langle x, y \rangle$, Equations (1.67) through (1.70), and (1.74) are proven by straightforward calculation.

1.7 injective (one-to-one): $s_1 \neq s_2 \Rightarrow as_1 \neq as_2$; because $as_1 = as_2 \Rightarrow s_1 = s_2$. surjective (onto): Given $s_2 \in G$, there exists $s_1 \in G$ with $as_1 = s_2$, namely, $s_1 = a^{-1}s_2$.

2.1 (a) Verify by direct computation.

(b) 1. Equation (2.47) implies $M(Tx_k) = \lambda_k M x_k = T(M x_k)$. Thus $M x_k$ is an eigenvector of T corresponding to the eigenvalue λ_k.

2. Since all eigenspaces E_{λ_k} of T are 1-dimensional, it follows that $M x_k = \mu_k x_k$.

$$\mu_k = \sum_{\ell=1}^{n} a_\ell \lambda_k^{n+1-\ell} \quad (= \text{ first component of } M x_k)$$

(c) $z_i = i$th row, $k_i = i$th column of M.

$$TM: \quad \text{permutation of rows} \quad \left| \begin{array}{cccc} z_1 & z_2 & \ldots & z_n \\ z_n & z_1 & \ldots & z_{n-1} \end{array} \right.$$

$$MT: \quad \text{permutation of columns} \quad \left| \begin{array}{cccc} k_1 & k_2 & \ldots & k_n \\ k_2 & k_3 & \ldots & k_1 \end{array} \right.$$

2.2 (a) $M = \begin{bmatrix} a_1 & a_2 \\ a_2 & a_1 \end{bmatrix}$ Eigenvalues of M: $\mu_1 = a_1 - a_2$, $\mu_2 = a_1 + a_2$ (see solution 2.1);

$$p = -2a_1 = -(\mu_1 + \mu_2), \quad q = \mu_1\mu_2 \Rightarrow \mu_{1,2} = -\frac{p}{2} \pm \sqrt{\frac{p^2}{4} - q}$$

(b) $M = \begin{bmatrix} 0 & a_2 & a_3 \\ a_3 & 0 & a_2 \\ a_2 & a_3 & 0 \end{bmatrix}$ with $\omega = e^{i2\pi/3}$.

Eigenvalues of M (see solution 2.1):

$$\begin{array}{llll} \mu_1 & = a_2\omega^2 & + a_3\omega \\ \mu_2 & = a_2\omega & + a_3\omega^2 \\ \mu_3 & = a_2 & + a_3 \end{array}$$

The characteristic polynomial is $(\lambda - \mu_1)(\lambda - \mu_2)(\lambda - \mu_3)$. By equating its coefficients with those of $\lambda^3 + p\lambda + q$, we obtain

$$p = -3a_2a_3 \qquad q = -(a_2^3 + a_3^3)$$

$$a_2 = \sqrt[3]{-\frac{q}{2} + \sqrt{\left(\frac{q}{2}\right)^2 + \left(\frac{p}{3}\right)^3}} \qquad a_3 = \sqrt[3]{-\frac{q}{2} - \sqrt{\left(\frac{q}{2}\right)^2 + \left(\frac{p}{3}\right)^3}}$$

2.3 M is a cyclic matrix with $a_1 = -2$, $a_2 = a_n = 1$, all other $a_i = 0$ (see solution 2.1). M has the eigenvalues $\mu_k = -4\sin^2(k\frac{\pi}{n})$, where $k = 0, 1, 2, \ldots, n-1$.

2.4 Let $D(\varphi_0)$ be the circle rotation through angle φ_0. The list of irreducible representations of G is given by

$$\vartheta_k : D(\varphi_0) \to e^{ik\varphi_0} \text{ where } k \in \{\ldots, -2, -1, 0, 1, 2, \ldots\}$$

ϑ_k and $\vartheta_{k'}$ are inequivalent for $k \neq k'$.

4.1 (a) The rotation axis is the eigenvector of the matrix (4.44) corresponding to the eigenvalue $\lambda = 1$.

(b) Trace $= 4\alpha_1^2 - 1 = 1 + 2\cos\varphi$.

4.2 $y_1 = \frac{1}{2}((a_3^2 - a_1^2)(-x_1 + x_4) + 2x_2(a_3a_4 - a_1a_2) + (a_4^2 - a_2^2)(x_1 + x_4))$, etc. To show: $y_1^2 + y_2^2 - y_4^2 = x_1^2 + x_2^2 - x_4^2$.

4.3 (b) $y_1 = (\alpha_1 - i\beta_1)x_1 + (-\beta_2 - i\alpha_2)x_2$,
$\quad y_2 = (\beta_2 - i\alpha_2)x_1 + (\alpha_1 + i\beta_1)x_2$,
$\quad \frac{1}{2}(x_1^2 - x_2^2) \to \frac{1}{2}(y_1^2 - y_2^2) = a\frac{1}{2}(x_1^2 - x_2^2) + b\frac{1}{2i}(x_1^2 + x_2^2) + c(-x_1x_2)$.
The coefficients a, b, c form the first column of (4.44), etc.

4.4 $x_k u_\ell$ is a homogeneous polynomial of degree $\ell + 1$. Equations (4.73), (4.94), (4.95), (4.96), and $\ell + 1$ in place of ℓ imply that on the unit sphere S^2 ($z_k = x_k$):

$$z_k Y_\ell = \alpha Y_{\ell+1} + \beta Y_{\ell-1} + \gamma Y_{\ell-3} + \cdots \qquad (*)$$

Let $\langle f, g \rangle = \int f \cdot \overline{g} \cdot dF$, with integration over S^2. By $\langle Y_m, Y_n \rangle = 0$ for $m \neq n$ and $(*)$, we have that $\langle z_k Y_\ell, Y_j \rangle \neq 0$ implies $j \in \{\ell + 1, \ell - 1, \ell - 3, \ldots\}$.
On the other hand, $\langle z_k Y_\ell, Y_j \rangle = \overline{\langle z_k Y_j, Y_\ell \rangle} \neq 0$ and $(*)$ imply $\ell \in \{j + 1, j - 1, j - 3, \ldots\}$. Thus $z_k Y_\ell = \alpha Y_{\ell+1} + \beta Y_{\ell-1}$.

4.5 (a) Monomials: $e_1 = x_1 y_1$, $e_2 = x_1 y_2$, $e_3 = x_2 y_1$, $e_4 = x_2 y_2$. Vector v has weight 0, thus $v = \alpha e_2 + \beta e_3$. If v' denotes the function v transplanted under $\vartheta_{1/2} \otimes \vartheta_{1/2}$, then by (4.47) and (4.8):

$$v' = \alpha e_2' + \beta e_3', \text{ where } \left\{ \begin{array}{l} e_2' = \overline{a}\overline{b}e_1 + a\overline{a}e_2 - b\overline{b}e_3 - abe_4 \\ e_3' = \overline{a}\overline{b}e_1 - b\overline{b}e_2 + a\overline{a}e_3 - abe_4 \end{array} \right\}$$

Hence $v = e_2 - e_3$.

(b) $\vartheta_0 : Z_0^0 = e_2 - e_3 = v$
$\quad \vartheta_1 : Z_1^1 = \sqrt{2}e_1$, $Z_0^1 = e_2 + e_3$, $Z_{-1}^1 = \sqrt{2}e_4$ (see (4.83), (4.84), (4.85)).

4.6 In the notation (4.85),

$$Z_{3/2}^{3/2} = \sqrt{3}X_1Y_{1/2} \qquad Z_{1/2}^{3/2} = X_1Y_{-1/2} + \sqrt{2}X_0Y_{1/2}$$

$$Z_{-1/2}^{3/2} = \sqrt{2}X_0Y_{-1/2} + X_{-1}Y_{1/2} \qquad Z_{-3/2}^{3/2} = \sqrt{3}X_{-1}Y_{-1/2}$$

4.8 (a) Let $A \in GL(n, \mathbf{R})$ have $\det A < 0$. If the group were connected, then by continuity of the determinant function and because $\det E = 1$, there would exist A_1 with $\det A_1 = 0$.

(b) Conclusion similar to that of Example 4.2. In the case SU(n), $\widetilde{A_0}$ is diagonal; in the case SO(n), $\widetilde{A_0}$ is a matrix with rotation boxes along the main diagonal (compare, e.g., (9.25)); and in the case GL(n, C), $\widetilde{A_0}$ is the Jordan canonical form.

4.9 Use Theorem 2.1:

(a) ϑ 3-dimensional:

Eigenvalues of M: $\quad \lambda_1 = 0$ (double) $\qquad \mid \lambda_2 = 4$ (simple)

Eigenspaces: $\qquad e_1 = x_1^2 - x_2^2,\ e_2 = x_1 x_2 \mid e_3 = x_1^2 + x_2^2$

Representation ϑ in the 2-dimensional invariant subspace; Formula (4.47) implies that

$$e_1' = x_1'^2 - x_2'^2 = \cos 2\varphi\, e_1 + 2\sin 2\varphi\, e_2$$
$$e_2' = x_1' x_2' = -\tfrac{1}{2}\sin 2\varphi\, e_1 + \cos 2\varphi\, e_2$$

Complete reduction: $\lambda_{1,2} = e^{\pm 2i\varphi}$; eigenvectors $(e_1 \mp 2ie_2)$ $\forall \varphi$; thus $\vartheta = \vartheta_2 \oplus \vartheta_0 \oplus \vartheta_{-2}$ (see solution 2.4).

(b) ϑ 4-dimensional:

Eigenvalues of M: $\quad \lambda_1 = 0$ (double) $\mid \lambda_2 = 8$ (double)

Eigenspaces: $\qquad u_1 = x_1^3 - 3x_1 x_2^2 \quad \mid v_1 = x_1^3 + x_1 x_2^2$
$$u_2 = 3x_1^2 x_2 - x_2^3 \mid v_2 = x_1^2 x_2 + x_2^3$$

Proceed as in (a):

for u_1, u_2 : $\lambda_{1,2} = e^{\pm 3i\varphi}$; eigenvectors $(u_1 \mp iu_2)$ $\forall \varphi$.

for v_1, v_2 : $\lambda_{1,2} = e^{\pm i\varphi}$; eigenvectors $(v_1 \mp iv_2)$ $\forall \varphi$.

Thus $\vartheta = \vartheta_3 \oplus \vartheta_1 \oplus \vartheta_{-1} \oplus \vartheta_{-3}$ (see solution 2.4).

4.10 Action of elements (4.104) on the monomials (4.102), together with (4.105) and summation over k, yields the character $\chi_\ell = e^{(\tau + it)} \dfrac{\sinh(\ell + 1/2)}{\sinh 1/2}$, thus $\overline{\chi_\ell} \neq \chi_\ell$.

5.1 (a) 1. $\displaystyle\sum_{i=1}^{N} \alpha_i = \begin{cases} \text{even} & \text{if the permutation is odd} \\ \text{odd} & \text{if the permutation is even} \end{cases}$

2. $(i\ j) = (i\ i{+}1)(i{+}1\ i{+}2)\ldots(j{-}1\ j)\ (j{-}2\ j{-}1)\ldots(i\ i{+}1)$

3. $(k+1\ k+2) = (12\ldots N)^k (12)(12\ldots N)^{-k}$,
$k = 0, 1, 2, \ldots, N - 2$

(b) Choose a fixed permutation of type $[\alpha_1, \alpha_2, \ldots, \alpha_N]$. The number of possible ways to occupy by $1, 2, \ldots, N$ the N places of this permutation is given by the following facts:

1. The number sets of the α_i cycles of length i can be permuted: $\alpha_i!$

2. In a cycle of length i the first number is arbitrary: i^{α_i}

Thus the total number of possible arrangements is $\Pi \alpha_i! i^{\alpha_i}$.

5.2   (a) h even : $\varphi = \frac{2\pi}{h}$, see table below.

Similarity classes	Iden-tity	\multicolumn{4}{c}{Rotations through angle $k\varphi$}	Reflec-tions	
		$\pm\varphi$	$\pm 2\varphi$	$\pm 3\varphi$	... (180°)		
Number of elements	1	2	2	2	... 1	$\frac{h}{2}$	$\frac{h}{2}$
χ_1^1	1	1	1	1	... 1	1	1
χ_2^1	1	1	1	1	... 1	-1	-1
(χ_3^1)	1	-1	1	-1	... -1	1	-1
(χ_4^1)	1	-1	1	-1	... -1	-1	1
χ_1^2	2	\multicolumn{4}{c}{$2\cdot\cos(k\varphi)$}	0	0
χ_2^2	2	\multicolumn{4}{c}{$2\cdot\cos(2k\varphi)$}	0	0
$\vdots$	$\vdots$	\multicolumn{4}{c}{$\vdots$}	$\vdots$	$\vdots$
$\chi_{h/2-1}^2$	2	\multicolumn{4}{c}{$2\cdot\cos((\frac{h}{2}-1)k\varphi)$}	0	0

(b) h odd : Table of (a) is modified as follows: The h reflections form a single similarity class, so that the characters χ_3^1, χ_4^1 are suppressed. Moreover, the 180° rotation is suppressed, and in the last row, $\frac{h}{2} - 1$ is replaced by $\frac{h-1}{2}$.

5.3   The objective is to compute some of the columns of the matrix $P^{(j)}$ in (5.160). The kth column of $D(s)$ can be read from (5.77). For example, $D(1\ 2\ 4\ 3)$ maps x_1 into $-y_2$; that is to say, the first column of $D(1\ 2\ 4\ 3)$ contains just one nonzero element in the sixth row, namely -1. The first five columns contain the basis vectors of the desired subspace. They are linear combinations of the w_i of (5.88).

5.4   Theorem 5.8, Equation (5.142).

5.5

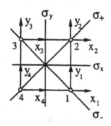

Proceed in a fashion similar to that of Example 5.3. By choosing the coordinate systems as in Fig. A.1, one obtains an 8-dimensional representation ϑ, and (5.148) yields

$$\vartheta = \vartheta_1^1 \oplus \vartheta_2^1 \oplus \vartheta_3^1 \oplus \vartheta_4^1 \oplus 2\vartheta_1^2$$

Fig. A.1

The algorithm (5.44) through (5.48) produces

$$\vartheta_1^1 : u = \tfrac{1}{\sqrt{8}}(1 \quad -1 \quad 1 \quad 1 \quad -1 \quad 1 \quad -1 \quad -1)^T$$
$$\vartheta_2^1 : v = \tfrac{1}{\sqrt{8}}(1 \quad -1 \quad -1 \quad -1 \quad -1 \quad 1 \quad 1 \quad 1)^T$$
$$\vartheta_3^1 : w = \tfrac{1}{\sqrt{8}}(1 \quad 1 \quad 1 \quad -1 \quad -1 \quad -1 \quad -1 \quad 1)^T$$
$$\vartheta_4^1 : x = \tfrac{1}{\sqrt{8}}(1 \quad 1 \quad -1 \quad 1 \quad -1 \quad -1 \quad 1 \quad -1)^T$$
$$\vartheta_1^2 : y_1 = \tfrac{1}{2}(1 \quad 0 \quad 1 \quad 0 \quad 1 \quad 0 \quad 1 \quad 0)^T$$
$$y_2 = \tfrac{1}{2}(0 \quad 1 \quad 0 \quad 1 \quad 0 \quad 1 \quad 0 \quad 1)^T$$
$$z_1 = \tfrac{1}{2}(0 \quad 1 \quad 0 \quad -1 \quad 0 \quad 1 \quad 0 \quad -1)^T$$
$$z_2 = \tfrac{1}{2}(1 \quad 0 \quad -1 \quad 0 \quad 1 \quad 0 \quad -1 \quad 0)^T$$

$$F = \begin{bmatrix} A & B & C & D & E & F & G & -D \\ & A & -D & G & F & E & D & C \\ & & A & -B & G & D & E & -F \\ & & & A & -D & C & -F & E \\ & & & & A & B & C & D \\ \text{symmetric} & & & & & A & -D & G \\ & & & & & & A & -B \\ & & & & & & & A \end{bmatrix}$$

A, B, C, D, E, F, G are functions of f, g (Fig. 5.9); u, v, w, x are eigenvectors of F corresponding to the eigenvalues

$$\lambda_1 = A - B + C + 2D - E + F - G$$
$$\lambda_2 = A - B - C - 2D - E + F + G$$
$$\lambda_3 = A + B + C - 2D - E - F - G$$
$$\lambda_4 = A + B - C + 2D - E - F + G$$

The quadratic equation for y_i, z_i gives the two following eigenvalues, each of which appears twice:

$$\lambda_{5,6} = (A + C + E + G) \pm (B + F)$$

5.6 Assume that there exists another irreducible continuous representation ϑ with character $\chi(\varphi)$. By an argument similar to that given at the end of Chapter 5, Equations (5.141) and (5.207) imply that $\chi = 0$. Contradiction.

7.1 $A(t) = e^t \begin{bmatrix} 1 & t \\ 0 & 1 \end{bmatrix}$ $\quad t \in \mathbf{R};$ $\quad A(t_1)A(t_2) = A(t_1 + t_2)$

7.2 $J^3 = -(u_1^2 + u_2^2 + u_3^2)J$. Normalization $u_1^2 + u_2^2 + u_3^2 = 1$, as t is an additional parameter. $e^{tJ} = E + \sin t \cdot J + (1 - \cos t)J^2$.

7.3 (a) A, B symplectic implies $(AB)^T S(AB) = S$.

(b) $SJ = (SJ)^T \Leftrightarrow J = \left[\begin{array}{c|c} U & V \\ \hline W & -U^T \end{array} \right]$ where U, V, W are arbitrary $n \times n$ matrices with $V = V^T$, $W = W^T$.

(c) If J_1, J_2 are two infinitesimal operators of the form given in (b), then so is $[J_1, J_2]$.

7.4 (a) $J_1' : t = 1/2, r = s = 0;$ $\quad J_2' : s = -1/2, r = t = 0;$
$J_3' : r = 1/2, s = t = 0.$
To show: $[J_1', J_2'] = J_3'$ and cyclically (see (7.39), (7.35)).

(b) The weights are the eigenvalues of $L' = iJ_3' : \lambda_{1,2} = \pm 1/2$ (each double). $\theta = 2 \cdot \theta_{1/2}$.

(c) The symmetry adapted basis vectors are

$$y_1 = (1 \quad i \quad 0 \quad 0)^T \qquad z_1 = (0 \quad 0 \quad i \quad 1)^T$$
$$y_2 = (0 \quad 0 \quad 1 \quad i)^T \qquad z_2 = (-i \quad -1 \quad 0 \quad 0)^T$$

7.5 (a) $D(a) : J \rightarrow J' = AJA^{-1}$. Let J be the infinitesimal operator of $B(t) \subset G$. The family $AB(t)A^{-1}$ with fixed A has J' as an infinitesimal operator, hence $J' \in \mathcal{L}$.

Let A, B be fixed. $D(AB) : J \rightarrow (AB)J(AB)^{-1} = A(BJB^{-1})A^{-1}$, thus $D(AB) = D(A)D(B)$. Furthermore, $D(E_n) = E_d$.

(b) Let K be an infinitesimal operator of the family $A(t)$:

$$\frac{d}{dt}\left(A(t)JA(t)^{-1}\right)\Big|_{t=0} = [K, J]$$

Use the Jacobi identity to show that
$D([K_1, K_2])J = [D(K_1), D(K_2)]J$.

7.6 In contrast to Problem 7.5(b), one can no longer use the equation $[K, L] = KL - LK$, but rather the structure relations (7.33) and the Jacobi identity for the basis vectors $J_1, J_2, \ldots, J_d$. The representation is faithful if and only if $[K, J] = 0 \; \forall J$ implies $K = 0$.

7.7 The $D(J_k')$ coincide with the J_k' of (7.38). For instance, one obtains the third column of $D(J_1')$ as follows: $[J_1', J_3'] = -J_2'$ by (7.35). The vector $-J_2'$ has the components $0, -1, 0$ with respect to the basis J_1', J_2', J_3'.

7.9 For J in the solution e^{tJ} to Problem 7.2, we may substitute, by virtue of (7.101), the special matrices $D(J_k')$ of solution 7.7.

7.11 According to (7.81), we have $U_k = J_k' \otimes E_2 \oplus E_2 \otimes J_k'$ with J_k' from (7.50), $k = 1, 2, 3$. The vector v, spanning the subspace to ϑ_0, has weight 0 (= eigenvalue of iU_3).
$e_1 = x_1 y_1 \qquad e_2 = x_1 y_2 \qquad e_3 = x_2 y_1 \qquad e_4 = x_2 y_2$
$U_3 e_1 = -ie_1 \quad U_3 e_2 = U_3 e_3 = 0 \quad U_3 e_4 = ie_4$, thus $v = \alpha e_2 + \beta e_3$
$L_+' = iU_1 - U_2$ is the raising operator as described in (7.72).
$L_+' v = L_+'(\alpha e_2 + \beta e_3) = 0$, thus $v = e_2 - e_3$.

7.12 Let $\ell \geq m$.

(a) The multiplicity of an eigenvalue of L_3 is equal to the number of points on the decreasing 45° line in Fig. 4.5. Thus we obtain the following table:

Eigenvalue of L_3 (up to factor $\hbar$)	Multiplicity
$\ell + m$	1
$\ell + m - 1$	2
$\ell + m - 2$	3
$\vdots$	$\vdots$
$\ell - m$	2m+1
$\ell - m - 1$	2m+1
$\vdots$	$\vdots$
$-(\ell - m)$	2m+1
$-(\ell - m) - 1$	2m
$\vdots$	$\vdots$
$-\ell - m + 1$	2
$-\ell - m$	1

(b) Equation (4.81) and considerations similar to those of Section 7.5.2 imply

Eigenvalue of L^2	Multiplicity
$\hbar^2 j(j+1)$	$2j+1$

$(j = \ell+m, \ell+m\text{-}1, \ldots, \ell\text{-}m)$

8.1

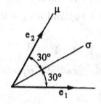

Let μ, σ be reflections about the corresponding lines (generating set) (see Fig. A.2).

$$D(\sigma) = \begin{bmatrix} 0 & 1 \\ 1 & 0 \end{bmatrix}, \qquad D(\mu) = \begin{bmatrix} -1 & 0 \\ 1 & 1 \end{bmatrix}$$

All representing matrices are integral.

Fig. A.2

8.2 Let, for example,

$$e_1 = (1 \quad 0 \quad 1)^T, \qquad e_2 = (1 \quad 1 \quad 0)^T, \qquad e_3 = (0 \quad 1 \quad 1)^T$$

be the lattice basis in a cartesian coordinate system. Equation (8.7) implies

$$\epsilon_1 = \tfrac{1}{2}(1 \quad -1 \quad 1)^T, \quad \epsilon_2 = \tfrac{1}{2}(1 \quad 1 \quad -1)^T, \quad \epsilon_3 = \tfrac{1}{2}(-1 \quad 1 \quad 1)^T$$

8.3 Consider Fig. 1.11.

(a) In the group T, let s_1 be the 120° rotation about the interior diagonal through vertex 1 and s_2 be the 180° rotation about the z-axis. $\{s_1, s_2\}$ is a generating set. 1-dimensional representations are found from (8.10).

$$\vartheta_1^3 : D(s_1) = \begin{bmatrix} 0 & -1 & 0 \\ 0 & 0 & -1 \\ 1 & 0 & 0 \end{bmatrix} \quad D(s_2) = \begin{bmatrix} -1 & 0 & 0 \\ 0 & -1 & 0 \\ 0 & 0 & 1 \end{bmatrix} \text{(Orientation of axes)}$$

(b) In the group O, s_1 is the 180° rotation about the line $x = -y, z = 0$ and s_2 the 90° rotation about the y-axis. $\{s_1, s_2\}$ is a generating set. According to (8.11),

$$s_1 = \begin{vmatrix} 1 & 2 & 3 & 4 \\ 2 & 1 & 3 & 4 \end{vmatrix} \qquad s_2 = \begin{vmatrix} 1 & 2 & 3 & 4 \\ 2 & 4 & 1 & 3 \end{vmatrix}$$

for some given labeling of the interior diagonals. The representing matrices are read from Table (5.77), in the columns corresponding to these two permutations.

8.4 (a) For $C_3 + ZC_3$, compare (1.52) and a table similar to (8.16).

(b) Character table of solution 5.2(a) with $h = 3$ is applicable here, because $(C_3, D_3) \cong$ plane group $D_3 \cong$ spatial group D_3. See Theorem 8.7, Fig. 9.2, and Section 8.2.1.

8.5 Ritz space (see Fig. 9.5): $f_1 = 1$, $f_2 = e^{2\pi i y_1}$, $f_3 = e^{2\pi i y_2}$, $f_4 = \overline{f_2}$, $f_5 = \overline{f_3}$. Let $f = H \sin 2\pi y_1 = \left(4\pi^2 \frac{\hbar^2}{2m} + v(y)\right) \sin 2\pi y_1$. Projection: Substitute f into (8.26). Computation of the various inner products by use of (8.52), (8.63), (8.70) gives

$$Pf = \frac{1}{2i}(\delta - \gamma)(f_2 - f_4) = (\delta - \gamma)\sin 2\pi y_1$$

8.6 See (8.59). Construct the symmetry adapted basis by (5.44) through (5.48).

(a) ϑ_1^1: The two eigenvectors are of the form $au_1 + bu_2$ and $a'u_1 + b'u_2$ where $a, b, a', b' \in \mathbf{C}$ and

$$u_1 = (2 \quad 0 \quad 0 \quad 0 \quad 0)^T, \qquad u_2 = (0 \quad 1 \quad 1 \quad 1 \quad 1)^T$$

Eigenvalues λ_1, λ_2 are obtained from solving the quadratic equation

$$\begin{vmatrix} \epsilon - \lambda & 2\alpha \\ 2\alpha & (2\beta + \gamma + \delta) - \lambda \end{vmatrix} = 0$$

(b) All other invariant subspaces are spanned by the following eigenvectors, here listed with the corresponding eigenvalues:

$$\begin{array}{llll} \vartheta_3^1: & v = (0 & 1 & -1 & 1 & -1)^T & \lambda_3 = -2\beta + \gamma + \delta \\ \vartheta_1^2: & w_1 = (0 & 1 & 0 & -1 & 0)^T & \lambda_4 = \lambda_5 = (\delta - \gamma) \\ & w_2 = (0 & 0 & 1 & 0 & -1)^T \end{array}$$

Here, by (8.52),

$$\alpha = \int_0^1 \int_0^1 v(y)e^{2\pi i y_1} \, dy_1 \, dy_2, \quad \text{where} \quad p = \begin{bmatrix} 0 \\ 0 \end{bmatrix}, q = \begin{bmatrix} 1 \\ 0 \end{bmatrix}$$

$$\beta = \int_0^1 \int_0^1 v(y)e^{2\pi i y_2} \, dy_1 \, dy_2, \quad \text{where} \quad p = \begin{bmatrix} 0 \\ 0 \end{bmatrix}, q = \begin{bmatrix} 0 \\ 1 \end{bmatrix}$$

See Fig. 8.5. Matrix entries γ, δ, ϵ as in (8.68), (8.69).

8.7 Equation (3.11) implies $\vartheta_{\text{per}} = 3\vartheta_1^1 \oplus \vartheta_3^1 \oplus \vartheta_4^1 \oplus 2\vartheta_1^2$.
Ritz operator M: symmetric 9×9 matrix with 11 free parameters (5.101). The symmetry adapted functions are linear combinations of the functions (8.49) with the following coefficients taken from Table (3.7) (dots stand for zero components):

$$\vartheta_1^1: \quad r_1 = \begin{array}{ccc} \cdot & \cdot & \cdot \\ \cdot & 4 & \cdot \\ \cdot & \cdot & \cdot \end{array} \quad r_2 = \begin{array}{ccc} 1 & \cdot & 1 \\ \cdot & \cdot & \cdot \\ 1 & \cdot & 1 \end{array} \quad r_3 = \begin{array}{ccc} \cdot & 1 & \cdot \\ 1 & \cdot & 1 \\ \cdot & 1 & \cdot \end{array}$$

$$\vartheta_3^1: \quad z = \begin{array}{ccc} \cdot & 1 & \cdot \\ -1 & \cdot & -1 \\ \cdot & 1 & \cdot \end{array} \qquad \vartheta_4^1: \quad u = \begin{array}{ccc} -1 & \cdot & 1 \\ \cdot & \cdot & \cdot \\ 1 & \cdot & -1 \end{array}$$

Table (3.7) and Equation (3.13) imply (up to a the factor $\sqrt{2}/2$):

$$\vartheta_1^2: v_1 = \begin{array}{ccc} \cdot & \sqrt{2} & \cdot \\ \cdot & \cdot & \cdot \\ \cdot & -\sqrt{2} & \cdot \end{array} \qquad v_1' = \begin{array}{ccc} \cdot & \cdot & \cdot \\ -\sqrt{2} & \cdot & \sqrt{2} \\ \cdot & \cdot & \cdot \end{array}$$

$$v_2 = \begin{matrix} -1 & \cdot & -1 \\ \cdot & \cdot & \cdot \\ 1 & \cdot & 1 \end{matrix} \qquad v_2' = \begin{matrix} 1 & \cdot & -1 \\ \cdot & \cdot & \cdot \\ 1 & \cdot & -1 \end{matrix}$$

$$M = \begin{bmatrix} \alpha_0 & \alpha_1 & \alpha_2 & \alpha_1 & \alpha_2 & \alpha_1 & \alpha_2 & \alpha_1 & \alpha_2 \\ & \alpha_3 & \alpha_5 & \alpha_6 & \alpha_7 & \alpha_8 & \alpha_7 & \alpha_6 & \alpha_5 \\ & & \alpha_4 & \alpha_5 & \alpha_{10} & \alpha_7 & \alpha_9 & \alpha_7 & \alpha_{10} \\ & & & \alpha_3 & \alpha_5 & \alpha_6 & \alpha_7 & \alpha_8 & \alpha_7 \\ & & & & \alpha_4 & \alpha_5 & \alpha_{10} & \alpha_7 & \alpha_9 \\ & \text{symmetric} & & & & \alpha_3 & \alpha_5 & \alpha_6 & \alpha_7 \\ & & & & & & \alpha_4 & \alpha_5 & \alpha_{10} \\ & & & & & & & \alpha_3 & \alpha_5 \\ & & & & & & & & \alpha_4 \end{bmatrix}$$

Calculate α_i $(i = 0, 1, 2, \ldots, 11)$ by use of (8.52) and the vector star of Problem 8.7 (see solution 8.6).

To compute the eigenvalues, we need to solve a cubic and a quadratic equation:

$$\begin{vmatrix} \alpha_0 - \lambda & 4\alpha_2 & 4\alpha_1 \\ 4\alpha_2 & (\alpha_4 + \alpha_9 + 2\alpha_{10}) - \lambda & 2(\alpha_5 + \alpha_7) \\ 4\alpha_1 & 2(\alpha_5 + \alpha_7) & (\alpha_3 + 2\alpha_6 + \alpha_8) - \lambda \end{vmatrix} = 0$$

$$\begin{vmatrix} (\alpha_3 - \alpha_8) - \lambda & \sqrt{2}(-\alpha_5 + \alpha_7) \\ \sqrt{2}(-\alpha_5 + \alpha_7) & (\alpha_4 - \alpha_9) - \lambda \end{vmatrix} = 0$$

z and u are eigenvectors corresponding to the eigenvalues $(\alpha_3 - 2\alpha_6 + \alpha_8)$ and $(-\alpha_4 - \alpha_9 + 2\alpha_{10})$, respectively.

8.8 The small group of k is D_2.
Equation (3.11) implies: $\vartheta_{\text{per}} = 2\vartheta_1^1 \oplus \vartheta_2^1 \oplus \vartheta_3^1 \oplus 2\vartheta_4^1$.

$$M = \begin{bmatrix} \alpha_1 & \alpha_3 & \alpha_4 & \alpha_5 & \alpha_4 & \alpha_3 \\ & \alpha_2 & \alpha_8 & \alpha_4 & \alpha_6 & \alpha_7 \\ & & \alpha_2 & \alpha_3 & \alpha_7 & \alpha_6 \\ & \text{symmetric} & & \alpha_1 & \alpha_3 & \alpha_4 \\ & & & & \alpha_2 & \alpha_8 \\ & & & & & \alpha_2 \end{bmatrix}$$

The Ritz matrix M has 8 free parameters (5.101). Symmetry adapted functions from (3.7), (3.13); compare solution 8.7. In the subspaces corresponding to ϑ_1^1 (upper sign) and ϑ_4^1 (lower sign), the operator M produces the two quadratic equations given by

$$\begin{vmatrix} (\alpha_1 \pm \alpha_5) - \lambda & \sqrt{2}(\alpha_3 \pm \alpha_4) \\ \sqrt{2}(\alpha_3 \pm \alpha_4) & (\alpha_2 \pm \alpha_6 + \alpha_7 \pm \alpha_8) - \lambda \end{vmatrix} = 0 \quad \begin{matrix} \text{with solutions} \\ \lambda_1, \lambda_2, \lambda_3, \lambda_4 \end{matrix}$$

The remaining two invariant subspaces are eigenspaces to the eigenvalues $\lambda_{5,6} = \alpha_2 \mp \alpha_6 - \alpha_7 \pm \alpha_8$. In contrast to (8.52), we now have

$$m_{pq} = \frac{h^2}{2m} |q + k|^2 \delta_{pq} + \int_0^1 \int_0^1 v(y) e(q - p, y) \, dy_1 \, dy_2$$

8.9 (3.11) implies: $\vartheta_{per} = 2\vartheta_1^1 \oplus \vartheta_1^2$.

$$M = \begin{bmatrix} \alpha_1 & \alpha_3 & \alpha_3 & \alpha_3 \\ & \alpha_2 & \alpha_4 & \alpha_4 \\ \text{symm.} & & \alpha_2 & \alpha_4 \\ & & & \alpha_2 \end{bmatrix}$$

The Ritz matrix M has 4 free parameters. In the subspace corresponding to ϑ_1^1, the operator M produces the quadratic equation

$$\begin{vmatrix} \alpha_1 - \lambda & \sqrt{3}\alpha_3 \\ \sqrt{3}\alpha_3 & (\alpha_2 + 2\alpha_4) - \lambda \end{vmatrix} = 0 \qquad \text{with solutions } \lambda_1, \lambda_2$$

The invariant subspace to ϑ_2^1 is an eigenspace with eigenvalue $\lambda_3 = \lambda_4 = \alpha_2 - \alpha_4$.

8.10 Let A be a proper or improper rotation that leaves Γ invariant and is described by the matrix A (A') relative to the lattice basis of Γ (Γ^{-1}). By (8.7),

$$(\epsilon_i, e_k) = (A\epsilon_i, Ae_k) = \delta_{ik} = \sum_{\rho=1}^{n} a'_{\rho i} a_{\rho k}$$

For fixed i, the last equation describes an inhomogeneous system of integer equations for the $a'_{\rho i}$. As A^{-1} is integral (see text after (8.6)), so is $(A^T)^{-1}$, and hence $a'_{\rho i}$ are integers $\forall \rho, i$.

8.11 By virtue of (8.96), the representing matrices of the elements (8.89), (8.90), (8.91) of D_2 are

$$D(s_{180^\circ}) = \begin{bmatrix} 1 & 0 & 0 & 0 & 0 \\ 0 & 0 & 0 & 1 & 0 \\ 0 & 0 & 0 & 0 & 1 \\ 0 & 1 & 0 & 0 & 0 \\ 0 & 0 & 1 & 0 & 0 \end{bmatrix} \qquad D(s_V) = \begin{bmatrix} 1 & 0 & 0 & 0 & 0 \\ 0 & 0 & 0 & -1 & 0 \\ 0 & 0 & 1 & 0 & 0 \\ 0 & -1 & 0 & 0 & 0 \\ 0 & 0 & 0 & 0 & 1 \end{bmatrix}$$

$$D(s_H) = \begin{bmatrix} 1 & 0 & 0 & 0 & 0 \\ 0 & -1 & 0 & 0 & 0 \\ 0 & 0 & 0 & 0 & 1 \\ 0 & 0 & 0 & -1 & 0 \\ 0 & 0 & 1 & 0 & 0 \end{bmatrix}$$

Equation (5.148) implies: $\vartheta_{per} = 2\vartheta_1^1 \oplus 2\vartheta_3^1 \oplus \vartheta_2^1$. By (5.101), the matrix M has 7 free parameters. Symmetry adapted basis:

$$\begin{array}{ll} \vartheta_1^1: & (\sqrt{2} \quad 0 \quad 0 \quad 0 \quad 0)^T \\ & (0 \quad 0 \quad 1 \quad 0 \quad 1)^T \\ \vartheta_3^1: & (0 \quad 1 \quad 0 \quad -1 \quad 0)^T \\ & (0 \quad 0 \quad 1 \quad 0 \quad -1)^T \\ \vartheta_2^1: & (0 \quad 1 \quad 0 \quad 1 \quad 0)^T \end{array} \qquad M = \begin{bmatrix} \alpha_1 & 0 & \alpha_2 & 0 & \alpha_2 \\ & \alpha_3 & \alpha_5 & \alpha_6 & -\alpha_5 \\ & & \alpha_4 & -\alpha_5 & \alpha_7 \\ \text{symm.} & & & \alpha_3 & \alpha_5 \\ & & & & \alpha_4 \end{bmatrix}$$

Group theory reduces the 5-dimensional eigenvalue problem to two 2-dimensional and a 1-dimensional problem. Compute the α_i by means of (8.52), where p, q are vectors of the reciprocal lattice.

8.12

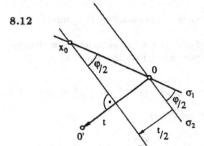

Fig. A.3

(a) Let σ_i be reflections in the lines plotted in Fig. A.3.

x_0 is the fixed point of the rotation taking O into O'. Proper isometries preserve distances and orientations.

(b) Translations (E/t) do not belong to the class of the elements (R/t) with $R \neq E$. Two translations $(E/t_1), (E/t_2)$ are similar if and only if $|t_1| = |t_2|$. Two elements (R_1/t_1), (R_2/t_2) are similar if and only if the oriented rotation angles φ_1, φ_2 satisfy $\varphi_1 = \varphi_2$ (i.e., $R_1 = R_2$).

9.1 Notation as in (9.3).

U_2 : Theorem 9.1 and (9.10) imply $\ell_1 > \ell_2$. Let $\ell_1 - \ell_2 = d$. The character $\chi_{\ell_1 \ell_2}$ is given by

$$e((\ell_1 - 1)x_1 + \ell_2 x_2) + e(\ell_2 x_1 + (\ell_1 - 1)x_2)$$
$$+ e((\ell_1 - 2)x_1 + (\ell_2 - 1)x_2) + e((\ell_2 - 1)x_1 + (\ell_1 - 2)x_2) + \ldots$$
$$+ \begin{cases} e\left(\frac{1}{2}(2\ell_2 + d + 1)(x_1 + x_2)\right) & \text{if } d \text{ is odd} \\ e\left(\frac{1}{2}[(2\ell_2 + d)x_1 + (2\ell_2 + d - 2)x_2]\right) \\ \quad + e\left(\frac{1}{2}[(2\ell_2 + d - 2)x_1 + (2\ell_2 + d)x_2]\right) & \text{if } d \text{ is even} \end{cases}$$

$SU(2)$: Theorem 9.2 and (9.10) imply that the character $\chi_{\ell_1 0}$ equals

$$e((\ell_1 - 1)x_1) + e((\ell_1 - 1)x_2) + e((\ell_1 - 2)x_1 + x_2)$$
$$+ e(x_1 + (\ell_1 - 2)x_2) + \ldots$$
$$+ \begin{cases} e\left(\frac{1}{2}(\ell_1 - 1)(x_1 + x_2)\right) & \ell_1 \text{ odd} \\ e\left(\frac{1}{2}[\ell_1 x_1 + (\ell_1 - 2)x_2]\right) + e\left(\frac{1}{2}[(\ell_1 - 2)x_1 + \ell_1 x_2]\right) & \ell_1 \text{ even} \end{cases}$$

With $x_1 + x_2 = 0$, one obtains (5.201) and (5.204). For odd (even) ℓ_1 the subscript ℓ of ϑ_ℓ becomes an integer (half integer); see Thm. 4.3.

9.2 Equations (9.29), (9.35), (9.36) imply $\ell_1 > 0$.

$$\chi_{\ell_1} = 2\left[\cos\left(\frac{\ell_1 - 1}{2}\right)\varphi_1 + \cos\left(\frac{\ell_1 - 3}{2}\right)\varphi_1 + \ldots + \begin{cases} 1/2 & \ell_1 \text{ odd} \\ \cos\frac{\varphi_1}{2} & \ell_1 \text{ even} \end{cases} \right]$$

9.3 Equations (9.29), (9.35), (9.36) imply

(a) $\chi_{40} = e^{i\varphi_1} + e^{i\varphi_2} + e^{-i\varphi_1} + e^{-i\varphi_2} = 2(\cos\varphi_1 + \cos\varphi_2)$.

(b) $SO(n)$ is a compact Lie group, thus ϑ_ℓ is equivalent to a unitary representation.
$\chi_{3\pm1} = e^{\frac{1}{2}(\varphi_1 \pm \varphi_2)} + e^{-\frac{1}{2}(\varphi_1 \pm \varphi_2)}$. The representation is multiple-valued. Determinants of the representing matrices equal 1.

9.4 Similar to solution 9.3.

$$\chi_{531} = e^{\frac{1}{2}(\varphi_1 + \varphi_2 + \varphi_3)} + e^{\frac{1}{2}(-\varphi_1 + \varphi_2 - \varphi_3)} + e^{\frac{1}{2}(-\varphi_1 - \varphi_2 + \varphi_3)} + e^{\frac{1}{2}(\varphi_1 - \varphi_2 - \varphi_3)}$$

9.5 Theorem 9.2 and (9.10) imply (notation (9.3)):

(a) $\chi_{4210} = e(\varphi_1) + e(\varphi_2) + e(\varphi_3) + e(\varphi_4), \quad \sum\limits_{i=1}^{4} \varphi_i = 0$

(b) $\chi_{4310} = e(\psi_1) + e(\psi_2) + e(\psi_3) + e(-\psi_1) + e(-\psi_2) + e(-\psi_3)$, where $\psi_1 = \varphi_1 + \varphi_2$, $\psi_2 = \varphi_2 + \varphi_3$, $\psi_3 = \varphi_1 + \varphi_3$. This is the character of the representation of $SO(6)$ by itself. (ψ_i are rotation angles).

(c) $\chi_{4320} = e(\varphi_1 + \varphi_2 + \varphi_3) + e(\varphi_1 + \varphi_2 + \varphi_4) + e(\varphi_1 + \varphi_3 + \varphi_4 + e(\varphi_2 + \varphi_3 + \varphi_4 = \overline{\chi}_{4210}$, as a comparison with (a) shows.

9.6 (a) $3 \otimes 3 \otimes \overline{3} = 15 \oplus 3 \oplus 3 \oplus 3 \oplus \overline{6}$

(b) Let b_1, b_2, b_3 and e_1, e_2, e_3 be the respective basis vectors of $\overline{3}$ and 3. $\overline{3} \otimes 3 = 8 \oplus 1$.
Octet(8): $e_1b_1, e_1b_2, e_2b_1, e_3b_2, e_2b_3, e_3b_3, e_3b_1 + e_2b_2,$
$\qquad\qquad e_2b_2 - e_3b_1 - 2e_1b_3,$
Singlet(1): $e_1b_3 - e_3b_1 + e_2b_2$

Symbol References

$\forall$ for all
$\exists$ there exists
$\alpha \Rightarrow \beta$ statement α implies statement β
$\alpha \Leftrightarrow \beta$ statements α and β are equivalent,
i.e., α implies and is implied by β

Let A, B be two sets:

$A \subset B$ A is a subset of B
$A \cup B$ union of A and B
$A \cap B$ intersection of A and B
$A \setminus B$ complement of B relative to A,
i.e., set of elements in A that are not in B
$a \in A$ a is an element of A
$a \notin A$ a is not an element of A
$\mathbf{R}$ set of real numbers
$\mathbf{C}$ set of complex numbers
$x \to y$ element y is assigned to element x
$\overline{c}$ complex conjugate of c

The transposed matrix of A is denoted by A^T. In particular, a vector v with the coordinates v_i is regarded as a one-column matrix. Thus $v = (v_1, v_2, \ldots, v_n)^T$.

$G_1 \cong G_2$	The two groups G_1 and G_2 are isomorphic.
$G_1 \times G_2$	direct product of the two groups G_1 and G_2
$V_1 \oplus V_2$	direct sum of the two vector spaces V_1 and V_2

If H is a normal subgroup of G, then G/H is the factor group of G modulo H.

Bibliography

[1] Allgower E.L., Böhmer K., Georg K., Miranda R. *Exploiting symmetry in boundary element methods*, to appear in SIAM J. Numer. Anal.

[2] Armstrong M.A. *Groups and Symmetry*, Springer, 1988.

[3] Birkhoff G., MacLane S. *A Survey of Modern Algebra*, 4th ed., Macmillan Publ. Co., 1976.

[4] Barut A., Raczka R. *Theory of Group Representations and Applications*, PWN-Polish Scientific Publishers, Warszawa, 1980.

[5] Bluman G.W., Kumei S. *Symmetries and Differential Equations*, Springer, 1989.

[6] Bossavit A. *Symmetry Groups and Boundary Value Problems. A Progressive Introduction to Noncommutative Harmonic Analysis of Partial Differential Equations in Domains with Geometrical Symmetry*, Computer Meth. in Appl. Mech. and Eng. Nr. 56, 167-215, 1986.

[7] Bourbaki N. *Groupes et algèbres de Lie*, Vol. 1285, Ch. IV, V, VI, Hermann Paris, 1968.

[8] Bradley C.J., Cracknell A.P. *The Mathematical Theory of Symmetry in Solids*, Oxford University Press, 1972.

[9] Breden G.E. *Introduction to Compact Transformation Groups*, Academic Press, 1972.

[10] Burzlaff H., Thiele G. *Kristallographie, Grundlagen und Anwendungen*, Band 1, Thieme Stuttgart, 1977.

[11] Chan Sang-Lung *Application de la théorie des représentations dans certains domaines sans symétrie classique*, C.R. Acad. Sc. Paris, t. 280, Série A, 389-391, 1975.

[12] Chevalley C. *Theory of Lie Groups*, Princeton University Press, 1946.

[13] Cicogna G. *Symmetry Breakdown from Bifurcations*, Lettere al Nuovo Cimento, Vol. 31, 600-602, 1981.

[14] Condon E.U., Shortley G.H. *The Theory of Atomic Spectra*, Cambridge University Press, 1959.

[15] Cooley J.W., Tukey J.W. *An Algorithm for the Machine Calculation of Complex Fourier Series*, Math. Comp. 19, 297-301, 1965.

[16] Coppersmith, D. and Winograd, S. *Matrix multiplication via arithmetic progressions*, Proceedings of the 19th ACM Symposium on the Theory of Computing (STOC), 1-6, 1987.

[17] Courant R., Hilbert D. *Methods of Mathematical Physics*, Vol. I, Interscience Publ. Inc. New York, 1953.

[18] Dellnitz M, Werner B. *Computational Methods for Bifurcation Problems with Symmetries—With Special Attention to Steady State and Hopf Bifurcation Points*, J. of Comp. and Applied Math., Nr. 26, 97-123, 1989.

[19] Dellnitz M. *Hopf-Verzweigung in Systemen mit Symmetrie und deren numerische Behandlung*, Dissertation Universität Hamburg, Verlag an der Lottbek, 1989.

[20] Diaconis P., Graham R.L. *The Radon Transform on Z_2^k*, Pacific Journal of Mathematics, Vol. 118, No. 2, 323-346, 1985.

[21] Diaconis P., Shahshahani M. *Time to Reach Stationarity in the Bernoulli-Laplace Diffusion Model*, SIAM J. Math. Anal., Vol. 18, No. 1, 208-218, 1987.

[22] Diaconis P., Rockmore D. *Efficient Computation of the Fourier Transform on Finite Groups*, Technical Report No. 292, Dept. of Statistics, Stanford University, California, 1988.

[23] Diaconis P., Graham R.L., Morrison J.A. *Asymptotic Analysis of a Random Walk on a Hypercube with many Dimensions*, in *Random Structures and Algorithms*, Vol. 1, No. 1, Wiley, 1990.

[24] Divisia F. *Les problèmes de l'indice général des prix*, Revue générale des sciences, 1-8, septembre 1927.

[25] Eichhorn W. *What is an Economic Index?*, in *Theory and Applications of Economic Indices*, Eichhorn W., Henn R., Opitz D., and Shephard R.W., Physica Würzburg, 1978.

[26] Fässler A., Mäder R. *Symmetriegerechte Basisvektoren für Permutationsdarstellungen aller endlichen Punktgruppen der Dimensionen 3 und 2*, Zeitschrift für angewandte Mathematik und Physik (ZAMP), Vol. 31, 1980.

[27] Fässler A., Schwarzenbach H. *Symmetriegerechte Polynome von Punktgruppen*, Zeitschrift für angewandte Mathematik und Physik (ZAMP), Vol. 30, 1979.

[28] Fässler A. *Application of Group Theory to the Method of Finite Elements for Solving Boundary Value Problems*, Thesis ETH Zürich No. 5696, 1976.

[29] Fisher I. *The Making of Index Numbers*, Third Edition, (Reprinted by Kelley A.M., 1967), Mifflin Boston, 1927.

[30] Flöscher K. *Zur Berechnung der Gewichte und Charaktere halbeinfacher Lie-Gruppen*, Dissertation ETH Zürich Nr. 6253, 1978.

[31] Freudenthal H. *Zur Berechnung der Charaktere der halbeinfachen Lieschen Gruppen*, Indag. Math. Vol. 16, 1954.

[32] Funke H., Voeller J. *A Note on the Characterization of Fisher's "ideal index"*, in (Eichhorn, 1978).

[33] Gatermann K. *Symbolic Solution of Polynomial Equation Systems with Symmetry*, Proceedings of ISSAC 90 (Tokyo, Aug. 20-24, 1990) ACM, New York, 1990.

[34] Georgi H. *Lie Algebras in Particle Physics*, Benjamin, 1982.

[35] Goldstine H. *A History of Numerical Analysis From the 16th Through the 19th Century*, Springer, 1977.

[36] Golubitsky M., Schaeffer D.G., Stewart I. *Singularities and Groups in Bifurcation Theory*, Vol. I, Springer, 1988.

[37] Golubitsky M., Schaeffer D.G., Stewart I. *Singularities and Groups in Bifurcation Theory*, Vol. II, Springer, 1988.

[38] Good I.J. *The Interaction Algorithm and Practical Fourier Series*, Jour. Royal Statist. Soc. B20, 361-372, 1958.

[39] Grossmann I., Magnus W. *Groups and Their Graphs*, Mathematical Association of America MAA, 1964.

[40] Guy J., Mangeot B. *Use of Group Theory in Various Integral Equations*, SIAM Journal of Appl. Math., Vol. 40, No. 3, 1981.

[41] Hall M. *The Theory of Groups*, 2nd ed., Chelsea Publ. Co., 1976.

[42] Hoel P.G., Port S.C., Stone C.J. *Introduction to Stochastic Processes*, 1972.

[43] Hargittai I. *Symmetry, Unifying Human Understanding*, Pergamon Press, 1986.

[44] Henrici P. *Essentials in Numerical Analysis*, John Wiley and Sons, 1982.

[45] Hersch J. *On Harmonic Measures, Conformal Moduli and Some Elementary Symmetry Methods*, Journal d'analyse mathématique, Vol. 42, 211-228, 1982-83.

[46] Hersch J. *Sur les fonctions propres des membranes vibrantes couvrant un secteur symétrique de polygone régulier ou de domaine périodique*, Comment. Math. Helv., Vol. 41, 222-236, 1966-67.

[47] James G.D., Kerber A. *The Representation Theory of the Symmetric Group*, Addison-Wesley, 1981.

[48] James G.D. *Representation of the Symmetric Groups*, Lecture Notes in Mathematics 682, Springer, 1978.

[49] Janssen T. *Crystallographic Groups*, North-Holland/American Elsevier, 1973.

[50] Kemeny J.G., Snell J.L. *Finite Markov Chains*, van Nostrand, 1965.

[51] Kostrikin A.I., Shafarevich I.R. *Algebra I, Basic Notions of Algebra*, Translated from the Russian by Reid M., Springer, 1990.

[52] Leech J.W., Newman D.J. *How to Use Groups*, Methuen London, 1969.

[53] Lenz R. *Group Theoretical Methods in Image Processing*, Springer, 1990.

[54] Lichtenberg D.B. *Unitary Symmetry and Elementary Particles*, Academic Press, 1978.

[55] Mackey G.W. *Unitary Group Representations in Physics, Probability, and Number Theory*, Benjamin, 1968.

[56] Madelung E. *Die mathematischen Hilfsmittel des Physikers*, Bd. IV, Springer, 1964

[57] Mäder R. *Programming in MATHEMATICA*, Addison-Wesley, 1990

[58] Magnus W., Karrass A., Solitar D. *Combinatorial Group Theory*, Wiley, 1966.

[59] Martin G.E. *Transformation Geometry, An Introduction to Symmetry*, Springer, 1982.

[60] Messiah A. *Quantum Mechanics*, Vol.2, North Holland, 1962.

[61] Mathiak K., Stingl P. *Gruppentheorie für Chemiker, Physiko-Chemiker, Mineralogen*, Vieweg, 1969.

[62] Miller W., Jr. *Symmetry Groups and Their Applications*, Academic Press, 1972.

[63] Montgomery C.G. *Principles of Microwave Circuits*, Dover NY, 1965.

[64] Onishchik A.L., Vinberg E.B. *Lie Groups and Algebraic Groups*, Translation from the Russian by Leites D., Springer, 1990.

[65] Ovsiannikov L.V. *Group Analysis of Differential Equations*, Translation edited by Ames W.F., Academic Press, 1982.

[66] Peterson W. W. *Error Correcting Codes*, M.I.T. Press, 1961.

[67] Pontrjagin L. *Topological Groups*, Princeton University Press, 1946.

[68] Prigogine I., Lefever R. *Stability and Self-Organization in Open Systems*, in *Membranes, Dissipative Structures and Evolution*, Nicolis G. and Lefever R. eds. Wiley, 1974.

[69] Qui C., Zhong W. *Analysis of Symmetric or Partially Symmetric Structures*, Computer Methods in Applied Mechanics and Engineering 38, 1-18, 1983.

[70] Rotenberg M., Bivins R., Metropolis N., Wooten J.K. *The 3-j and 6-j Symbols*, The Technology Press, Massachusetts Institute of Technology, Cambridge, 1959.

[71] Sagle A.A., Walde R.E. *Introduction to Lie Groups and Lie Algebras*, Academic Press, 1973.

[72] Sattinger D.H., Weaver O.L. *Lie Groups and Algebras with Applications to Physics, Geometry, and Mechanics*, Springer, 1986.

[73] Schwarz H.R. *Numerical Analysis, A Comprehensive Introduction*, Wiley, 1989.

[74] Schwarz H.R. *Finite Element Methods*, Academic Press, 1988.

[75] Schwarzenbach H. *Anwendung der darstellungstheoretischen Methoden auf die Lösung von Randwertproblemen partieller Differentialgleichungen mittels Probierfunktionen*, Dissertation ETH Zürich Nr. 6338, 1979.

[76] Serre J.P. *Linear Representations of Finite Groups*, Springer, 1977

[77] Serre J.P. *Lie Algebras and Lie Groups*, Benjamin, 1965

[78] Shubnikov A.V., Belov N.V. *Colored Symmetry*, Oxford University Press, 1964

[79] Stanton D. *Invariant Theory and Tableau*, Springer (IMA Volume 19), 1989.

[80] Stanton D. *Orthogonal Polynomials and Chevalley Groups* in *Special Functions: Group Theoretical Aspects and Applications*, Askey R., Koornwinder T.H., Schempp W., eds., Reidel Dordrecht, 1984.

[81] Stiefel E. *Kristallographische Bestimmung der Charaktere der geschlossenen Lieschen Gruppen*, Comm. Math. Helv. Vol. 17, 165-200, 1944.

[82] Stiefel E. *Ueber eine Beziehung zwischen geschlossenen Lieschen Gruppen und diskontinuierlichen Bewegungsgruppen euklidscher Räume und ihre Anwendung der einfachen Lieschen Gruppen*, Comm. Math. Helv. Vol. 14, 350-380, 1942.

[83] Strang G. *Linear Algebra and its Applications*, 2nd ed., Academic Press, 1976.

[84] Streitwolf H.W. *Group Theory in Solid State Physics*, Beekman, 1971.

[85] Thompson T.M. *From Error-correcting Codes Through Sphere Packings To Simple Groups*, Mathematical Association of America, 1983.

[86] Ungricht H. *Anwendungen der Darstellungstheorie endlicher Gruppen auf Probleme der numerischen linearen Algebra*, Thesis ETH Zürich No. 6915, 1982.

[87] Vanderbauwhede A. *Local Bifurcation and Symmetry*, Research Notes in Mathematics, Vol.75, Pitman, 1982.

[88] van der Waerden B. L. *Group Theory and Quantum Mechanics*, Springer, 1974.

[89] Vinberg E.B. *Linear Representations of Groups*, Translated from the Russian by Iacob A., Birkhäuser, 1989.

[90] Vogt A. *Rapide historique du problème des indices et études de la solution de Divisia*, journal de la société de statistique de Paris, 1989.

[91] Vogt A. *Some Suggestions Concerning an Axiom System for Statistical Price and Quality Indices*, Communications in Statistics, Decker M. New York, 1987.

[92] Weaver H.J. *Theory of Discrete and Continuous Fourier Analysis*, Wiley, 1988.

[93] Weil A. *L'intégration dans les groupes topologiques et ses applications*, 2 ième éd., Hermann Paris, 1951.

[94] Werner B. *Computational Methods for Bifurcation Problems with Symmetries and Applications to Steady States of n-box Reaction-Diffusion-Models*, in *Numerical Analysis 1987* Griffiths D.F. and Watson G.A. eds., 279-293, Pitman 1988.

[95] Werner B. *Eigenvalue Problems with the Symmetry of a Group and Bifurcations*, in *Continuation and Bifurcations: Numerical Techniques and Applications*, ed. Roose D., Nato Asi Series, 1990.

[96] Weyl H. *The Classical Groups*, 2nd ed., Princeton University Press, 1946.

[97] Weyl H. *Symmetry*, Princeton University Press, 1952.

[98] Weyl H. *Symmetrie*, Uebersetzung der Originalausgabe durch Bechtolsheim L. *Symmetry*, Birkhäuser Basel, 1955.

[99] Weyl H. *The Theory of Groups and Quantum Mechanics*, Translated from the 2nd German ed. *Gruppentheorie und Quantenmechanik* by Robertson H.P., Dover, 1950.

[100] Wigner E. *Group Theory and Its Application to the Quantum Mechanics of Atomic Spectra*, Academic Press, 1964.

[101] Wolfram S. *MATHEMATICA, A System for Doing Mathematics by Computer*, Addison-Wesley, 1988.

Index

abelian group 4, 67, 145
    irreducible representation 36
    reducible representation 36
abelian subgroup 82, 86, 90, 99,
    239
abstract Lie algebra 188
abstract structure of the symmet-
    ric group 9
addition of angular momenta 205
adjoint representation 207
algebraic multiplicity 167
algorithm
    complete reduction, $SO(3)$, $SU(2)$
      89
    complete reduction, $U(n)$, $SU(n)$
      248
    computation of multiplicities
      144
    computation of weights 253
    division 244
    eigenvalue calculations 117
    symmetry adapted basis 113,
      155, 198, 260
    using orthogonality 115
alternating
    elementary sum 242
    group 13
      similarity classes 136
    representation 22
angular momentum 204
annihilation operator 261
approximation 13, 44
array 37
    symmetry adapted 37, 41
associative law 3
automorphic
    elementary sum 245
    Fourier polynomial 242

    function 242
autonomous system of ordinary dif-
    ferential equations 160
auxiliary file 244
average of continuous functions 157
axiom 5
    commensurability 5
    commodity reversal 5
    identity 5
    proportionality 6
azimuthal quantum number 203

Baker-Campbell-Hausdorff-formu-
    la 194
band form 62
basic cell 209
basis 25
    of a function space 26
    orthonormal 29, 56
    symmetry adapted 43, 54, 166
basis generator 115, 129, 261
Bernoulli-Laplace diffusion model
    176
bi-invariant probability 172
bifurcation 159, 166
    classification 167
    diagram 160, 162
    Hopf 160, 170
    linear problem 161
    nonlinear problem 161
    parameter 160
    Pitchfork 160
    point 160
      G-simple 167
      Hopf 167
      simple 167
      static 160
      steady-state 167

problems with symmetries 159
steady-state 169, 170
subgroup 167
Bloch wave 230
block diagonalization 41
boundary condition 63
boundary value 59
brain current 70
Brillouin zone 212
symmetry of the lattice 213
Brusselator 168

Cartan
group 242
method of weights 89, 239
Casimir operator 196
Cayley parametrization 83, 90
Cayley-Hamilton equation 206
central force field 202
character 139, 141, 145, 251
decomposition 140
finite group 145
irreducible representation of
U(n) 243
Kronecker product 140
on similarity classes 140
regular representation 148
representation by permutation
141
representation of U(3) 240
representation of U(n) 242
character matrix 149
character table 145, 236
$D_3 + ZD_3$ 217
cyclic group 215
dihedral group 144, 157, 270
octahedral group 215
symmetric group 142, 158
tetrahedral group 215
circle group 157
Clebsch-Gordan
coefficients 94, 201
series 93, 158
combination
law 1
of permutations 10
commensurability axiom 5

commodity reversal axiom 5
commutative law 4
commutator 187
properties 188
compact Lie group 105, 151, 155,
163
irreducible representation 155
complete reducibility of a representation 19
unitary representation 28
matrices 19
of a finite group 28
of an abelian group 36
complete reduction of a representation 99
of cyclic group 22
of Kronecker product 101
SO(3) 93
SU(n) 248
U(n) 248
of SL(2, C) 99
of SO(3) and SU(2) 86, 89
of SU(3) 254
of SU(n) 247
of U(n) 247
completeness
of spherical functions 98
of the representations of SU(2)
155
complex
conjugate representation 99,
258
free parameter 47
composition 3
of transformations 4
computational complexity 67
condition number 63
conglomerate 36
congruence map of a lattice 219
conjugate transpose 173
connected 75
construction, real orthonormal basis 56
continuous matrix group 74, 181,
188
free real parameters of 74
convolution 66, 173

coordinate transformation 15
coordinate-independent represen-
    tation 139
cross operator 44
crystal class 219, 220
crystal system 220
crystallographic point group 219,
    233
crystallographic space group 232
cycle notation 137
cyclic group 4, 22, 68, 214
    character table 145, 215
    complete reduction of a rep-
        resentation 22
    representation 22
cyclic matrix 48

data 69
decomposition
    of a character 140
    of a representation space 36
    of an isotypic component 36
    of an operator 41
degree
    of a representation 11
determining invariant subspaces 17
diagram
    of SU(3) 242
    of SO(n) 253
differential operator, discrete 48
diffusion
    model, Bernoulli-Laplace 170,
        176
    process 170
digital image processing 70
dihedral group 7, 13, 23, 144, 157,
    214
    character table 157
    plane and spatial 214
    representation by permutations
        52, 226
    set of generators of 23
    similarity classes 157
dimension
    of a Lie algebra 190
    of a representation 11, 140,
        167

of a representation of a Lie
    algebra 190
of an irreducible representa-
    tion 246, 253
discretization 13, 44, 63
    of a differential operator 48
distribution
    hyper-geometric 171
    stationary 170, 171, 179
division algorithm 244
dominant vector 246
dominant weight 88, 101
double-valued 85, 86, 93, 99
dual Hahn polynomial 176
duplication property 69

earthquake 70
economics 5
eigenvalue 35, 43, 62, 97, 222
    G-simple 167
    computation 45
    problem 43
eigenvector 43, 97
electroencephalography 70
element 1
    generating 4
    identity 2
    neutral 2
    unit 2
elementary particle physics 254,
    261
elliptic equation 59
equation
    autonomous system of ordi-
        nary differential 160
    Cayley-Hamilton 206
    elliptic 59
    evolution 160
    partial differential 13
equilibrium 163
equivalent
    point 212
    representation 15
        of finite groups 145
equivariant 163
    branching lemma 165
error correcting codes 179

evaluation of polynomials 70
evolution equation 160
exceptional group 242
exponential map 185

factor reversal test 6
faithful representation 11, 14, 163
family, one-parameter 181
fast Fourier transform 65
finite abelian group
    representation 36
finite element method 59
finite group 108
    complete reducibility of a rep-
        resentation 28
    irreducible representation 145
    representation 28, 145
finite point group
    symmetry adapted basis vec-
        tors 57
fixed point space 164
fixed point subspace 165
force matrix 122
formula, complete reduction of Kro-
        necker product 102
four group 23, 30, 231
Fourier
    inversion formula 65
    operator 69
    transform 65
free
    complex parameter 47
    real parameter 74, 132
full permutation group 8, 138
function
    defined on a set of points 52,
        54
    lattice 13
    odd 163
    spherical 90, 175
    spherical surface 91, 98
function space, basis 26
fundamental domain 51, 212, 241,
        252

Gelfand pair 174
general linear group 74, 183

generating symmetry adapted ba-
        sis 113, 116, 155, 198, 260
generating element 4
generating set 4, 273
    of the symmetric group 157
GL(n, **C**) and GL(n, **R**) 188
    infinitesimal operator 183
group 3
    ($\pm$)- 214
    (+)- 214, 218
    abelian 67
    alternating 13
    Cartan 242
    circle 157
    continuous matrix 74
    crystallographic point 219, 233
    crystallographic space 232
    cyclic 22, 68, 214
    dihedral 7, 13, 23, 157, 214
    exceptional 242
    finite 108
    full permutation 138
    general linear 74
    hyper-octahedral 252
    icosahedral 215
    kaleidoscope 242
    Klein four 23, 30, 231
    Lorentz 80, 98
    matrix 74, 75
    octahedral 135, 215
    of permutations 7
    of proper isometries 237
    of rotations 73
    orthogonal 73, 74
    point 134
    proper orthogonal 74
    real unimodular 74
    small 213
    space 232
    special linear 74, 76
    special orthogonal 74
    special unitary 74, 75
    symmetric 8, 67, 138
    symplectic 242
    tetrahedral 135, 214
    unimodular 74, 98
    unitary 74, 75, 239

Weyl 241, 252
group table 9
group theory method 43, 64, 97, 159

Hahn polynomial, dual 176
Hamilton operator 223
harmonic polynomial 90
hermitian
    conjugate 173
    matrix 77
    metric 27
    operator 204
higher 246
holohedral crystal class 220
homogeneous polynomial 84, 90, 95
homomorphism 10, 79
Hopf bifurcation 160
    point 167
hyper-octahedral group 252

icosahedral group 215
    irreducible representations 216
identity axiom 5
identity element 2
induced representation 232
infinitesimal operator 182, 183, 186, 187
    in matrix groups 186
    in $O(n)$ 184
    in $SO(n)$ 184
    in $SU(n)$ 184
    in $U(n)$ 184
    of Kronecker product 201
inner product 27
invariant subspace 16, 20, 113, 195
    explicit determination 17
inverse element 2
inversion 216
irreducible representation 16, 17, 24, 144, 243, 253
    1-dimensional 22
    of a compact Lie group 155
    of an abelian group 36
    of $SL(2, C)$ 99
    of $SO(3)$ 85, 86

of $SU(2)$ 84, 86, 155
of the icosahedral group 216
of the octahedral group 215
of the symmetric group 22
of $U(n)$ 243
    space 16
irreducible subspace 93, 114, 147, 257
isometry 219, 232
isomorphism 10, 216
    Lie algebra 189
isotropy subgroup 164
isotypic component 36, 38, 109, 116, 167
    decomposition 36
    non-unique decomposition 36, 115

Jacobian matrix 160, 166

kaleidoscope group 242
kernel 34, 79
Klein four group 23, 30, 231
Kronecker product 26
    character 140
    complete reduction of
        $SL(2,C)$ and Lorentz group 101
        $SO(3)$ 93
        $SU(n)$ 248
        $U(n)$ 248
    infinitesimal operator 201
    unitary 93

Laguerre circular geometry 103
Laplace operator 44, 60, 96, 203
lattice 209
    congruence map of 219
    function 13, 44
    reciprocal 211
    vector 209
        numbering 224, 231
law
    associative 3
    commutative 4
    of combination 1
left translation 31, 152
length of a vector 27

Lie algebra 188, 191, 193
 abelian 188
 abstract 188
 dimension 190
 isomorphism 189
 of SO(3) 198
 of SU(2) 198
 of SU(3) 254
  symmetry adapted basis 260
 representation 190, 191
Lie group, compact 105, 151, 155,
  163
Lie subalgebra 193
limit cycle, stable 162
linear operator 43, 47
linear transformation 80
Lorentz group 80, 98
 complete reduction of the Kronecker product 101
Lorentz transformation 78
lowering operator 197

mapping 232
Markov chain 171, 172
 rate of convergence 71
MATHEMATICA 94
matrix
 cyclic 48
 force 122
 hermitian 77
 Jacobian 160, 166
 of a completely reducible representation 20
 orthogonal 73
 Pauli spin 190
 permutation 12
 reduced form 16
 representing 19
 scattering 118
 skew-hermitian 184
 symplectic 206
 unitary 29
matrix group 74, 75
 continuous 181, 188
 infinitesimal operators 186
maximal torus 240
 SO(5) 250

membrane problem 43, 120
methane 123
method of weights 89
metric, hermitian 27
Minkowski cone 79
molecular oscillation 121
monomial 84, 95
multiple products of operators 3
multiplication
 of integers 70
 of polynomials 70
 theorem 143
multiplicity 36

natural representation 12
 permutation matrix 52
nearest neighbor random walk 171
neutral element 2
Newton principle 122
norm of a vector 27
number
 of irreducible inequivalent representations
  of a finite group 151
  of a regular representation
   149
 of similarity classes
  alternating group 136
  octahedral group 135
  orthogonal group 134
  permutation group 138
  SU(2) 154
  symmetric group 139
  tetrahedral group 136
  U(3) 240
numbering lattice vector 224, 231

octahedral group 135, 215
 character table 215
 irreducible representations 215
 similarity classes 135
odd function 163
one-parameter family 181
one-parameter subgroup 186
one-to-one 3
onto 3
operator 40

annihilation 261
Casimir 196
cross 44
decomposition 41
differential 48
Hamilton 223
Laplace 44, 60, 96, 203
linear 43, 47
lowering 197
raising 198
Ritz 220
symmetry 40
orbit 52
    of a group 163
    type 52
orbital quantum number 203
order 5
orthogonal
    basis 54
    group 73, 74, 188
        infinitesimal operator 184
        similarity classes 134
    matrix 73
    subspace 28
    vector 27
orthogonality
    of characters 142
    of irreducible representations
        108
    relation 142
orthonormal basis 29

(+)-group 214, 218
(±)-group 214
    first kind 216, 218
    second kind 218
parameter
    free complex 47
    free real 74, 132
partial differential equation 13
path of an eigenvalue 169
Pauli spin matrices 190
permutation 8
    assigned to a group element
        9
    even or odd 12
    group 7, 8

similarity classes 138
matrix 12
    natural representation 52
    representation 51, 52, 54
    type 137
perturbation theory 64
phase portrait 162
Pitchfork bifurcation 160
point group 134
price index 5
price reversal test 6
primitive cell 209
probability
    asymptotic 170
    bi-invariant 172
    distribution 170, 173
    transition 173
product
    inner 27
    Kronecker 26
projection
    onto a subspace 112, 146
    operator 147
projector 147
proper
    linear subspace 16
    orthogonal group 74
    rotation 2, 82
proportionality axiom 6

quantity reversal test 6

raising operator 198
random walk 172
    nearest neighbor 171
range 3
    finite 8
rank of basis generator 129
real unimodular group 74
realization 10
reciprocal lattice 211
reciprocity theorem 174
reduced form of a matrix 16
reducible representation 16
reduction of a representation 18
    of SO(3) by homogeneous poly-
        nomials 96

of the cyclic group 22
reflection 13
regular representation 148
  character 148
  number 149
  of the symmetric group on 3
    elements 148
relationship between
  continuous groups 76
  $SL(2, C)$ and Lorentz group
    77
  $SU(2)$ and $SO(3)$ 81
relativity theory 79
representation 11, 12, 84
  1-dimensional 16
  adjoint 207
  alternating 22
  by diagonals 87
  by permutations 51, 52, 54,
    141, 227
  complete reducibility 19, 36,
    155
  complex conjugate 99, 258
  construction formula 193
  coordinate-independent 139
  degree or dimension 11
  equivalent 15, 16
  faithful 11, 14
  induced 232
  irreducible 16, 17
    1-dimensional 22
    of an abelian group 36
  matrix
    completely reducible 20
    reducible 16
  natural 12
  not completely reducible 19
  of cyclic groups 22
  of finite abelian groups 36
  of finite groups 28, 145
  of Lie algebras 190, 191
    dimension 190
  of matrix groups 83
  of $SO(3)$ 88, 91, 95
  of $SO(3)$ by spherical func-
    tions 91
  of $SU(2)$ 91

of the symmetric group on 3
  elements 11, 13, 22
of the symmetric group on 4
  elements 30, 126
property 84
reducible 16, 18
regular 148
space 11
  decomposition 36
  irreducible 16
  unit 11
  unitary 27, 29
representation theory
  of continuous groups 86
  of the circle group $SO(2)$ 158
  of the symmetric group on 4
    elements 30, 126
representing matrix 19, 41, 109
right translation 152
Ritz
  matrix 222
  method 222
  operator 220
  space 222
rotation 2, 13, 73, 87
  matrix 13, 82
  proper 2, 82
rotational part 232

scattering matrix 118
Schur's lemma 33
  generalization of 39
  on Lie algebras 196
set of generators 4
  of the dihedral group 23
similar elements 133
similarity classes 133, 135, 145
  alternating group 136
  character 140
  dihedral group 157
  octahedral group 135
  permutation groups 138
  $SU(2)$ 154
  subgroups of orthogonal groups
    134
  symmetric group 139
  tetrahedral group 136

U(3) 240
simultaneous time and factor reversal test 7
simultaneously and sharply measurable 204
skew-hermitian matrix 184
skew-symmetric matrix 184
SL(2, C) 99, 104, 188
  complete reduction 99
  complete reduction of the Kronecker product 101
  continuous irreducible representation 99
  irreducible representation 99
SL(2, R) 103, 188
SL(n, C) and SL(n, R) 183
  infinitesimal operator 183
small group 213
SO(2) 157
SO(3) 88, 91, 135
  complete reduction of representations 86
  irreducible representation 85, 86
  reduction of representation by homogeneous polynomials 96
  representation by spherical functions 91
  representation of 95
  subgroup of 134
  weights of Kronecker product 92
SO(5)
  maximal torus 250
  Weyl group 250
SO(n) 188
  diagram 253
  infinitesimal operator 184
  torus 252
  weights 253
solution
  set 160
  steady-state 160, 163
space group 232
special linear group 74, 76
special orthogonal group 74

special unitary group 74, 75
spectrum 65
spherical functions 90, 175
  completeness 98
spherical polar coordinates 91
spherical surface function 91, 98
spin angular momentum 204
spontaneous symmetry-breaking 165
stabilizer 213
stable limit cycle 162
state variable 160
stationary distribution 170, 179
statistics 71
steady-state solution 160, 161, 163
structure constants 188
structure relations 189, 255
SU(2) 88, 91
  complete reduction of representations 86
  completeness of representations 155
  irreducible representation 84, 86, 155
  similarity class 154
  weights of Kronecker product 92
SU(3) 239, 256
  complete reduction 254
  diagram 242
  generating symmetry adapted basis 260
  Lie algebra 254
  Weyl group for 241
SU(n) 184, 188, 244, 248
  complete reduction of a representation 247
  complete reduction of the Kronecker product 248
  dimension of an irreducible representation 246
  infinitesimal operator 184
subgroup 3
  of SO(3) 134
  of the symmetric group 8, 136
subspace
  invariant 16, 20, 113
  irreducible 114

orthogonal 28
proper linear 16
summation over a group 28, 105
symmetric group 8, 67, 138
    abstract structure 9
    character table 142
    generating set 157
    on 3 elements 11, 13, 22
    on 4 elements 30, 126
    on 5 elements 138
    regular representation 148
    similarity classes 139
    subgroups 136
symmetry 60, 64, 96
    -breaking, spontaneous 165
    adapted array 37
    adapted basis 43, 54, 56, 57,
        61, 116, 166, 169
    amount of 164
    maximal 164
    of a representation 163
    of an operator 35, 40
    operation 35
symmorphic group 233
symplectic group 242
symplectic matrix 206

test 5
    factor reversal 6
    price reversal 6
    quantity reversal 6
    simultaneous time and factor
        reversal 7
    time reversal 6
tetrahedral group 135, 214
    character table 215
    similarity classes 136
time reversal test 6
time series 70
torus 87
    maximal 240
    of SO(5) 250
    of SO(n) 252
tower of subgroups 67
trace 79, 81, 139
transformation 3, 11, 61, 79
    group 4

    complete 4
law 81
linear 80
Lorentz 78
of finite range 8
translation
    left 31, 152
    right 152
translative part 232
transplantation of a function 83,
    235
transposition 9
trapezoidal rule 63
tridiagonal matrix 62
type of a permutation 137
type table of the symmetric group
    on 5 elements 138

U(3) 239
    character 240
    similarity classes 240
U(n) 184, 188, 248
    character of a representation
        242, 243
    complete reduction of a rep-
        resentation 247
    complete reduction of the Kro-
        necker product 248
    dimension of an irreducible rep-
        resentation 246
    infinitesimal operator 184
    irreducible inequivalent con-
        tinuous representation 243
unimodular group 74, 98
unimodular matrix 80
unit element 2
unit representation 11
unitary 93
    group 74, 75, 239
    matrix 29
    representation 27, 29
        complete reduction 28

value index 6
variation distance 171, 173, 175
vector
    dominant 246

length 27
norm 27
orthogonal 27
space of complex-valued class
    functions 151
star 225

waveguide junction 118
Weierstrass approximation theo-
    rem 98
weight 88, 100, 247
    diagram 92, 100, 256, 257
    dominant 88
    of the Kronecker product of
        SO(3), SU(2) 92
    of the representation
        of SO(3) 88
        of SO(n) 253
    vector 253
Weyl group 240, 241, 252
    of SO(5) 250
    of SU(3) 241

Young tableau 67, 244